litterae textuales

A SERIES ON MANUSCRIPTS AND THEIR TEXTS

EDITED BY

J.P. GUMBERT
M.J.M. DE HAAN
A. GRUYS

THE BRUSSELS HORLOGE DE SAPIENCE

La .vii.e hystoire.

Le second chapitre. Comment on peut venir a la congnoissance de la diuinite par la passion de Jhucrist et quelle forme dieu prist par la ditte passion de Jhucrist.

The Brussels Horloge de Sapience

Iconography and Text
of Brussels, Bibliothèque Royale, MS. IV 111

Peter Rolfe Monks

With an Edition of the Déclaration des Hystoires
and a Translation by K.V. Sinclair

E.J. BRILL
LEIDEN • NEW YORK • KØBENHAVN • KÖLN
1990

Library of Congress Cataloging-in-Publication Data

Monks, Peter Rolfe, 1941-
The Brussels Horloge de Sapience: iconography and text of Brussels, Bibliothèque royale, MS. IV 111 / Peter Rolfe Monks; with an edition of the Declaration des hystoires; and a translation by K.V. Sinclair.
p. cm.—(Litterae textuales, ISSN 0169-8702)
Includes bibliographical references (p.) and index.
ISBN 90-04-09088-6 (pbk.)
1. Bibliothèque royale de Belgique. Manuscript. IV 111.
2. Suso, Heinrich, 130?-1366. Horologium sapientiae.
3. Illumination of books and manuscripts, Medieval—Belgium—Brussels. 4. Mysticism—History—Middle Ages, 800-1500—Sources—Manuscripts. 5. Christian art and symbolism Medieval, 500-1500.
I. Bibliothèque royale de Belgique. Manuscript. IV 111.
II. Title III. Series.
Z105.5.B53M85 1990
091—dc20
90-2215
CIP

Frontispiece: Brussels, Bibliothèque Royale,
MS. IV 111, fol. 23 (slightly reduced).

ISSN 0169-8702
ISBN 90 04 09088 6

PRINTED IN THE NETHERLANDS

Contents

Acknowledgements

The research for this study was commenced in 1982. I was privileged to be received on several occasions by Professor Eleanor P. Spencer at her home in Paris. It is a pleasure to recall now these visits and to record Professor Spencer's kindness and her support for the project.

During my sojourns in New York, Professor John H. Plummer, Curator of Medieval and Renaissance Manuscripts at the Pierpont Morgan Library most generously made time in a busy schedule to discuss with me aspects of the Rolin Master's works.

Encouraging and helpful suggestions about the contents of this study were made to me by Dr Pierre Cockshaw, Conservateur, Cabinet des Manuscrits, Bibliothèque Royale, Brussels and Professor of Art History at the University of Brussels. To him I extend my sincere thanks.

Discussions with other scholars in the field of manuscript illumination have been beneficial to my understanding of the parameters of iconographical studies. In particular, I mention Professor Margaret M. Manion, Herald Professor of Fine Arts in the University of Melbourne, Dr Lilian M.C. Randall, Curator, Department of Manuscripts, Walters Art Gallery, Baltimore; the late Professor James D. Breckenridge of the Department of Art History, Northwestern University, Evanston; Dr Sandra Hindman, formerly of Johns Hopkins University, now in the Department of Art History at Northwestern University.

For assisting me to consult miniatures by the Rolin Master and works relevant to this study, I am indebted to the following: Mr Roger S. Wieck, Houghton Library, Harvard University; Mr Jan Fontein, Director, Museum of Fine Arts, Boston; Mr William Cuffe, Yale University Art Gallery; Mr William Voelkle, Pierpont Morgan Library; Dr W.B. Rayward, Dean of the Graduate Library School, University of Chicago; Ms Janet Backhouse, British Library; Mr Michael Bott, Keeper of Archives and Manuscripts, and Catherine I. Reynolds, Department of the History of Art, both of Reading University; M. Georges Dogaer, Conservateur en Chef, and Madame Marguerite Debae and Madame Claudine Lemaire, all of the Bibliothèque Royale, Brussels; M.J. Vervaet, Hoofdconservator, Koninklijk Museum voor Schone Kunsten, Antwerp; Madame Amélie Lefèbvre, Conservateur des Collections, Musée Condé, Chantilly; M. Matthieu Pinette, Conservateur du Musée Rolin, Autun; Madame Marie-Josette Perrat, Conservateur, Bibliothèque Municipale, Autun; M. François Avril, Conservateur, Département des Manuscrits, Bibliothèque Nationale, Paris; Madame Anne-Françoise Labie, Madame Anne-Marie Gènevois and M.A. Vernet of the Institut de Recherche et d'Histoire des Textes, Paris; Madame C. Gleyze, Conservateur en Chef,

Bibliothèque de la Ville, Lyons; Madame C.A. Chavannes-Mazel, Koninklijke Bibliotheek, The Hague; Madame Klara Garas, Director, Museum of Fine Arts, Budapest; Dr. Otto Mazal and Madame Eva Irblich, of the Österreichische Nationalbibliothek, Vienna; Mr P. Singleton, University of Melbourne Library; Dr N.A.C. Radford, Chief Librarian, University of Sydney Library; Mr M.J. Walkley, Department of French Studies, University of Sydney.

Archbishop Simeon, Metropolitan for Western Europe of the Bulgarian Patriarchate, provided me with introductions to scholars and art historians in several libraries and museums.

I feel that the observations I make from time to time in the study about the Order of Preachers would have been much less informative if I had not received the unstinting help of Fr. Bernard Montagnes O.P. of the Convento S. Sabina (Aventino), Rome.

The transcriptions of extracts from the French *Horloge de Sapience*, which appear in the footnotes of chapter V, and the edition of the French text of the *Déclaration des Hystoires*, with an English translation and glossary, are the work of Professor K.V. Sinclair.

The comments and advice of Dr J.P. Gumbert of Leiden University have been most generous and perceptive. His suggestions about the composition of the Brussels codex and the number of hands are gratefully acknowledged.

The production of this volume has been in the skilled hands of the staff of Messrs E.J. Brill of Leiden. I have welcomed their guidance in many matters, not least of which are page lay-outs and photographic reproductions. An anonymous donor has generously met the costs of the plates.

Peter Rolfe Monks
March, 1987

Introduction

Since the Second World War the field of study embracing iconography and text has attracted art historians and literary scholars to the sides of manuscript scholars whose domain it was so long. Members of the new mix seem united in their desire to throw light not merely on the manuscript as a portable art museum but on the textual concepts within its confines as potential *points de départ* or catalysts for the art work on the folios. Put another way, one may be able to appreciate the idea that when miniaturists employ familiar modes of representing abstract ideas, they become important purveyors of key concepts, some rooted in a common culture and others special to particular artists.

Studies of the visual definitions of essential or minor ideas have been numerous. The ones now mentioned constitute a mere sample and have been selected purely because of the relevance of their technical approach to our research interests. In 1967 Sister Marie Brisson attempted, unsuccessfully I believe, to discover in pictorial imagery of the Passion the source or the projection of an obscure detail mentioned in the Crucifixion narrative of Robert the Carthusian's *Chastel Périlleux*.[1] Adopting a chronological span of some eight centuries of Western Art, Meyer Schapiro studied several iconographic models that purport to depict the narrative of Exodus XVII, 9-13, wherein Moses and the Israelites engage the Amalekites at Rephidim.[2] Claire Sherman's primary focus in her paper on the illustrations of two codices of Oresme's translation of Aristotle was "on the texts and images of the two manuscripts as interdependent rather than as separate entities."[3] Donal Byrne studied iconographic models in two contemporary copies of the famous *Livre des propriétés des choses*. In one instance, innovations in the series of images were observed to reflect the interests of a patron.[4] In a richly illustrated monograph which appeared in 1979, J.H. Marrow explored the association of literary conflations of the Passion narrative of the Gospels with actual images executed in Northern European Art of the late Middle Ages and Early Renaissance.[5]

[1] Sister Marie Brisson, "An Unpublished Detail of the Iconography of the Passion in *Le Chastel Périlleux*," *Journal of the Warburg and Courtauld Institutes*, XXX (1967), pp. 398-401.

[2] Cf. M. Schapiro, *Words and Pictures. On the Literal and the Symbolic in the Illustration of a Text*, The Hague, 1973.

[3] Claire R. Sherman, "Some Visual Definitions in the Illustrations of Aristotle's *Nicomachean Ethics* and *Politics* in the French Translation of Nicole Oresme," *Art Bulletin*, LIX (1977), pp. 320-330.

[4] D. Byrne, "The Boucicaut Master and the Iconographical Tradition of the *Livre des Propriétés des Choses*," *Gazette des Beaux-Arts*, XCI (1978), pp. 149-64.

[5] J. H. Marrow, *Passion Iconography in Northern European Art of the Late Middle Ages and Early Renaissance*, Courtrai, 1979.

Elizabeth Salter and Derek Pearsall published a joint paper about, *inter alia*, the role of the frontispiece in the illustrated book. The positioning was seen to be potentially influential, it creates expectations in the reader's mind, and it may display an artist's intellectual affinity with the textual content.[6] The increased functional emphasis on miniatures in late medieval secular texts was the subject of Lesley Lawton's article in 1981. She chose the illustrations of Lydgate's *Troy Book* as her case study.[7] The thematic content of the new cyclic programme devised around 1400 for copies of Raoul de Presle's translation of St. Augustine's *City of God* occupied the endeavours of Sharon Smith in a 1982 paper. She observed that the programme "generally follows the text of the *City of God* with illustrations that tend to be naively literal in a lively and vivacious vein." She was also concerned to demonstrate the "nature of the relationship between the intellectual conception of the subject matter and its pictorial relationship."[8]

The manuscript catalogued as IV 111 in the Bibliothèque Royale, Brussels, and containing the *Horloge de Sapience* as its principal text, may appear at first sight to be just another richly decorated volume in the manner of so many codices of the great collections amassed between 1370 and 1435 by Charles V, the Duke of Berry, the Duke of Bedford, and by Philippe le Hardi and Philippe le Bon. Like these earlier volumes, the Brussels codex was provided with cyclic programmes of illustrations. But there is much more of interest to state. The illuminations in the *Horloge de Sapience* constitute a challenge to art historians, by any intellectual or artistic standard. For, as Schiller observed, and we concur with her viewpoint, "artistic realisation of an intellectual idea depends upon the expressional potentialities of the artistic means of a given period."[9] This study attempts therefore to lay bare the concordances and discordances between the visual statements and the textual passages of the Brussels manuscript. The symbols employed, the manner of their incorporation by recognisable and traditional icons, the significance of the Dominican foundations, will each be examined in due course.

The organisation of the wealth of material at our disposal is conducted along traditional scholarly lines. The opening chapter of the study introduces the reader to the mid-fifteenth-century volume, Brussels, Bibliothèque Royale, MS. IV 111, its history, its texts and its art work. There follows a chapter about the craftsmanship of the two artists concerned, the Rolin Master and the Bedford Master's Chief Associate.

This manuscript is the only known *Horloge* to contain a cyclic approach to the iconography. The unusual textual matter from the pen of the Dominican mystic, Heinrich Seuse, *alias* Suso, no doubt contributes to the intriguing aspects of the pictorial narratives. Then there is a contemporary commentary of the illustrations, designated as the *Déclaration des Hystoires*; it commands the reader's attention, not only because it interposes itself between reader and viewer, but because it is unique in *Horloge* iconography.

Understandably, the most developed discussion is reserved for the final chapter which treats in turn the iconography and text of the *Horloge* series of pictures and of the three illustrations that decorate Gerson works. The commentary for each miniature opens with the relevant textual passage from Suso or Gerson, as the case may be. The quotation is a modern English rendering of the French original. Whenever the French *Horloge de Sapience* departs noticeably from the Latin of Suso's *Horologium*, attention will be drawn to the problem.

[6] Cf. Elizabeth Salter and D. Pearsall, "Pictorial Illustration of Late Medieval Poetic Texts: the Role of the Frontispiece or Prefatory Picture," in *Medieval Iconography and Narrative. A Symposium*, Odense, 1980, pp. 100-23.

[7] Lesley Lawton, "The Illustration of Late Medieval Secular Texts, with Special Reference to Lydgate's *Troy Book*," in *Manuscripts and Readers in Fifteenth-Century England* (ed. D. Pearsall), Cambridge, 1983, pp. 41-69.

[8] See Sharon D. Smith, "New Themes for the *City of God* around 1400: the Illustrations of Raoul de Presle's Translation," *Scriptorium*, XXXVI (1982), pp. 68-82, pls. 5-9.

[9] Gertrud Schiller, *Iconography of Christian Art*, Greenwich, Conn., 1971-1972, I, p. 1.

Appropriate passages of the *Déclaration* are next adduced, also in a modern English translation. Since the commentator was at all times at liberty to offer an exegesis based on whatever took his fancy, one should not be surprised if he departs occasionally from the source text of Suso.

The final component of the evaluation of the iconography and text of each visual structure will be a twentieth-century interpretation.

There follow the Plates and the French text of the *Déclaration* and an English translation with Glossary. A Selected Bibliography lists the more important publications that were consulted for this study. All Biblical citations derive from the *Vulgata Clementina*, recorded in the Bibliography under *Biblia Sacra*. The General Index contains not only references to persons and places, but to themes and symbols and to each title of the pictorial units.

CHAPTER ONE

The Manuscript

1. DESCRIPTION OF BRUSSELS, BIBLIOTHÈQUE ROYALE, MS. IV 111

The Brussels manuscript is bound in a binding of blind-stamped brown calf over wooden boards, possibly more or less contemporary with the manuscript itself. The brass clasp may be contemporary as well; but the heavy brass cornerpieces do not look authentic and may be later. The back is restored. There are modern vellum pastedowns and flyleaves at both ends.

The manuscript itself consists of 273 vellum leaves,[1] numbered by a contemporary hand (certainly one of the hands originally involved in producing the book) by placing Roman numerals in the centre of the top margins. Although the first surviving leaf acts as a fly-leaf, it counts as *j*, the second folio being numbered *ij*. The numbering is then consecutive to *lvj*; the leaf that follows it is *lviij*; there are no other gaps until the last numbered leaf which carries *cclxxiiij*; on its verso the text breaks off.

The book is composed in quires, normally of eight leaves; they bear catchwords on the last verso, except for three quires: the first (preliminary), the penultimate (where they are presumably trimmed off) and the last (incomplete). The book is, however, not quite homogeneous: it is executed in five sections. Irregularities in quiring (shorter or longer quires to fit the text as precisely as possible) mark the ends of these sections; other evidence, which we shall note, confirms the divisions. The construction of the book is as follows:

quire 1^{12}, fols. 1-12: preliminary matter.
quires 2^{8}, 3^{6}, $4\text{-}7^{8}$, 8^{8} lacks one, $9\text{-}19^{8}$, 20^{2}, fols. 13-156: main section, *Horloge de Sapience* etc. It is not apparent why quire 3 is shorter. From quire 8, fols. 51-58, fol. 57 has been removed (see below pp. 6, 101-2, 178-9).
quires $21\text{-}24^{8}$, 25^{10}, fols. 157-198: a group of minor texts.
quires $26\text{-}28^{8}$, 29^{6}, fols. 199-228: another group of minor texts.
quires $30\text{-}34^{8}$, 35^{8} lacks two, fols. 229-274: another group of minor texts. The last two leaves of the last quire (and, possibly, one or more further quires) are lost.

[1] Cf. *Bulletin de la Bibliothèque Royale*, 5 (1961), p. 48; *Bruxelles, Bibliothèque Royale Albert Ier. Quinze années d'acquisitions*, Brussels, 1969, p. 84; Eleanor P. Spencer, "L'*Horloge de Sapience*, Bruxelles, Bibliothèque Royale, MS. IV. 111," *Scriptorium*, XVII (1963), p. 277, n. 1.

The leaves measure approximately 370×255 mm. and have been ruled to carry two columns of text, each with 40 lines. The ruling is executed in pale-brown ink.

The script is a French Gothic *bâtarde*, neat and roundish. More than one hand was engaged on the work; the question of the division of labour will be reviewed below.

The minor decoration is uniform throughout the volume. Apart from line fill-ins and red and blue paragraph markers with penwork flourishes attached, there are two types of initials. The major initials are large, burnished-gold initials of high quality, accompanied in the margins by graceful embellishments in the form of ivy-leaf and floral designs. In the *Horloge de Sapience* these open chapters; in the minor works by Gerson and others, they mark the beginnings of what the scribes considered to be separate works. Minor initials alternate gold with dark blue penwork and blue with red penwork. They open the chapters of the minor works; in the *Horloge* they serve to open the sections immediately following the miniatures.

Finally, there are miniatures; these will be treated fully below. It should only be mentioned here that the placing of the last three miniatures (by a second painter) is not accidental: on fols. 157, 199v (199 being a sort of chapter list) and 229, they open the last three sections of the volume as analysed above.

Some open-weave curtains of a faded burgundy colour survive, which were originally glued between leaves carrying miniatures and folios facing them. Most are now detached.

2. TEXTS

The list of contents supplied by the Belgian scholar librarians was deliberately cursory, while Spencer's statement about the texts was incomplete in places.[2] It has been necessary to examine the transcriptions afresh, folio by folio. We print a new list of the contents (using > to mark the divisions into sections of the volume, as described above; two folio numbers are given in those cases where the rubric is found at the end of one page and the text itself begins on the next leaf).

(Fol. 1 — originally pastedown, now lifted; ruled, but not used.)

Fol. 2 — Table of Contents for the whole codex, with text titles, text openings and numbers of the folios on which each text commences.

Fol. 2v — (originally not used; later added: Croy family arms, motto, ducal coronet, Collar of the Order of the Golden Fleece, and the date 1618.)

Fols. 3-11 — *Declaration des hystoires de l'Orloge de Sapience*: Premierement au commencement du livre est dame Sapience en forme et figure de femme ...

Fols. 11-11v. — *Les Declarations de troys hystoires qui ne sont pas de l'Orloge de Sapience et est la premiere des .x. Commandemens*: En ce mesme volume y a troys...

Fols. 12-12v. — Table of sixteen chapters for Book I of the *Horloge de Sapience*.

>

Fols. 13-89v. — Book I of the *Horloge de Sapience*,[3] the French translation by a Lorraine Fran-

[2] See Spencer, *Horloge*, p. 277, n. 1 where titles and folio numbers are reproduced from the fifteenth-century Table of Contents, but not from the folios themselves. The Table, however, omits mention of four texts: the *Meditations* by St. Augustine, the sermon *Beati qui lugent* by Jean Gerson, the poem *Chansonnette amoureuse* which has been transcribed in the codex as part of another work, and, lastly, the *Piteuse complainte* attached to the end of the *Mendicité spirituelle* and attributed to Jean Gerson. In a second article on the manuscript entitled, "Gerson, Ciboule and the Bedford Master's Shop," *Scriptorium*, XIX (1965), pp. 104-5, Spencer printed the Table from fols. 2-2v. and supplied references to modern critical comment on many of the texts, in particular, those by Chancellor Gerson. The prelate's complete works, edited by P. Glorieux, had only just commenced appearing in print at the time of the *Ciboule* paper. Glorieux, *Gerson, Oeuvres*, VII, pp. xl-xli mentions only thirteen texts.

[3] There is no modern critical edition of this medieval French text.

ciscan of Henry Suso's Latin work, *Horologium Sapientiae*:[4] Salmon en son livre de Sapience ou premier chapitre dit Sentite de domino in bonitate... There is a gap in the text of chapter X because of the removal of fol. 57.

Fol. 90 — Table of eight chapters of Book II of the *Horloge de Sapience.*

Fols. 90v.-133 — Book II of the *Horloge de Sapience.*

Fols. 133-155v. — St. Augustine, *Soliloques*:[5] Sire Dieu, je desire que je congnoisse toy qui es celui qui me congnois...

Fol. 155v. — St. Augustine, *Méditations*:[6] Sire, je ay a dire une secrete parole...

Fol. 156 — Ruled, but blank; the verso side ruled and col. *a* is blank; col. *b* has the title of the following text.

>

Fols. (156 v.) 157-161v. — Jean Gerson, *Dix Commandements de Dieu*:[7] Gloire soit a Dieu [en] qui nom pour le salut des ames du simple peuple crestien...

Fols. 161v.-170 — Jean Gerson, *Le Prouffit de sçavoir que est pechié mortel ou veniel*:[8] Qui bien considere la bonté de Dieu envers nous il a cause de grandement le regracier...

Fols. 170v.-172v. — Jean Gerson, *Examen de conscience*:[9] Qui se veult mettre de l'estat de pechié en l'estat de grace et de salut...

Fols. 172v.-176 — Jean Gerson, *Sermon*:[10] Memento finis. Ouvrons maintenant les yeulx et les oreilles de nostre entendement...

Fols. 176v.-181 — Jean Gerson, *Sermon*:[11] Beati qui lugent mathei quinto dit ce theume Ceulx ycy sont bieneureulx qui les cuers ont douloureux...

Fols. 181-194v. — Robert Ciboule, *Douaires de la gloire pardurable*:[12] Absterget Deus omnem lacrimam ab oculis sanctorum iam non erit amplius neque luctus neque clamor neque ullus dolor etc. Apoc. vij°. En ces paroles qui sont escriptes en l'Apocalypse...

Fols. 194v.-198 — Pierre d'Ailly, *Jardin amoureux de l'ame*:[13] En l'abbaye de devote religion fondee en mondain desert est le jardin de vertueuse consolacion...

Fol. 198 — Pierre d'Ailly, *Chansonnette amoureuse*:[14] Pour la couronne amoureuse gaignier dont aux amans fait amoureux present...

Fol. 198v. — Ruled but blank except for part of col. *b* where is copied the long rubric of the following text.

>

Fols. (198v.) 199-226 — Jean Gerson, *Mendicité spirituelle*:[15] Ma povre, ma malade, ma chartriere, ma miserable ame...

[4] This Latin original has since been recently published: *Heinrich Seuses Horologium Sapientiae. Erste kritische Ausgabe unter Benützung der Vorarbeiten von Dominikus Planzer O.P.*, ed. P. Künzle O.P., Freiburg, 1977.

[5] Partial ed. by Geneviève Esnos in *Ecole française de Rome. Mélanges d'archéologie et d'histoire*, LXXIX (1967), pp. 299-370. See also J. Sonet, *Répertoire d'incipit de prières en ancien français*, Geneva, 1956, p. 341., no. 1944, and additions in K.V. Sinclair, *Prières en ancien français*, Hamden, Conn. 1978, pp. 138-9, no. 1944.

[6] Cf. Geneviève Esnos in *Ecole française de Rome. Mélanges d'archeologie et d'histoire*, LXXIX (1967), pp. 311-18; Sonet, *Répertoire*, p. 354, no. 2028; corrections and additions in Sinclair, *Prières*, pp. 142-3, no. 2028.

[7] The copy ends: "...pecher et trespasser le commendement." which means the scribe has copied only the first major part of the longer tract entitled *Miroir de l'âme.* Ed. Glorieux, *Gerson, Oeuvres*, VII, pp. 193-203.

[8] Ed. Glorieux, *Gerson, Oeuvres*, VII, pp. 370-89.

[9] Ed. Glorieux, *Gerson, Oeuvres*, VII, pp. 393-99. The Latin *Forma absolucionis* has not been copied at the end of the text which in this Brussels codex concludes: "...oublier negligemment ses pechiez."

[10] Ed. Glorieux, *Gerson, Oeuvres*, VII, pp. 690-8.

[11] Ed. Glorieux, *Gerson, Oeuvres*, VII, pp. 549-60.

[12] The work is an extract of a longer text entitled *Le Livre de la sainte méditation.* There is no attribution in the Brussels copy's rubric or explicit. Spencer, *Ciboule*, pp. 104-5 correctly identified the author and the parent text.

[13] Spencer's statement (*Ciboule*, p. 105) on the authorship question is most unsubtle. The work still continues to be associated with the name of Pierre d'Ailly or Jean Gerson. An edition was printed in Glorieux, *Gerson, Oeuvres*, VII, pp. 144-54. See also the discussion about the authorship and manuscripts in P. Y. Badel, "Pierre d'Ailly, auteur du *Jardin amoureux*," *Romania*, XCVII (1976), pp. 369-81.

[14] The poem has here been copied as prose as part of the *Jardin amoureux.* Editions and manuscripts are listed in K. V. Sinclair, *French Devotional Texts of the Middle Ages*, 3 vols, London, 1979-1988, II, p. 149, no. 4995.

[15] Ed. Glorieux, *Gerson, Oeuvres*, VII, pp. 220-80.

Fols. 226-227 — Jean Gerson, *Piteuse Complainte*:[16] Jhesus, vray espous de virginité, Jhesu de chasteté loyal ami...

Fols. 227v.-228v. — Ruled but blank except for col. *b* of 228v. where appear the title and opening words of the theme for the next sermon.

>
Fols. (228v.) 229-258v. — Jean Gerson, *Sermon*:[17] Ad Deum vadit Joh^{is} XIII°. Ideo penitemini et credite evvangelio marci quarto decimo (*sic*) A Dieu s'en va...

Fols. 258v.-274v. — Celestine Brother unidentified, *Fruits de tribulation*:[18] Si comme dit saint Pol l'apostre nous ne avons pas en ce monde cité ou habitation...

Fol. 274v. — Anonymous, *Huit Béatitudes*:[19] Benois soient ceulx qui sont povres d'esperit car le royaulme des cieulx est a eulx... (breaks off at the bottom of col. *b* at the words: doulceurs espirituelles que sans comparaison sont plus. — It is clear that this treatise was always the last in the codex, since it figures as the last entry in the original Table of Contents on fol. 2).

3. SCRIBES AND DATE

The review of the contents makes it quite clear that the manuscript is a collection of mystical treatises, theological tracts and sermons, many by prominent churchmen of fifteenth-century France. The descriptions of the codex by the staff of the Bibliothèque Royale in Brussels, already mentioned above, characterised the script as "une gothique bâtarde de module moyen, extrêmement soignée," and there was no suggestion that more than one hand could have been involved. Spencer in her turn ventured this comment: "The original format, however, seems to be that of two different fifteenth-century books, probably from the same atelier but written and illuminated separately before being assembled and prefaced by a third fifteenth-century scribe who added a table of contents for both parts, a *déclaration des hystoires*, or descriptive index to all the miniatures, a marginal numbering of the miniatures of the first manuscript and a complete pagination."[20] The distribution of the labour was given by Spencer as fols. 2-12, 13-133v., 133v.-274v. To this detailed comment one should juxtapose her statement made in the *Ciboule* paper, "Although the format of the book suggests that it was written by two, possibly three scribes or at different times, there is nothing to contradict the impression that the work was all done in one shop within a relatively short time."[21]

The distinction of hands, in a type of script as formal as the *bâtarde* of this manuscript, is notoriously difficult. Still, it is possible to find some differences, mainly by studying the distribution of two characteristics: the predominant use of a *g* with a 'weak' tail, dangling downward in a thin line, contrasting with a *g* which has a firm, more or less horizontal tail; and the use of numerous, few, or very few looped ascenders on *b h l*. These characteristics, together with the general aspect of the pages, lead to the distinguishing of three hands (although a close study might reveal that one or more of these ought to be further subdivided). Somewhat tentatively, the result is as follows:

[16] The copyist transcribes this text as part of the *Mendicité*. Ed. Glorieux, *Gerson, Oeuvres*, VII, pp. 213-16. The text is also attributed at times by scribes to Pierre d'Ailly, see Sonet, *Répertoire*, p. 180, no. 998; additions and corrections are in Sinclair, *Prières*, pp. 86-7, no. 998.

[17] Ed. Glorieux, *Gerson, Oeuvres*, VII, pp. 449-519. It has not entirely superseded an earlier scholarly edition by D. H. Carnahan, *The Ad Deum Vadit of Jean Gerson*, Urbana, Ill., 1917.

[18] The author has not been identified, as Spencer, *Ciboule*, p. 105, n. 14 observes, while noting the date 1445 for the text in another codex. On this Celestine version of an already popular tract, see Kathleen Chesney, "Notes on some Treatises of Devotion intended for Margaret of York," *Medium Aevum*, XX (1951), p. 24.

[19] Cf. Rosalie Vermette, "The *Huit Béatitudes* in Old French Prose," *Manuscripta*, XVIII (1974), pp. 105-110 who does not mention this manuscript or the version it contains.

[20] Spencer, *Horloge*, p. 277.

[21] Spencer, *Ciboule*, p. 107.

preliminary section

quire 1, fols. 1-12: hand A, with a firm tail to the *g* and very few loops.

main section

quires 2-15, fols. 13-122: hand B, with a 'weak' *g* and many loops.

quires 16-18, fols. 123-146: hand C, with a 'weak' *g* and few loops. At the end of this stint it becomes more difficult to distinguish the hand(s).

quires 19-20, fols. 147-156: hand A.

third section

quires 21-22, fols. 157-172: hand C.

quires 23-25, fols. 173-198: hand A.

fourth section

quires 26-28, fols. 199-222: hand C.

quire 29, fols. 223-228: hand B.

fifth section

quires 30-35, fols. 229-274: hand A.

This shows that the work was produced in planned co-operation by several scribes. It also seems to show that scribe A had the main responsibility for the execution of the work: twice he finished sections begun by other scribes, and the preliminary section, doubtlessly written last, is by him (he also wrote the rubrics for the three minor sections, which are on the last pages of the preceding sections; and he may have been responsible for the foliation).

A further conclusion is possible. The *Déclaration* was copied by scribe A, and therefore still during the original process of production of the book. On the other hand its text was not drawn up (as will be shown below) before the miniatures had been painted. This means that the *Déclaration* is strictly contemporary with, and intimately connected with, the whole planning and execution of the volume.

The *Douaires de la gloire pardurable* on fols. 181-194v., an extract from Robert Ciboule's *Livre de la sainte Méditation*, was considered by Spencer to be of some assistance in the dating of the whole manuscript.[22] She referred her readers to Combes' dating of Ciboule's work contained in a long essay on the preacher, entitled *Un Témoin du socratisme chrétien au XVe siècle: Robert Ciboule 1403-1458.* Combes alluded to the *Livre de la sainte Méditation* in these terms: (Ciboule) "paraît mériter l'attention par la rigueur systématique avec laquelle, dans une œuvre de pure vulgarisation théologique écrite en français vers 1450, il s'est appliqué à tirer du thème traditionnel de la connaissance de soi les conséquences qu'il implique, selon lui, pour la formation spirituelle du chrétien."[23] In a tabulated summary of the dates of the works by the preacher, Combes had a rubric 'Dates probables' under which appears a cautious modification of the above date 'vers 1450'. He wrote, "Peut-être après 1451, et certainement dans l'ordre suivant: *Le Livre de la sainte Méditation...*" The reason for the revised dating is nowhere supplied. Even in a short notice on Ciboule published some twenty years later, Combes was silent about a specific date for the text in question.[24] Given the uncertainty surrounding the date of the work and of its extract, *Les Douaires*, it is hazardous to use its presence in the Brussels volume as the sole means of dating the transcription of the manuscript.

The only credible dating at this juncture for the scribal work is 'mid-fifteenth-century', however vague that appears to be.

[22] *Ibid.*

[23] The study by A. Combes appeared on pp. 93-259 of vol. VIII (1933) of the *Archives d'histoire doctrinale et littéraire du moyen âge.* The quotation is from p. 93.

[24] Cf. A. Combes, "Ciboule (Robert)," in *Dictionnaire de spiritualité ascétique et mystique*, ed. M. Villet, Paris, 1953, II, cols. 887-90.

4. TWO MEDIEVAL OWNERS

The Brussels volume has the distinction of displaying two different fifteenth-century escutcheons. The first coat of arms is that of a member of the Gouffier family: *d'or à trois jumelles de sable*. It has been painted four times on fol. 13, that is, thrice in the margins and once on a surcoat of a kneeling figure in the illumination of *Sapientia in Majesty*. The identification was made in 1961 by the Section d'héraldique of the Institut de Recherche et d'Histoire des Textes in Paris and conveyed to the Bibliothèque Royale in Brussels. Its librarians went further, they named the family member and offered an explanation for the acquisition of the volume, even suggesting that Gouffier ordered it for himself: "...parmi les membres de la famille Gouffier, le seul dont l'importance puisse justifier l'exécution d'un livre aussi luxueux est Guillaume, baron de Roannais, sieur de Boisy. Sénéchal de Saintonge en 1451, il fut conseiller et premier chambellan du roi Charles VII en 1454. On peut même supposer sans difficulté que seule l'accession à une charge aussi élevée a pu inspirer à notre personnage l'idée de commander un manuscrit aussi précieux. Cette hypothèse permettrait de situer l'exécution du manuscrit en 1454 ou peu après."[25]

The hypothesis fails to explain why this particular owner should order a copy of a text such as the mystical *Horloge de Sapience*, or for that matter, why he could not have received it as a gift.

The above suggestion and identification by the Belgians find no acknowledgement in Spencer's comments although she produced the same Guillaume Gouffier as bearer of the arms in the manuscript: "The figure of the kneeling knight is contemporary with the rest of the miniature but is ignored by the writer of the *déclaration*. He once bore on his tunic the arms of the Gouffier family of France... Of course, one would not discount the possibility that the text was illuminated for an even earlier patron whose arms were never filled in, or were well erased when it came into the hands of a Gouffier."[26] She went on to identify the former owner as Guillaume Gouffier (*c.* 1420-1495): "who spent his whole life in the service of one court or another..."[27] In a biographical sketch she traced the principal high and low points of his long life, including his disgrace and banishment, then his return to favour under Louis XI, but she did not speak about his period of service in the West of France. Above all, she offered a very plausible argument to explain how the Gouffier manuscript came later into the possession of a duc de Bretagne, and it will be necessary to discuss this at an appropriate moment. In presenting for the reader's attention the following observations on Gouffier's life and times, reference has been made to detailed statements in works by earlier scholars.[28]

Guillaume held positions of trust at court in the reigns of Charles VII (1422-1461), Louis XI (1461-1483) and Charles VIII (1483-1498). Beaucourt reports[29] that in 1436 and 1437 he was under the guardianship of his uncles, Guillaume and Jean, one of whom had been in 1418 an *écuyer* and *chambellan* to the Dauphin, the future Charles VII. As the young Guillaume matures, he remains in a circle of royal retainers close to the person of the monarch. His presence is recorded in 1442-1444 in the service of Charles d'Anjou, comte du Maine and younger brother of Marie d'Anjou, the king's wife. In the last of these three years he received money from the royal purse for

[25] *Bulletin de la Bibliothèque Royale de Belgique*, V (1961), p. 57. The coat of arms (?) painted on the top edge of the volume is indistinguishable.

[26] Spencer, *Horloge*, p. 296.

[27] *Ibid.*, p. 297.

[28] I have consulted the standard genealogical compendia, by P. de G. Anselme de Sainte-Marie, *Histoire généalogique et chronologique de la maison royale de France*, 9 vols, 3rd ed., Paris, 1726-1733, V, pp. 607-8; and E. Henri, E. Beauchet-Filleau, C. de Chergé, *Dictionnaire historique et généalogique des familles du Poitou*, 2nd ed. by J. Beauchet-Filleau, 6 vols, A-L, Poitiers and Paris, 1891-1978, IV, pp. 251-2; F. Galantino, *I Conti del Forese ed i Gouffier de Boysi*, Milan, 2 parts, 1880-1881, *passim*.

[29] G. du Fresne de Beaucourt, *Histoire de Charles VII*, 6 vols, Paris, 1881-1891, IV, p. 178, n. 3.

attendance at Angers, with the rank of *écuyer*. Other receipts and accounts in the immediate years that follow show him being reimbursed for expenses incurred on the king's behalf and for such purposes as "the upkeep of his station" and "payment towards a new robe". The term *valet de chambre* now accompanies *écuyer*.[30]

In the year 1445, during the monarch's sojourn at Razilly, two leagues from Chinon, he placed Gouffier in the household of the royal mistress, Agnès Sorel.[31] Two of her brothers, Charles and Jean, were already *écuyers d'honneur* and another two, André and Louis, were *hommes d'armes* in the royal bodyguard of archers.[32] Gouffier was clearly no stranger to the Sorel family. He accompanied Agnès on a pilgrimage to Sainte-Geneviève in 1449,[33] and attended her needs when she set up her small court at Loches the same year.[34] Meanwhile, Charles VII was absent in Normandy, following the progress of his captain, Dunois, as he routed the English at Rouen and Harfleur.[35] The king spent part of the winter of 1449-1450 at the Benedictine Abbey of Jumièges and was surprised to receive a visit from his pregnant mistress. Obviously she could not remain in the monastery nor in the encampments, so the Abbot placed at her disposal his villa named Le Mesnil, situated to the South-East of the liberty. A short illness carried her off on 11 February, and at her deathbed was Gouffier.[36] He must also have known personally her three testamentary executors[37] as a result of his privileged station. These were Agnès's doctor, Maître Robert Poitevin; the unofficial court banker, shrewd entrepreneur and self-made magnate, Jacques Coeur;[38] and the royal Maître des Comptes, Etienne Chevalier.

Chevalier was born *c.* 1410 at Melun and in mid-life was entrusted with royal financial matters. He moved from the rank of Maître des Comptes to the top position of Royal Treasurer in 1451 and held the office for the remainder of his life. It is fitting to recall that in the year Agnès died Chevalier commissioned Jean Fouquet to paint a diptych, one panel of which was the Virgin and Child, for his favourite church, Notre-Dame-de-Melun. It is widely held by modern art critics, following much moralising among nineteenth-century predecessors, that the portrait of the Virgin enshrines the features of Agnès Sorel.[39] Chevalier also commissioned Fouquet to illustrate *c.* 1452-1456 a celebrated Book of Hours[40] in memory, it is thought, of his own cherished wife, Catherine Budé, killed in an accident in 1452. One should also draw attention to the theory that in the manuscript's *Magi Adoring the Virgin and Child* Fouquet gave the three kings the features of Charles VII, the Dauphin Louis, and the monarch's favourite son, Charles de France, born 28 December 1446.[41]

[30] *Ibid.*, IV, p. 178, notes 3 and 4.

[31] *Ibid.*, IV, pp. 170-7.

[32] *Ibid.*, IV, p. 174, n. 4.

[33] P. Champion, *La Dame de beauté, Agnès Sorel*, Paris, 1931, p. 53.

[34] Beaucourt, *Charles VII*, IV, p. 217, note.

[35] Champion, *Dame de beauté*, pp. 59-61.

[36] She died of dysentery, according to a contemporary chronicler, Thomas Basin, Bishop of Lisieux; see his *Histoire de Charles VII*, ed. and translated by C. Samaran and H. de Surirey de Saint-Rémy, Paris, 1933 and 1944, II, pp. 281-3; Champion, *Dame de beauté*, pp. 61-2.

[37] See P. Cavailler, "Le Compte des éxécuteurs testamentaires d'Agnès Sorel," *Bibliothèque de l'Ecole des Chartes*, CXIV (1956), pp. 97-114.

[38] See P. Clément, *Jacques Coeur et Charles VII*, Paris, 1873; Louise Guiraud, *Recherches et conclusions nouvelles sur le prétendu rôle de Jacques Coeur*, Paris, 1900; C. Marinesco, "Jacques Coeur et ses affaires aragonaises, catalanes et napolitaines," *Revue historique*, CCVI (1951), pp. 224-37; and *Les Affaires de Jacques Coeur. Journal du procureur Dauvet*, ed. M. Mollat, Paris, 1952-1953.

[39] The panel is now in the Koninklijk Museum voor Schone Kunsten, Antwerp; cf. K. G. Perls, *Jean Fouquet*, London, 1940, p. 19 and pls. VII and VIIa; P. Wescher, *Jean Fouquet and his Time*, London, 1947, pp. 48 and 51; C. Schaefer, "Le Diptyque de Melun de Jean Fouquet conservé à Anvers et à Berlin," *Jaarboek van het Koninklijk Museum voor Schone Kunsten, Antwerpen*, 1975, pp. 7-100. See a bibliography for the painting in Nicole Reynaud, *Jean Fouquet*, Paris, 1981, pp. 46-55, 91-2.

[40] Colour reproductions of the surviving paintings can be viewed in C. Sterling and C. Schaefer, *The Hours of Etienne Chevalier*, London, 1972, while very clear black and white ones are a feature of Wescher, *Fouquet*, pls. 1-27.

[41] Cf. Sterling and Schaefer, *Etienne Chevalier*, p. 18 and pl. 2.

In the Spring of 1450 Gouffier was granted lands from the estate of Agnès Sorel, probably as a wedding gift. His marriage to Louise d'Amboise was celebrated that year at Tours on 8 April.[42] The King appointed him Sénéschal of Saintonge on 20 June 1451. While the monarch was in the South-West of France that Summer news came of the capture of Bordeaux and the defeat of the English there. On 31 July at Taillebourg Castle (now Charente-Maritime), Gouffier, Dunois the Bâtard d'Orléans and Dammartin were ordered to arrest Jacques Coeur.[43]

In 1454 Gouffier's close friend, André de Villequier died; he was the monarch's *premier chambellan.* André's widow, Antoinette de Maignelais, became the royal mistress and Gouffier was appointed to the top position in the Royal Household. Fortune then turned her Wheel and cast down the High Chamberlain. Gouffier became the victim of intrigue was arrested in 1457 on a charge of practising sympathetic magic on his Lord and Master, signed a false confession and was stripped of his possessions in May 1458 by the *Haut Conseil.* He even suffered the ignominy of banishment to beyond thirty leagues from the king's person. Spencer reports that "during his period of banishment Gouffier is said to have gone to the service of Pierre duc de Bourbon where he would have known the young Pierre de Beaujeu and his bride, Anne de France, Louis XI's daughter."[44]

Perhaps the next most significant event in Gouffier's life was his return to royal favour *c.* 1465 with lands and income commensurate with his new station: to serve Charles de France, now aged about nineteen.[45] The courtier, being a familiar of the prince, would most certainly have seen the Book of Hours[46] that had been prepared at Mehun-sur-Yèvre for Charles, *c.* 1461-1465, and decorated by an *enlumineur*, possibly named Jean de Laval, but also containing one miniature in Fouquet's style.[47]

In 1472 Gouffier married for a second time, his bride being Philippe de Montmorency, widow of Charles de Melun. Within two years he suffered the loss of an old friend, Etienne Chevalier, who was interred in the church of Notre-Dame-de-Melun. Gouffier's pattern of court service did not change noticeably as he aged: in 1484 he acted on behalf of the new boy king, Charles VIII, then only thirteen, by heading a delegation to the Estates-General at Tours. Before his death in May 1495, Gouffier had become guardian to Charles VIII's own offspring, born in 1492, at least this is the import of part of the inscription on his tomb at Amboise.[48]

Spencer's views must now be examined about the date Gouffier acquired the Brussels Manuscript and his reasons for doing so. In her article on the *Horloge* she wrote "The arms of Gouffier in

[42] Her father and mother were Pierre d'Amboise and Anne-Marie de Bueil, cf. Anselme, VII, pp. 183-4. Pierre became chamberlain to Charles VII in 1460 and continued in the Royal Household when the Dauphin Louis became king the following year, cf. Galantino, *Conti del Forese*, part II, p. 143. One of Louise's brothers, Charles, was Sénéschal of Poitou in 1476; another, Pierre, was Bishop of Poitiers (1481-1505), cf. F. Villard, "Pierre d'Amboise, évêque de Poitiers," in *Mélanges offerts à René Crozet*, Poitiers, 1966, II, p. 1382.

[43] The chronicler Thomas Basin discusses the capture, imprisonment and subsequent escape of Coeur in his *Histoire de Charles VII*, see II, pp. 283-7. On the arrest, see also A. Castelot and A. Decaux, *Histoire de la France et des Français au jour le jour*, 8 vols, Paris, 1976-1977, III, pp. 172 and 185.

[44] Spencer, *Horloge*, p. 298.

[45] *Ibid.* The standard biography for Charles is by H. Stein, *Charles de France, frère de Louis XI*, Paris, 1921. Stein does not mention Gouffier.

[46] Now Paris, Bibliothèque Mazarine, MS. 473. On the history of this and its black velvet *chemise* lined with black satin, see Stein, *Charles de France*, pp. 11 and 507-8. For other codices owned by Charles, see L. Delisle, *Le Cabinet des Manuscrits de la Bibliothèque Impériale*, 2 vols, Paris, 1868-1874, I, pp. 84-6; and J. Plummer, *The Last Flowering. French Manuscript Painting 1420-1530*, New York, 1982, pp. 47-8.

[47] A *Betrayal* scene on fol. 13; cf. Perls, *Fouquet*, p. 35 and fig. 204; Wescher, *Fouquet*, p. 40. On the painters who worked in this manuscript, see a more penetrating analysis by Margaret B. Freeman, "The *Annunciation* from a Book of Hours for Charles de France," *Bulletin of the Metropolitan Museum of Art*, XIX (1960), pp. 105-118; C. Schaefer, "Deux enluminures du Maître de Jouvenel des Ursins à la Biblioteca Nacional à Lisbonne," *Arquivos do Centro Cultural Português*, VII (1973-1974), p. 138, n. 46; Sterling and Schaefer, *Etienne Chevalier*, pp. 13, 122, n. 16; Reynaud, *Fouquet*, pp. 56-8, 92.

[48] Spencer, *Horloge*, p. 298.

the *Horloge de Sapience*, if attributable to Guillaume, would be before his first marriage in May 1450."[49] In her second article where she established to her way of thinking that the manuscript's transcription was not completed before 1451, she did not renounce her earlier hypothesis; rather, she appears to withdraw from the problem altogether: "There is no trace in this book of either of his two marriages in 1450 to Louise d'Amboise, in 1472 to Philippe de Montmorency."[50] As for a possible explanation of his acquisition of such a richly illustrated codex in the first place, she advances no theory, but is content to observe: "There is no record of his concern with religious, intellectual or artistic matters."[51]

It is a fact that Gouffier was a courtier who moved in the inner circle of royalty's domestic and private life. There is little in his biography to suggest he was entrusted with important affairs of State, such as embassies to foreign powers or missions to allies or foes.[52] No document has yet been published, it would appear, that points to his being an ambitious collector of manuscripts, even Books of Hours.[53] One notes with interest, however, that a relative of his friend André de Villequier, who died, it will be recalled, in 1454, owned six years later her own copy of the French *Horloge de Sapience.* The manuscript, which does not contain any illuminations, is dated 1460; the owner was Thomine de Villequier, comtesse de Villars, of Saint-Aignan (Loir-et-Cher).[54]

Nothing is documented about Gouffier's religious contacts and friendships, apart, that is, from his meeting in the course of duty the legates, bishops and cardinals who waited on the king from time to time. The most pertinent question remains unanswered at the moment: why should a royal familiar become interested in or wish to own a copy of a Dominican's mystical work in French? Spencer had a relevant comment: "The fact that the second part of the book contains some of the works which Jean Gerson had recommended in 1417 for the education of a dauphin, the future Charles VII, is hardly sufficient a base for suggesting that by putting the book into Gouffier's hands someone was hoping to reach a young mind in his charge."[55] But the answer could also lie in another direction altogether: Gouffier did not commission the manuscript or buy it, he received it as a gift from a high-placed friend or confidant. In this context the most likely person to be the donor would be the monarch himself. He could afford to pay for the transcription and extensive illustration programme of the Brussels volume. The choice of the principal text could have been suggested to him by his Dominican confessor, Robert Baygard,[56] who had advised the king since 1448 on matters of conscience, morality and education. As for the motive of the presentation, it seems logical to assume it was to celebrate the wedding in 1450 and the king's increased trust in his worthy servant. Not only did Gouffier receive lands from Agnès Sorel's estate, as already mentioned, but a royal boon in the form of an illustrated manuscript.

Yet, apart from the manuscript, there is another Dominican connection in Gouffier's life. What was not noticed by Spencer is the fact that his third child, Louise, became a Dominican nun. Her birth-date is not on record. Assuming a birth to his wife each year after their marriage in 1450, the earliest birth-date for the infant Louise would be *c.* 1453. If we then allow eighteen years for her to reach maturity and make up her mind to become a novice[57] in the Order of Preachers, the

[49] *Ibid.*, p. 296, n. 48.

[50] Spencer, *Ciboule*, p. 108.

[51] Spencer, *Horloge*, p. 298.

[52] Cf. P. Bernus, "Le Rôle politique de Pierre de Brezé," *Bibliothèque de l'Ecole des Chartes*, LXIX (1911), p. 319.

[53] The Guillaume Gouffier, seigneur de Bonnivet, whose signature appears in Chantilly, Musée Condé, MS. 698 (483), an early sixteenth-century volume of geographical localities in Latin, is a relative of the courtier Guillaume Gouffier; on this ms., see Samaran and Marichal, *Catalogue*, I, Text vol., p. 393.

[54] Now Paris, Bibliothèque de l'Arsenal, MS. 2315. See Jeanne Ancelet-Hustache, "Quelques indications sur les manuscrits de l'*Horloge de Sapience*," in *Heinrich Seuse. Studien zum 600 Todestag (1366-1966)*, ed. F. M. Filthaut, Cologne, 1966, p. 166.

[55] Spencer, *Horloge*, pp. 298-99.

[56] I should like to thank Fr. Bernard Montagnes O.P. of the Convento S. Sabina (Aventino), Rome, for identifying the confessor.

[57] Cf. Mandonnet, *Order of Preachers*, p. 356: "Each convent received novices, these, according to the Constitutions, must be at least eighteen years of age, but this rule was not strictly observed."

earliest years for her entry into the Order would be *c.* 1472. Even her year of death in the Dominican Convent at Poissy is subject to contention, Anselme stating it occurred on 12 May 1487, while the *Dictionnaire historique et généalogique des familles du Poitou* gives the date as 12 May 1496. These bare facts, such as they are, and the many assumptions cause one to believe that Gouffier's *Horloge* volume could not have been acquired with Louise's future in view. The manuscript had been completed before *c.* 1453, if that year was indeed the one in which Louise was born.

The only soundly based observation that seems possible is that in the life-time of Gouffier's first wife, Louise d'Amboise, the Dominican influence in their family circle was sufficiently strong for their daughter Louise to decide to become a nun in the Order of Preachers.

Be that as it may, the arms that cover the Gouffier ones on fol. 13 of the Brussels *Horloge* are clearly the black and white ermine coat of the Dukes of Brittany. Nearby on the same folio are stylised *cordelières*, they were first granted to the dukedom in 1440. The notice of the manuscript published by the staff of the Bibliothèque Royale de Belgique in 1961 included the following form of identification: "... le duc de Bretagne qui acquit le livre après Gouffier doit être recherché entre 1454 et 1458. C'est soit Pierre II (1450-1457)... soit son oncle et successeur Arthur III (1457-1458), soit François II (1458-1488)."[58] No ideas, suggestions or hypotheses were advanced to explain why the volume passed to such a ducal family. Scholars are indebted to Spencer for a plausible theory in this regard: "At the court Guillaume Gouffier was closely associated with Arthur comte de Richemont, *connétable de France* for thirty-three years, as well as with the king's mignon, François, then comte d'Etampes.[59] On the basis of what we can discover about Guillaume Gouffier's interests, and assuming always that he was once the owner, let us propose a working hypothesis to the effect that when Gouffier was stripped of his possessions in 1456-57 (*sic*) he handed it to his friend Antoinette de Maignelais, who then presented it to François II de Bretagne when she joined him the next year."[60] One wonders if Antoinette is needed as a transmission agent, since Gouffier was demonstrably a close friend of François. One further observes that the possibility of a Dominican connection with a duc de Bretagne has not been addressed at all by earlier critics. Before looking at that problem, it may help to supply more background information about the Breton family alliances[61] and their relationships with Charles VII and his court, and later, with Louis XI and his entourage.

François I, born 1414, married firstly Yolande d'Anjou (1412-1440) and secondly, Isabella Stuart (*c.* 1427-*c.* 1494).[62] From the second union he had two daughters, Marie and Marguerite. There being no surviving male heirs, the dukedom passed in 1450 to his brother, Pierre de Montfort, born 1418, who took the title Pierre II.[63] His wife was Françoise d'Amboise (1427-1485);[64] none of the sons she bore him survived, so on his death in 1457 the succession passed to his uncle Arthur. His rule in turn was brief, he died at the family seat in Nantes on Christmas Day 1458.[65]

[58] *Bulletin de la Bibliothèque Royale de Belgique*, V (1961), p. 57.

[59] This François d'Etampes will become the last Duke of Brittany in 1458, taking the style François II.

[60] Spencer, *Horloge*, p. 299.

[61] A genealogical table of the House of Brittany in the fifteenth century is readily accessible in Harthan, *Book of Hours*, p. 190. Portraits of the ducal families I mention are all represented in a Missal commissioned by Pierre II for the Carmelites at Nantes, now Princeton, Princeton University Library, MS. Garrett 40; see Plummer, *Last Flowering*, pp. 23-4.

[62] A Book of Hours bearing her Scottish arms impaling those of Brittany is now Cambridge, Fitzwilliam Museum, MS. Fitzwilliam 62. Its five hundred or so pictures are the product of the Rohan Master's workshop for an earlier owner *c.* 1417-1418 who was perhaps her predecessor in the ducal marriage-bed, Yolande d'Anjou. See Harthan, *Book of Hours*, pp. 114-17.

[63] Pierre II owned a sumptuous Book of Hours, richly decorated with some fifty paintings and vignettes, which was completed at Nantes *c.* 1455-1457. It is now in Paris, Bibliothèque Nationale, MS. latin 1159; see Harthan, *Book of Hours*, pp. 118-21.

[64] Cf. Harthan, *Book of Hours*, p. 190.

[65] An 'official' chronicle of Arthur's life and times was composed in Breton circles in the third quarter of the fifteenth century, see *Chronique d'Arthur de Richemont par Guillaume Gruel*, ed. A. Le Vavasseur, Paris, 1890. For a lost portrait of Arthur by Fouquet, known from a sixteenth-

The heir presumptive, François d'Etampes, born 1435, became the new Duke of Brittany as François II. His wife was a cousin, Marguerite de Bretagne, already mentioned above. Their formal union lasted until her death in 1469, although he took a mistress in *c.* 1458 in the person of Antoinette de Maignelais, widow of the king's late premier chambellan, André de Villequier, and formerly the royal mistress after Agnès Sorel. Two years after Marguerite's demise, François married again, this time for political reasons: to strengthen the alliance of the princes of the West and South-West of France against Louis XI. The new bride was Marguerite de Foix,[66] daughter of Gaston IV, comte de Foix. Once again no male issue survived the father in 1488, but a daughter did, by name Anne. Through her own two marriages to two French kings, Charles VIII and Louis XII, she united Brittany and the French realm.

To explain how a Gouffier manuscript passed into the hands of a duc de Bretagne, one must examine alliances and friendships that could have facilitated the transmission. Given Gouffier's marriage to Louise d'Amboise and the marriage of Françoise d'Ambroise to Pierre II de Bretagne, a family alliance existed between a high-placed courtier and a peer of the realm. Equally important, however, could have been the friendship between Gouffier as valet de chambre and later chambellan, and François d'Etampes when mignon and familiar of Charles VII.

Since François lived to 1488 and Gouffier to 1495, the former could have come into ownership of the manuscript any time before 1488. Spencer argued for a date close to that of his inheritance of the dukedom, that is 1458. There would seem to be valid reasons why Gouffier would not have held the volume until late in François's life, the most important being his own disgrace and absence from the royal court until 1465, and secondly, François's pursuit from about that time onwards of a hard line of antagonism towards Louis XI on both the political and military fronts.

As for the disgrace of Gouffier, it began in 1457, as was stated above, and ended in banishment in May 1458. François was still comte d'Etampes. Was the Brussels manuscript one of the possessions Gouffier was obliged to part with to comply with the Haut Conseil's judgment? If it were, the Crown may well have passed it to François d'Etampes or the Breton peer may have requested it as a royal boon.

Alternatively, when Gouffier was arrested in 1457 he could have offered the volume to his friend François; he would surely not have needed to use their mutual acquaintance, Antoinette de Maignelais, as an intermediary, as Spencer suggested. The gift would have preceded by many months the decision of the Haut Conseil to strip Gouffier of his possessions.

These theories suppose that François or a duc de Bretagne wanted to own the manuscript, for reasons other than pure avarice. It is not known if François was a book collector in his own right; no inventory of a library appears to have survived. Did he covet the volume because of its art work, or was he a Philistine in matters of taste and art appreciation? There seems no satisfactory answer to these questions, since factual evidence is lacking on which to base any cogent theories. But there is still one unexplored factor in this history of patronage—the Dominican connection. A link between a duc de Bretagne and the text of a Preaching Friar in the Brussels manuscript has not been established by earlier scholars, yet, it seems that one may have existed.

The great Dominican preacher Vincent Ferrer died at Vannes in the heart of Brittany in 1419. Born in Valencia in 1350, he had studied at the Universities of Barcelona and Toulouse. His entry

century drawing, see Reynaud, *Fouquet*, pp. 35-6, 91. Further information about his patronage of the arts is supplied by C. Schaefer, "Lemaire de Belges, Fouquet et maître Paoul Goybault: la peinture murale du Jugement Dernier de la Sainte-Chapelle de Châteaudun," *Bulletin de la Société Nationale des Antiquaires de France*, 1985, pp. 249-64.

[66] A splendid Book of Hours was prepared for her in the 1470's with twelve large miniatures in addition to the conventional Calendar illustrations. The manuscript is now in London, Victoria and Albert Museum, MS. Salting 1222. The painter is unidentified, but the style is Parisian and he may well have been a follower of Jean Fouquet. Cf. Harthan, *Book of Hours*, pp. 122-5.

into the Order of Preaching Friars occurred when he was only seventeen. In 1379 he joined the entourage of Cardinal Pedro de Luna who was elected antipope in 1394 and assumed the title of Benedict XIII. The latter appointed Vincent his confessor and made him an apostolic plenipotentiary. However, a period of sickness in 1399, during which Vincent beheld a vision of Christ with Saints Dominic and Francis, changed the course of his life from a papal diplomat to a practising preacher. Marguerite de Savoie and Bernardino da Siena were among his converts during the extensive travels he made in the next two decades through Germany, Italy, Spain and France.

Throughout Europe he became acknowledged as a divine orator.[67] His oratory and charismatic personality were capable of swaying vast throngs, and at gatherings miracles occurred in his presence. He delivered his popular sermons, usually in Catalan,[68] but seems to have had no difficulty in making himself understood wherever he was, be it Germany, Provence, Italy or Brittany. Jean V de Bretagne, duke from 1399 to 1442, summoned Vincent to Brittany in 1417 and it was while in Vannes in 1419 that he died. For the next thirty-five years, Jean V and his successors, François I and Pierre II, made every effort to have the Dominican canonised.[69] Their initiative was recognised when Calixtus III solemnly canonised Vincent Ferrer on 29 June 1455. On 4 June the following year Pierre II and his family organised a ceremony in Vannes during which the *dépouille mortelle* of the Dominican was elevated and translated to a fresh reliquary in the Cathedral of Saint-Pierre. It was then sealed with three locks whose keys were entrusted to the Bishop of Vannes, to the Papal Legate and to the Duke himself.

The duke who succeeded Pierre II, François II did not neglect or eschew Dominican connections either. His alliances and friendships in religious circles reach, moreover, into the private life of Charles de France in the period that Gouffier was in Charles' household, from *c.* 1465 to *c.* 1472. Shortly after succeeding to the duchy in 1458 François II sent a person he trusted implicitly as his envoy to the Papal Curia in Rome. His name was Roland Le Gozic, a Dominican and former prior of Saint-Martin at Lamballe.[70] About 1466 after five years of the reign of the Spider King Louis XI, François II succeeded in winning to his side the French heir apparent, Charles de France.

Charles was in fact a political refugee from the harassments and machinations of his elder brother, the French king. Vannes welcomed Charles and although he did return to the Kingdom of France for cautious raids or incognito visits, he remained until his death in 1472 a staunch ally of François II.[71] In the year of his move to Vannes, that is 1466, his accounts record a payment of 22 livres and 10 sous to a Dominican scribe, one Dominique de Millan, for translating from Latin into French a text called *Kakalien.*[72] At some period between 1466 and his death Charles appointed Roland Le Gozic as his personal confessor. One can only surmise it was at François II's suggestion. Charles' mistress in his closing years was Colette de Chambes, dame de Montsoreau and wife of Louis d'Amboise, vicomte de Thouars. Louis was a relative of Louise d'Amboise, wife of Guillaume Gouffier. Louis died in 1470 and the following year Colette contracted a pernicious and wasting sickness, thought by modern historians to be syphilis. At the time she dictated her last will and testament, Colette had as a witness her lover's confessor Le Gozic.[73]

[67] Cf. P. Fages, *Histoire de Saint Vincent Ferrier, apôtre de l'Europe*, Louvain and Paris, 2 vols., 1901 and 1904.

[68] See J. Sanchis Sivera, *Sant Vicent Ferrer, Sermons*, Barcelona, 2 vols., 1932 and 1934.

[69] P. Fages, *Procès de la Canonisation de saint Vincent Ferrier*, Louvain and Paris, 1904.

[70] Cf. Stein, *Charles de France*, p. 351, n. 1. Apart from his role as ducal envoy and later as confessor to Charles de France, Le Gozic was well known in French ecclesiastical circles. I am very grateful to Fr. B. Montagnes O.P. of the Convento S. Sabina (Aventino) in Rome for supplementing my knowledge of this Dominican. Le Gozic was named Inquisitor General of France in 1453 and still held the title in 1473. He was *difinitor* of the Province of France in 1459 and the Provincial of France in 1462, a post he resigned from in 1464.

[71] On the matter of political alliances, see Stein, *Charles de France*, pp. 184-276.

[72] Cf. Delisle, *Cabinet des manuscrits*, I, p. 84, n. 6.

[73] Cf. Stein, *Charles de France*, pp. 432-5.

When Charles de France died in 1472 from the same disease, he named the same Dominican as his testamentary executor.[74] The ties with the Preaching Order continued in a direct line of this royal branch into the reign of François I (1515-1547). Charles had two daughters by Colette, one of whom was known as Jeanne de Guyenne, her father's title at his death was duc de Guyenne. Jeanne entered the Dominican Order and later rose to become sub-prioress of the Convents of Blaye and Saint-Pardoux-la-Rivière in the Périgord. She was still alive in the reign of François I and regularly received sums from the royal purse.[75]

This may be the appropriate moment to attempt to draw together the threads of evidence left trailing on previous pages, bearing in mind the need to explain the transmission of the Brussels manuscript of a Dominican text from a courtier to a duke, perhaps occurring in a period (1455-1458) when the courtier is disgraced and three dukes in rapid succession could become the owners.

Given that the Breton dukes' interest in Vincent Ferrer had culminated in public witness at Vannes in 1456, it is possible that, if offered a manuscript depicting a Dominican preaching[76] in its first illustration, and containing a text by a celebrated Dominican preacher of the previous century, a duc de Bretagne would be pleased to acquire the volume. This pleasure would surely have been heightened if the offer had come through a wife who was related by marriage to the wife of the book's owner, Guillaume Gouffier. If Pierre II did acquire it, he could not have held it for long, since he died in September 1457. Receiving the volume as an inheritance, Arthur III would not have had very long to appreciate its illuminations. Within a year he was dead and François II would have received the volume from his estate.

Another line of reasoning assumes that Pierre II would have taken no initiative to acquire the volume before his death and that the sexagenarian Arthur III would have seen no cause to acquire it from a disgraced courtier, even if the latter had been a friend. François II, however, could be seen as more involved in the issues than his relatives. He had three good grounds for acquiring the manuscript: a long standing personal friendship with Gouffier reaching back into each's boyhood at court; a deeply felt sympathy for Gouffier's predicament, especially as the confession to the charges was written under duress; the Dominican connections his Breton family enjoyed historically and currently at the time. Whether this acquisition took the form of a purchase from Gouffier or of a royal favour on the forfeiture of the courtier's possessions, must remain a matter for conjecture.

The next owner of the Brussels manuscript after a duc de Bretagne cannot be identified. Assuming François II retained it until his death in 1488, then there follows a period of about one hundred and thirty years before another owner marks his possession.

5. HISTORY SINCE THE MIDDLE AGES

No document has come to light yet that supplies information on the owners and location of the volume between 1488 and 1618, the year in which a member of the influential Croy family placed on fol. 2v. his escutcheon *d'argent à trois fasces de gueules*, surmounted by a ducal coronet and surrounded by the Collar of the Order of the Golden Fleece.[77] A representation of two hands in

[74] *Ibid.*, p. 457. The testament was published in B. de Mandrot, *Le Journal de Jean de Roye*, II, pp. 281-7. Le Gozic's name is on p. 284.

[75] Stein, *Charles de France*, pp. 432-33.

[76] One of the component elements in the first illumination on fol. 13.

[77] The arms are merely announced as those of a Croy family member in the manuscript notice published in the *Bulletin de la Bibliothèque Royale*, 5 (1961), pp. 48-58.

the act of clasping is placed over the arms, and in a banderole one reads "W. J'AIME QUI M'AIME. CROY". Beneath the pendant jewel of the collar is a heart bearing the letter "G", transpierced by crossed arrows and surmounted by an imitation ducal coronet. Lower still is another stylised banderole reading "16. JE SOUSTIENDRAY CROY ET J'AIME QUI M'AIME. 18."[78] Spencer argued that the owner was Charles-Alexandre de Croy (1581-1624) who became Duc de Croy in 1613 and was appointed a member of the Order of the Golden Fleece in 1617.[79] This identification finds corroboration from another source. Five manuscripts bearing the same Golden Fleece insignia and collar around the same coat of arms with an identical motto and year 1618 were identified in 1955 by Marguerite Pecquer as belonging to the same nobleman, Charles-Alexandre de Croy.[80] However that may be, the *Horloge* volume is not included in the repertoire of codices owned by the Croy family which Bayot published in 1928.[81]

The manuscript remained in the Croy family until after World War II, when it was sold to a Belgian dealer in rare books, M. F. Miette. While in his hands it was seen by a collector of autographs and paintings, Dr Dujardin, but he did not acquire it immediately. A certain M. Simonson purchased the volume from M. Miette and then sold it to Dr Dujardin. These transactions took place before 1952.[82] By 1955 the manuscript was in the collection of the widow Dujardin, for in that year it appeared in an exhibition in Madrid with her ownership acknowledged.[83] In 1961 it was purchased by the Bibliothèque Royale in Brussels and the acquisition was announced in the Library's *Bulletin*.[84]

Scholars in many fields have had occasion to discuss the manuscript, its contents or its art work. The first was an international authority on horology, H. Michel, who published two papers on the time-pieces in the *Clock Chamber* miniature.[85] The distinguished art historian Eleanor P. Spencer was the first scholar to study the miniatures and their styles and themes.[86] The theme of Temperantia, the Virtue whose symbol is a clock, was discussed by Lynn White, Jr. in a paper published in 1969; he included reference to the image of Sapientia, the regulator of the clock in the *Clock Chamber*.[87] In the same year Philippe Verdier reviewed the painting of the *Seven Liberal Arts* and reproduced it in the course of his examination of the principal theme.[88] The volume has, naturally enough, been seen to constitute an important monument in the history of book illustration.[89] Musicologists also paid homage to the manuscript on account of the musical instruments depicted

[78] Cf. also *Bruxelles. Bibliothèque Royale Albert Ier. Quinze années d'acquisitions*, Brussels, 1969, pp. 84-6.

[79] Spencer, *Horloge*, p. 299.

[80] Marguerite Pecqueur, "Répertoire des manuscrits de la Bibliothèque de l'Arsenal peints aux armes de leur premier possesseur (XIIIe-XVIe siècles)," *Bulletin de l'Institut de Recherche et d'Histoire des Textes*, IV (1955), p. 123.

[81] Cf. A. Bayot, *Martin le Franc. L'Estrif de Fortune. Etude du manuscrit 9510 de la Bibliothèque Royale de Belgique*, Paris, 1928, Appendix, pp. 52-6.

[82] I paraphrase here the comments on the manuscript's post-1618 history conveyed to me in a letter dated 16 December 1982 by Mme Claudine Lemaire of the Bibliothèque Royale, Brussels.

[83] *L'Héritage de Bourgogne dans l'art international. Exposition de manuscrits à peintures*, Madrid, 1955, p. 38, no. 52. The miniature *Sapientia in Majesty* is reproduced on p. 39 and given the erroneous title *Vierge en gloire*.

[84] Cf. *Bulletin de la Bibliothèque Royale*, 5 (1961), pp. 48-58. It was later referenced in *Bruxelles, Bibliothèque Royale Albert Ier. Quinze années d'acquisitions*, Brussels, 1969, p. 84-6; and in *La Librairie de Bourgogne et quelques acquisitions récentes de la Bibliothèque Royale Albert Ier—cinquante miniatures*, Brussels, [1970], pls. 12-15 in colour (fols. 13v, 17r, 23r, 39r).

[85] H. Michel, "L'*Horloge de Sapience* et l'histoire de l'horlogerie," *Physis*, II (1960), pp. 291-8; and "Some New Documents in the History of Horology," *Antiquarian Horology*, III (1962), pp. 288-91.

[86] Cf. Spencer, *Horloge* and *Ciboule* articles, details of which are supplied in notes 1 and 2, above.

[87] Cf. L. White, Jr., "The Iconography of Temperantia and the Virtuousness of Technology," in *Action and Conviction in Early Modern Europe: Essays in Memory of E. H. Harbison*, Princeton, 1969, pp. 197-219.

[88] P. Verdier, "L'Iconographie des arts libéraux dans l'art du moyen âge jusqu'à la fin du quinzième siècle," in *Arts Libéraux et Philosophie au moyen âge*, Montréal, 1969, pp. 332-4 and pl. III.

[89] Cf. *Le Livre illustré en Occident du haut moyen âge à nos jours*, Brussels, 1977, p. 57, no. 44, pl. 36.

in its illustrations.[90] In 1985 an article appeared on affinities between the style of one of the artists and that of a miniaturist in the entourage of Philippe le Bon.[91] Another paper published in *Scriptorium* in 1986 raised questions about the methods of reading visual statements in the Brussels volume.[92] The art work of the whole manuscript was the subject of a doctoral dissertation in 1986.[93] The most recent publication concerns iconographic programmes found in pre-1450 copies of the *Horloge de Sapience.*[94]

[90] Cf. *L'Iconographie musicale dans les manuscrits de la Bibliothèque Royale Albert Ier*, ed. Isabelle Huttois, Brussels, 1982, pp. 120-3, no. 199, figs. 22 and 23.

[91] Cf. P. R. Monks, "The Influence of the Master of Jean Rolin II on a Master of the Vienna *Girart de Roussillon*," *Codices Manuscripti*, XI (1985), pp. 1-8.

[92] P. R. Monks, "Reading Fifteenth-Century Miniatures: the Experience of the *Horloge de Sapience* in Brussels, Bibliothèque Royale, MS. IV. 111," *Scriptorium*, XL (1986), pp. 242-8, pls. 13-16.

[93] P. R. Monks, *A Study of the Art Work of the Rolin Master in the Horloge de Sapience, Brussels, Bibliothèque Royale, MS. IV. 111, and of his other known surviving Works*, Melbourne, 1986.

[94] P. R. Monks, "Pictorial Programmes in Manuscripts of the French Version of Suso's *Horologium Sapientiae*," *Archivum Fratrum Praedicatorum*, LVII (1987), pp. 31-43, 2 pl.

CHAPTER TWO

The Artists

1. GENERAL

When originally completed, the art work comprised thirty-nine miniatures. The first thirty-six, by an artist we shall call the Rolin Master, illustrated the *Horloge de Sapience* and the remaining three, attributed to the Chief Associate of the Bedford Master,[1] accompanied texts by Gerson. The outstanding quality of the art work was first recognised by the authors of the notice of the volume, which appeared in the *Bulletin de la Bibliothèque Royale*;[2] they employed such dignified epithets as ''monument de l'enluminure française,'' ''mine de thèmes iconographiques'', and ''richesse de l'iconographie,'' while the major artist was thought to possess a ''virtuosité inégalée.'' Spencer made no reference to this encomium when she published her own seminal study,[3] however, in her turn, she judged the paintings and their creators as representative of some of the best that French ateliers produced in the middle of the fifteenth century.

The miniatures by the Rolin Master are simple monoscenes, pairs of monoscenes or diptychs, triptychs, and polyptychs. On the average, the full-page compositions are 260 × 170 mm.; fifteen of the other sizes measure 170 × 100, five are 170 × 150 and six are 95 × 75 mm.

Although there were originally thirty-six separate paintings by the Rolin Master in the Brussels *Horloge*, the reader will find in the discussion in chapter V, that mention is made of only thirty-two. What has happened to the other four? A religious wrote an account of all the paintings shortly after their completion and called it the *Déclaration des Hystoires*. On four occasions two illustrations are considered as one by the expositor. The rubricator indicates the relationship by *de* 'belonging to', 'part of'. For example, the diptych *Temptations of the Flesh* and *Repentant in his Cell*, on fol. 18, is numbered *la cinquième hystoire* and the *Epicureans* on fol. 18v. is characterised as *de la v*e *hystoire*. The other binary or two-part illustrations can be found on fols. 16v. and 17, 46v. and 47, 72 and 72v. The use of A and B in our numbers for the plates and the *Déclaration* will help to remind the viewer that both illustrations need to be studied as elements of one pictorial statement.

So much for the thirty-six paintings by the Rolin Master which will be examined as thirty-two integrated illustrations. Of these thirty-two, two were removed before the volume was acquired by

[1] Spencer, *Horloge*, p. 277 and *Ciboule*, p. 107.

[2] See vol. V (1961), pp. 48-58.

[3] Spencer, *Horloge*, pp. 277-99 and *Ciboule*, pp. 104-8.

the Bibliothèque Royale, Brussels, but after they had all been seen by the author of the *Déclaration*. The two were triptychs and positioned on the lost fol. 57,[4] probably one group of the recto side and the other on the verso.

The spatial area in which the Rolin Master has created his painting is confined in each case within thin, black, white and dark red lines, between which are polished-gold fill-ins. The artist's figures and compositional elements occasionally extend over these lines. One may deduce, therefore, that the frames were in place before he began his task. When a picture is positioned at the top of a column of text or of a folio, its top edge is customarily arched and serrated.

2. THE MASTER OF JEAN ROLIN II

The name of the Master of Jean Rolin II was first suggested by Eleanor P. Spencer in 1963. Writing about the workshop of Maître François, she commented: "In distinguishing the several hands within this prolific Paris atelier, I had already chosen for this painter the arbitrary and temporary title of the Master of Jean Rolin II, on the basis of important commissions executed for that patron. Now in this *Horloge de Sapience* we have a new and much more significant criterion for his work."[5]

Spencer repeated this name for the painter in her second paper on the *Horloge* that appeared in 1965.[6] Nevertheless, in a still later article published in 1974 she referred to the *Horloge* painter as the Master of the Brussels *Horloge de Sapience*, "an awkward title ...convenient to use only until his legal name is known."[7]

Scholars who have subsequently had occasion to mention the illuminator have taken their cue from Spencer's articles and have used either or both of the two appellations. The Master of Jean Rolin II is preferred by C. A. Chavannes-Mazel in 1980,[8] by J. Plummer in 1982,[9] and by R. S. Wieck in 1983.[10] Margaret Manion and Vera Vines in 1984 retained both titles.[11] In 1985 James Marrow chose to refer to the painter as the Master of Jean Rolin II.[12]

Jean Rolin II was a Cardinal who was a member of a very influential Burgundian family in the fifteenth century. Patronage of the arts by family members began with the Cardinal's father, Nicolas Rolin.

Nicolas (*c.* 1380-1461)[13] was appointed Chancellor of Burgundy by Philippe le Bon in 1422 and held the post for more than thirty years.[14] Although continually pressed by the burdens of State,

[4] Spencer, *Horloge*, p. 277, n. 1 and p. 281, n. 16.

[5] Spencer, *Horloge*, pp. 292-3.

[6] Spencer, *Ciboule*, p. 104.

[7] Spencer, "Dom Louis de Busco's Psalter," in *Gatherings in Honor of Dorothy E. Miner*, ed. Ursula E. McCracken, Lilian M. C. Randall and R. H. Randall, Jr., Baltimore, 1974, p. 237.

[8] In *Schatten van de Koninklijke Bibliotheek*, The Hague, 1980, p. 89.

[9] *The Last Flowering. French Painting in Manuscripts 1420-1530*, New York, 1982, p. 62.

[10] R. S. Wieck, *Late Medieval and Renaissance Illuminated Manuscripts*, Cambridge, Mass., 1983, p. 28.

[11] Margaret M. Manion and Vera F. Vines, *Medieval and Renaissance Illuminated Manuscripts in Australian Collections*, London, 1984, pp. 186-7. In their discussion of the miniaturist's work, one observes that the second name is employed more often than the first one.

[12] Cf. James H. Marrow, "Miniatures inédites de Jean Fouquet: les Heures de Simon de Varie," *Revue de l'Art*, LXVII (1985), pp. 7-32.

[13] The principal biographies and biographical sketches are those of: Ch. Bigarne, *Etude historique sur le chancelier Rolin et sur sa famille*, Beaune and Dijon, 1860; A. Perier, *Un Chancelier au XVe siècle: Nicolas Rolin 1380-1461*, Paris, 1904; G. Valat, "Nicolas Rolin, Chancelier de Bourgogne," *Mémoires de la Société Eduenne*, n.s. XL (1912). pp. 73-145; XLI (1913), pp. 1-73; XLII (1914), pp. 52-148; H. Pirenne, *Histoire de Belgique*, Brussels, 1900-1932, 7 vols., II, 4th ed., pp. 395-7; J. Bartier, *Légistes et gens de finances au XVe siècle*, Brussels, 1952, pp. 54, 433-4; P. Cockshaw, *Les Secrétaires de la Chancellerie de Flandre-Bourgogne sous Philippe le Bon*, Brussels, 1964, *passim* (I am indebted to M. Pierre Cockshaw for these last three references); R. Berger, *Nikolas Rolin, Kanzler der Zeitenwender im burgundisch-französischen Konflikt 1422-1461*, Fribourg, 1971.

[14] An account of this hegemony can be read in R. Vaughan, *Valois Burgundy*, London, 1975.

Rolin found time to exercise beneficial and often munificent patronage as well as to display initiatives in many areas of learning and the arts, both plastic and monumental.[15] For example, he founded a university at Dole in 1423. He made generous benefactions to his own parish church, Notre-Dame-d'Autun, and caused it to be raised to the status of a Collegiate Church in 1450. Many of its service-books, as will be seen, were donated by him. In Dijon likewise, capital of the dukedom,[16] he renovated and augmented the size of his residence, making it a fitting mansion for the duke's principal administrator.

Flemish builders and architects were summoned by Rolin to Beaune to design and construct a new hospice for the poor of the town. The completed edifice is not only one of the finest civic buildings of the Northern Renaissance, it is a monument to the Rolin family whose name, portraits, emblems, mottoes and escutcheons are everywhere. Thirty-one tapestry hangings were ordered by Nicolas *c.* 1450 for use as screens in the hospice. One of them portrays St. Anthony against a background or a *semé* composed of turtle-doves on branches, single stars, the word SEULLE and the monogram NG repeated. There are also representations of the arms of Nicolas and his second wife, Guigone de Salins.[17] In the hospice chapel, serving as a retable for the high altar was one of Rogier van der Weyden's finest paintings, the *Last Judgment*, commissioned *c.* 1442 by the Chancellor and completed between 1446 and 1449.[18]

Before that particular commission was initiated, however, Nicolas' features had been immortalised by another Flemish master, the great Jan van Eyck. In the *Rolin Madonna*, also known as the *Rolin Virgin*, the *Virgin of Autun* or the *Virgin with Chancellor Rolin*, completed *c.* 1432, Nicolas is seen kneeling at a prie-dieu on the left, facing towards the Virgin and Child.[19] The painting was commissioned specifically for his parish church at Autun by a member of the Rolin family, thought for a long time to have been Nicolas, but recent scholarship sees his son, Cardinal Jean Rolin II, as the real benefactor.[20]

Art historians do not seem to have appreciated that Nicolas commissioned liturgical books from local scribes in the town of Autun.[21] At his death in 1461 an inventory was made of the manuscripts he and his wife had so generously presented to the Collegiate Church in Autun. One finds listed five Missals, seventeen Antiphonals, seven Graduals, four Breviaries, eleven Psalters, one Gospel-Book, two books of Epistles, one Passionary, one Legendary in three volumes, two Bibles in five

[15] For what follows, see Perier, *Nicolas Rolin*, pp. 353-73.

[16] For life in Dijon in the fifteenth century, see W. R. Tyler, *Dijon and the Valois Dukes*, Norman, 1971.

[17] Cf. W. D. Wixom, *Treasures from Medieval France*, Cleveland, 1967, pp. 298-99.

[18] Reproduced in W. Schöne, *Die Grossen Meister der niederländischen Malerei des 15 Jahrhunderts*, Leipzig, 1939, p. 71; M. Davies, *Rogier van der Weyden*, London, 1972, pl. 40 (centre panel). The work is also discussed in detail in Nicole Veronée-Verhaegen, *L'Hôtel-Dieu de Beaune*, Brussels, 1973. I thank M. Pierre Cockshaw for this last reference.

[19] Elisabeth Dhanens, *Hubert and Jan van Eyck*, New York, 1980, p. 267. For a discussion of the portraits of Nicolas by painters of miniatures, see J. Desneux, "Nicolas Rolin, authentique donateur de la *Vierge* d'Autun," *Revue des Arts*, IV (1954), pp. 195-200.

[20] From among the more recent articles and monographs I have read on the subject of this panel, I select the following for mention: J. Lejeune, *Les van Eyck, peintres de Liège et de sa cathédrale*, Liège, 1956; J. Snyder, "Jan van Eyck and the *Madonna* of the Chancellor Nicolas Rolin," *Oud Holland*, LXXXII (1967), pp. 163-71; M. Felheim and F. W. Brownlow, "Jan van Eyck's *Chancellor Rolin and the Blessed Virgin*," *Art Journal*, XXVIII (1968), pp. 22-6, 58; H. Roosen-Runge, *Die Rolin-Madonna des Jan van Eyck*, Wiesbaden, 1972; Anne H. van Buren, "The Canonical Office in Renaissance Painting. Part II. More about the *Rolin Madonna*," *Art Bulletin*, LX (1978), pp. 617-33; Dhanens, *Van Eyck*, pp. 266-81; Purtle, *Marian Paintings*, pp. 59-97; Molly T. Smith, "On the Donor of the Van Eyck *Rolin Madonna*," *Gesta*, XX (1981), pp. 273-9, argued that "Jean Rolin, on leaving Autun for Chalon in 1431, commissioned this painting for his father's chapel in Notre-Dame as a permanent visual intercession for his father's soul, thus inspiring Van Eyck to create a rich, unique and very personal composition."

[21] A. de Charmasse, "Note sur l'inventaire des livres liturgiques donnés à l'église collégiale de Notre-Dame d'Autun par Nicolas Rolin, chancelier de Bourgogne," *Mémoires de la Société Eduenne*, n.s. XXXIII (1905), pp. 285-304; Marie-Josette Perrat, *Autun, Bibliothèque Municipale. Le Livre au siècle des Rolin*, Autun, 1985, pp. 8-10.

volumes, fifteen Processionals and two Prayer-Books. Of all these, five are described as "enluminé d'or et d'azur", but the words "image", "hystoire" or "hystorié" do not occur anywhere. As far as can be ascertained, no document published to date allows the modern historian of miniatures to link the name of Nicolas Rolin with any specific *enlumineur.* His motto was *Deum Time* and his arms were *d'azur à trois clefs d'or posées en pal 2 et 1.*[22] A manuscript that contains a representation of the Rolin family's escutcheon or motto is a Book of Hours in Madrid, in which the Rolin Master's hand is not represented.[23]

The great Chancellor had three sons from his first wife, Marie des Landes: Guillaume, Antoine and Jean, known also as Jean Rolin II.[24] Spencer devotes three lines to the third son without reference to any publication about his life. Her statement reads: "Jean Rolin II, third son of the chancellor, Nicolas Rolin, was appointed cardinal in 1449, died in 1483. Although he made great gifts to his cathedral in Autun, he had a house in Paris and was interested in the Sorbonne press established in 1471 as well as in the library of the University."[25] More needs to be said about his life, if only because of his patronage of the arts and of illuminators of manuscripts.

At his baptism in 1408 was Jean's god-father, Jean sans Peur.[26] After early studies which included Civil and Canon Law, he was launched on a career in the Church. He served two years as Bishop of Chalon before being elected in 1436 as pastoral leader of the See of Autun under the name of Bishop Jean Rolin II. Two years later he accompanied his father and Philippe le Bon to Bourges to participate in the preparation of Charles VII's Pragmatic Sanction.[27] He acted as Administrator of the important diocese of Lyons in 1443 during the minority of Charles de Bourbon, the Archbishop designate. The pallium, normally reserved for metropolitans,[28] was bestowed on him by Eugenius IV (1431-1447). At the age of 41 he was created a Cardinal and in 1452 he made a personal visit to Rome to meet Nicholas V (1447-1455). In 1473 Jean was at the right hand of Charles le Téméraire from the very moment the duke entered the gates of Dijon for the first time to the end of the celebrations. The Cardinal's good works for his diocese, its churches and towns, hardly abated throughout his life-time. He loved Beaune as much as he loved Autun and planned to be buried in its Collegiate Church of Notre-Dame. In 1481 he was greatly chagrined when the Pope removed the church from his control. Jean had clearly fallen foul of Louis XI's vengeful acts against Burgundy.

The Cardinal's patronage included panel-painters. He may have been acquainted with at least two, and possibly three, if credence is given to the theory mentioned above that he commissioned the *Rolin Madonna* from Jan van Eyck. During his numerous visits to Beaune Jean had every opportunity to admire in the 'family' hospice his father's commission, the *Last Judgment* by Rogier van der Weyden. And at Autun was hanging a splendid portrait of Jean executed *c.* 1480 by the Maître de Moulins.[29] The work is in fact a *Nativity* in which the artist places the Virgin, Joseph and the Holy Child in the centre and left foreground, and on the right, kneeling devoutly in adora-

[22] See Perier, *Nicolas Rolin*, p. 27.

[23] Madrid, Biblioteca Nacional, MS. Res. 149, cf. J. Dominguez Bordona, *Manuscritos con pinturas*, Madrid, 1933, I, pp. 324-7, no. 834, pls. 272-3. Another codex with the same arms in The Hague, Koninklijke Bibliotheek, MS. 76.E.8, a Psalter, displayed in the 1983 Exhibition in the Rijksmuseum Meermanno-Westreenianum, cf. A. S. Korteweg (ed.), *Liturgische Handschriften uit de Koninklijke Bibliotheek*, The Hague, 1983, front cover and p. 14, no. 36.

[24] On the close family likeness of Jean Rolin II to his father Nicolas, see Desneux, *Authentique donateur*, pp. 195-200.

[25] Spencer, *Horloge*, p. 295, n. 43.

[26] I rely at this point on the biographic sketch drawn by Bigarne, *Etude historique*, pp. 29-41.

[27] Cf. N. Valois, *Histoire de la Pragmatique Sanction de Bourges sous Charles VII*, Paris, 1906.

[28] For a description of the medieval pallium, see Joan Evans, *Dress in Medieval France*, Oxford, 1952, p. 74, pl. 79.

[29] Now in the Musée Rolin at Autun. I should like to thank the Director of the Museum, M. Matthieu Pinette, for allowing me to examine the painting.

tion is the Cardinal, bare-headed, wearing heavy scarlet robes with sculptured folds, with his distinctive hat and tassels suspended behind him from a nail above his escutcheon. The family motto *Deum Time* can be read in the upper right and left-hand corners of the panel.[30]

Jean died after a painful illness at Cravant (arr. d'Auxerre) at the end of June, 1483 and was laid to rest in his own cathedral in Autun. Four librarians from the Sorbonne visited his Paris residence for the purpose of assessing his estate. The inventory they took has fortunately been preserved and it makes interesting reading.[31] Setting aside manuscripts that belonged to a friend, Jean Monnet "chanoine de l'église de Paris", one finds that the Cardinal owned three Missals, one Psalter, a Chronicle of Popes and Emperors, a Commentary on Valerius Maximus, Letters of Pius II, a Boethius, and texts by St. Augustine and by St. Bernardino da Siena. None is described as *enluminé*. What is also worth noting is the fact that Jean Rolin II kept in his Paris *hôtel* a copy of Heinrich Suso's Latin work, the *Horologium Sapientiae*. The description in the inventory reads:

> Item, ung autre livre en parchemin, escript a la main en lectre bastarde, appellé *Orloge de Sapience*, commenceant ou second feuillet *de multorum*, et finissant ou penultime *unum*, relyé entre deux ays et couvert de cuyr vert, prisé xii. s. p.

Although the title is supplied in French, the Latin words that serve as identification clearly point to the language of the volume as being Latin and therefore to a Latin text, namely the *Horologium*.[32]

Documents attest that in 1448 he commissioned an Avignon painter, by name Jean des Plans from Uzès, to illuminate a Missal for use in his own private chapel at the Celestine Convent in that city. The manuscript seems to be no longer in existence, or if it is extant, it has not been identified.[33] In the same Rolin Chapel there was formerly a fresco, painted *c.* 1445, representing the *Sacrament of the Mass* and the *Assumption*. "An old copy of this work, lost to us today, authorises us to see in it a certain relationship to the Aix pictures and to the miniatures of the Master of King René," wrote Wescher.[34] Several other manuscripts are today preserved at Autun that were once in the Cardinal's possession. Some bear only the Rolin family motto *Deum Time*, others carry both the motto and his personal arms: *écartelé aux 1er et 4e d'azur à trois clefs d'or posées en pal, aux 2e et 3e d'argent à la bande d'azur chargée en chef d'une oie d'argent becquée et membrée de gueules.*[35] A careful examination *in situ* of these volumes at Autun revealed that the Rolin Master's hand was in two only, 108A and 114A, both Missals for the Use of the diocese of Autun. Another manuscript that bears the Cardinal's arms and contains the hand of our illuminator is Lyons 517, also a Missal for the Use of Autun. In her 1963 *Horloge* article, Spencer considered that these three Missals were "ordered" for the Cardinal.[36] To our way of thinking, this view cannot be convincingly substantiated. The presence of the prelate's escutcheon on the folios does not, it is believed, allow a conclusion that he commissioned them. He may have been a purchaser of stock copies, and his arms entered at the time of acquisition.

One attempt to determine the true name of the Rolin Master was made as far back as 1901 when

[30] The date 1480 has been argued in the light of the age of the Cardinal in his portrait. He was 72 in 1480. On the painter, see C. Sterling, "Du nouveau sur le Maître de Moulins," *L'Oeil*, CVII (1963), pp. 2-15, 65-8.

[31] Ed. by Charmasse, *Inventaire*, pp. 297-304. There is no specific statement that any of these items had been acquired by Jean from his father Nicolas.

[32] While on the subject of inventory, it is worth observing that in his residence in Paris the Cardinal had three timepieces, see chapter V, note 40 below.

[33] P. Ancona and E. Aeschlimann, *Dictionnaire des miniaturistes du moyen âge et de la Renaissance*, 2nd ed. Milan, 1949, p. 119; Samaran and Marichal, *Catalogue*, VI, Text vol., p. xiv.

[34] Cf. Wescher, *Fouquet*, p. 68.

[35] Cf. Samaran and Marichal, *Catalogue*, VI, Text vol., pp. 438-440. See also P. Liebaert, "Les Livres liturgiques du cardinal Rolin et d'Antoine de Chalon," *Revue de l'art chrétien*, 1912, pp. 442-7; Perrat, *Le livre au siècle des Rolin*, pp. 12-22.

[36] Spencer, *Horloge*, p. 295.

the Missal, Lyons 517, was the subject of a monograph by an archivist of that city, one Léon Galle. His description of the lower border decoration on fol. 8 mentions the arms of Cardinal Jean Rolin II, and continues: "surmontant l'écusson, le chapeau de cardinal, sur lequel on lit: BENIGNE JUYOT; les lettres en sont quelque peu effacées."[37]

Some twelve years later the respected art historian, F. de Mély, included the artist among his collection of titles and names known as *Les Primitifs et leurs signatures.* He read the surname as Guyot, not Juyot, as printed by Galle, and after mentioning other fourteenth and fifteenth-century illuminators in Lyons whose family name was Guyot, concluded that the small illustrations, at least, in Lyons 517 are the work of Bénigne Guyot.[38]

De Mély did not observe that the name Guyot and the name Bénigne had historical associations with Dijon, the capital of Burgundy. Guyot was and still is a Burgundian Christian name, for that matter. In the first third of the thirteenth century there was a celebrated poet called Guiot de Dijon.[39] Jean sans Peur christened one of his illegitimate sons Guyot. In 1425 Philippe le Bon dispatched this bastard brother on a pilgrimage to the Holy Sepulchre in Jerusalem at the head of a party of Burgundian nobles.[40] The same year, the duke's Chancellor, Nicolas Rolin, sent a certain Guiot Bourgoin from Dijon to Berry to effect the evacuation of select towns beset by marauding bands of armed men. Nicolas also entrusted in 1427 one of his counsellors, Guiot de Jaucourt, sire de Villernoult, with a letter for Arthur de Richemont, Connétable de France.[41] As for the name Bénigne, it was no doubt inspired by that of the patron saint to whom was dedicated Dijon's splendid principal church, now its cathedral.[42] It therefore seems quite possible that a painter named Bénigne Guyot or Guyot Bénigne could be a member of a Burgundian family who had ties with Dijon.

Shortly after the Second World War Bénigne Guyot's name was mentioned by two other art critics. Wescher spoke of him in the same breath as Maître François and Jacques de Besançon, even though the comment was an aside about apprentices whom the great Bedford Master might have trained: "Perhaps younger painters were trained in his workshop, such as the prolific Maître François or Bénigne Guyot and Jacques de Besançon, and possibly also Fouquet." When Paolo d'Ancona and Erhard Aeschlimann published their comprehensive *Dictionnaire des miniaturistes* in 1949, Guyot was given an entry, and his work was stated to be in Lyons 517, a reference clearly derived from de Mély's account.[43]

In her Master's thesis for the University of Melbourne in 1962 Margaret Manion had occasion to examine the relationships between two paintings in the Wharncliffe Hours by Maître François and two similar formats in Lyons 517.[44] She reported, on the authority of a communication from Dorothy Miner, that Spencer had questioned de Mély's "attribution and has suggested that the

[37] See L. Galle, *Note sur le Missel d'Autun de la Bibliothèque de la Ville de Lyon*, Paris, 1901, p. 5. J. W. Bradley had not recorded the name of Bénigne Guyot in his three-volume *Dictionary of Miniaturists*, published from 1887 to 1889.

[38] Cf. F. de Mély, *Les Primitifs et leurs signatures. Les Miniaturistes*, Paris, 1913, pp. 251-4. My gratitude for this reference is here expressed to Madame Anne-Marie Gènevois et M. André Vernet, of the Institut de Recherche et d'Histoire des Textes, Paris. In spite of de Mély's argument that Bénigne Guyot could have been from a Lyonnais family, the point was not taken up by M. Audin and E. Vial, *Dictionnaire des artistes et ouvriers d'art du Lyonnais*, Paris, 1918-1919. The name is not included in the B and G alphabetical listings of vol. I. M. Pierre Cockshaw advises me that Guiot is a Burgundian Christian name.

[39] See *Les Chansons attribuées à Guiot de Dijon et Jocelin*, ed. Elisabeth Nissen, Paris, 1928.

[40] Cf. L. de Laborde, *Les Ducs de Bourgogne. Etude sur les Lettres*, Paris, 3 vols. 1849-1852, I, p. 234.

[41] See G. Valat, "Nicolas Rolin, Chancelier de Bourgogne," *Mémoires de la Société Eduenne*, n.s. XVII (1914), p. 107 and 125.

[42] Cf. L. Chomton, *Saint-Bénigne de Dijon: les cinq basiliques*, Dijon, 1923; V. Flipo, *La Cathédrale de Dijon*, Paris, 1928; Tyler, *Dijon and the Valois Dukes*, pp. 12-13, 20.

[43] See Wescher, *Fouquet*, p. 23; and Ancona and Aeschlimann, *Dictionnaire*, pp. 103-4, pl. LIII.

[44] Margaret Manion, *A Study of the Wharncliffe Horae*, Master's Thesis in the University of Melbourne, 1962, pp. 142-151. The section on Guyot was not included in her 1972 study entitled *The Wharncliffe Hours.*

signature in the Missal may refer to the work of a border illuminator." Margaret Manion now continues, "It would then be possible to envisage the miniatures in the Missal as the work of the young Maître François himself. The American scholar, however, does not favour this conclusion and although her complete findings on the subject are yet to be published, already there are evident reasons for supporting the hypothesis that the artist of the Missal and Maître François are two different people."[45] The point that emerges here is that although Spencer knew about the name Bénigne Guyot in Lyons 517 before 1962, she omitted any mention of him in her two articles of 1963 and 1965 when she placed Lyons 517 in the canon of the Rolin Master.

Guyot's name was also omitted from Françoise Cotton's census of the illuminated manuscripts in the Bibliothèque de la Ville at Lyons.[46] To be sure, manuscript 517 is briefly described, but without reference to the artist's identity. She does, however, include Galle's monograph in her bibliography. Given the impossibility at the moment of establishing that Bénigne Guyot was the Rolin Master, it is proposed to retain only the alternate nomenclature, Master of Jean Rolin II, or Rolin Master.

The Rolin Master's *floruit* years appear to have been *c.* 1440 to *c.* 1465. A succinct account of his style was published by Spencer[47] in 1963; a more recent study, based on a considerable amount of new stylistic evidence, seems to exhaust the subject.[48] A sharper focus is now possible when one is required to identify his hand. His compositional structures may contain archaic traits such as diaphragm arches, and designs that recall the hands of the Boucicaut and Bedford Masters. Isolated elements may even be Eyckian in origin or derive from Rogier van der Weyden.

On most occasions the Rolin Master displays a consistent use and control of space. The figures are boldly separated from buildings or other inanimate objects, or figures from figures, and buildings from buildings. A volumetric structure may be balanced by a landscape, but one cannot go so far as to say that volumetric harmony was an artistic preoccupation with him. One also notes that spatial allocations for land masses as opposed to skies are consistently two to one, so that horizons, visual or false, are two thirds the way up the picture plane. In monoscene interiors, the volumes attributed to domestic objects such as beds, book-cases, tables and chairs tend to overwhelm the host cell or chamber, perhaps because the artist does not pay a lot of attention to perspective. Multiple vanishing points and neckless or faceless figures drawn from oblique angles are further signs that the Rolin Master had only a fragmented comprehension of perspective.

Terrains are quite realistic, the more so when city or ramparts or both are on skylines. Configurations of city circumvallations are accomplished. Landscapes are panoramic. Trees are characteristically bushy but stunted, with extended lower branches, and are often carefully positioned in rows. Quinquepartite fern-like clumps grow on grassy patches. When craggy outcrops are depicted, they always lean to the right in the middle ground. Paths are a feature of the artist's hillsides, but unlike Eyckian ones which enter thickets, the Rolin Master's wander away from the viewer and climb up and over hill-tops, to be lost from view.

A feature of the artist's architectonic forms is decorative detail. Further, his interiors may contain mystery elements, such as glimpses of sparsely lighted rear chambers, spiral stair-cases or hallways.

Individuality of portraiture is a distinctive trait of the Rolin Master's anthropomorphic types and attitudes. Sapientia's postures are supple and articulate when she is youthful, but staid and matronly when she represents a mature woman. The Disciple's figure too changes with age. For a traditional iconographic type, for example, Christ in the Passion scenes, there is little variation in the

[45] Manion, *A Study of the Wharncliffe Horae*, pp. 145-6.

[46] Françoise Cotton, "Les Manuscrits à peintures de la Bibliothèque de Lyon. Essai de catalogue," *Gazette des Beaux-Arts*, LXV (1965), pp. 294-5.

[47] Spencer, *Horloge*, pp. 292-5.

[48] For detailed discussions of the points made here, the reader is referred to P. R. Monks, *A Study of the Art Work of the Rolin Master*, Melbourne, 1986, pp. 201-445.

portraiture. St. John the Divine presents another model of constancy, he is endowed with a quality of adolescent androgyny that seems eternal, whether he is present at the Last Supper, Pentecost, the Crucifixion or the Court of Heaven.

Among the mannerisms expressed for less important people in the crowd scenes or social groups, one observes bony and elongated heads, prominent noses, short upper-lips, ears composed of thick ovals indented with two dots. High-born personages tend to present high foreheads and a hair-line receding on the cranium. The artist is unsuccessful when it comes to portraying his actors in profile, and has a predilection for three-quarter views.

Although his range is limited, the Rolin Master displays a capacity for sentiment. One meets brilliantly expressive vignettes for boredom, anger, terror, apprehension, paralysing grief and ecstasy.

Clothing, cloths and coverings, all provide evidence of the painter's fascination with detail. He perfects to his advantage a Bedford Master tone-on-tone design in his own whitish-blue, grey tone-on-tone lozenge patterns. Forceful drafting of the designs that one sees on cloths of honour and wall hangings are eminently distinctive. The artist is a master of cloth volume, the linear folds being inevitably tailored to the position of the torso.

The Rolin Master is most at ease with colour and enjoys rich and brilliant chromatic articulation. He understands the potential of colour for effect, but that effect responds more to a personal aesthetic than to the drama enacted or the setting evoked.

The miniaturist comprehends the pictorial function of light, witness its use to convey a mood, or to pick out figures or for perspectival purposes. Light always floods his tableaux from the left.

At the risk of over-simplification, one may epitomise the variety of elements that constitute his style, in these terms: resolute forms, monumental drapery, clear palette, and personal landscape conformations.

After close comparisons of these stylistic traits with those of hands alleged by art historians to be that of the Rolin Master, it has been possible to reject numerous attributions, to add new ones and to confirm earlier ones. Cast out from among works previously believed to be by the artist are illuminations in Chantilly 76 (1362), 122-123 (322-323), 282 (491); London, Yates Thompson 3; Paris, B. N. Latin 10545 and 17294; Vienna 1900.

As already mentioned his hand occurs exclusively in three Missals: Autun 108A and 114A, and Lyons 517. To these liturgical volumes must be added the Books of Hours, Walters 251 and The Hague 74. F. 1.

In London, Egerton 2019 and Additional 25695, the artist's hand occurs along with one that imitates the Bedford Master's style. In Vienna 1840, one finds the Rolin Master's hand, the Bedford Master's work and a third, as yet unidentified hand. In Walters 285 our painter exercised his skills, but two and possibly three other Parisian miniaturists, also unidentified, participated in the programme of illustrations.

Both Brussels IV. 111, containing the *Horloge de Sapience* and texts by religious writers of the fifteenth century, and the Hours of Simon de Varie (now fragmented in The Hague 74. G. 37, 74. G. 37A, and Malibu, J. Paul Getty Museum 7), contain the hands of the Rolin Master and of the Bedford Master's Chief Associate. We shall return to this painter below.

The presence of the Rolin Master's hand in the same manuscript as the hands of other illustrators indicates that he worked on the volume in the same place as other artists, but at a time that was different.

The fact that our miniaturist is the sole illustrator in some instances, and one among several painters on other occasions, suggests he was a Master of an atelier in his own right, and that he could have been invited to assist in the completion of illustrative programmes left unfinished by predecessors or contemporaries.

Certain stylistic elements that he employed are found in common with contemporaries and later artists, such as Maître François, the Master of Vienna s.n. 13237 and the Master of Wolfenbüttel 2326. These findings in turn lead one to believe that his fellow illuminators recognised his work as being at a pinnacle of respect and technical accomplishment. As such, he was important enough to imitate.

3. THE BEDFORD MASTER'S CHIEF ASSOCIATE

The existence of the three illustrations in the Brussels volume which are attributed to the Chief Associate of the Bedford Master was announced as early as 1955 when the manuscript was part of an exhibition in Madrid. The catalogue described them as "trois tableaux annoncés d'ailleurs dans la table des histoires et exécutés par un artiste moins habile que le peintre des tableaux de l'*Horloge de Sapience.*"[49]

The scholar librarians who prepared the notice of the manuscript on its entry into the Bibliothèque Royale, Brussels in 1961 at least indicated the subjects of the illuminations: "...f. 157r. histoire d'Aaron, de Moïse et de Josué; f. 199v. la Trinité, entourée des saints et des saintes, ainsi que d'un choeur d'anges; vision du paradis céleste; un homme exhorte son âme gisant sur un brancard; f. 229r. six scènes de la Passion de Jésus. Par un procédé extrêmement habile, les six tableaux paraissent à première vue fondus en un seul. En fait, un examen plus attentif de la peinture révèle une division de celle-ci en six compartiments rectangulaires. Le cloisonnement n'est pas indiqué par un encadrement, par exemple, mais obtenu par une disposition très subtile des personnages ou du décor, qui engendre une succession de traits rectilignes produisant l'effet recherché..."[50]

Attempts to identify the hand of the artist were made by Spencer in her 1963 *Scriptorium* article; it was characterised as the "late style of the Master of the Duke of Bedford",[51] but in her 1965 paper she bestowed on the painter the temporary name, the Bedford Master's Chief Associate.[52] Her views on the stylistic traits of this artist are worthy of close attention; the major points are excerpted and reproduced below.

Of the first painting she wrote: "The nearly bisymmetrical lay-out is rare for narratives in the Bedford Master's shop, but the brilliant vermilion vision of God is transcribed from a more familiar figure in blue (e.g., Manchester, J. Rylands Libr., MS. lat. 164, f. 13). The pairs of walking figures and elements in the two crowds have their parallels in certain border miniatures of the Salisbury Breviary (Paris, B.N., lat. 17294, ff. 343, 347, 473 *et al.*)."[53]

Concerning the six episodes from the Passion cycle, she believed that each are "derived from almost standardized patterns for the same sequence in liturgical books illuminated in this shop. The juxtaposition of the six scenes without their frames fails completely to create a single continuous space in which successive acts take place, although the painter has made minor adjustments of landscape and drapery."[54]

Spencer goes on to say that when viewed as illustrations for a manuscript, the three paintings "...are fairly effective in terms of the clarity of their content and the spread of colour upon the page." She also points out that the "...fumbling soft painterly brushwork contrasts sharply with that of the other painter in this book... the Master of Jean Rolin II."[55]

[49] *Héritage de Bourgogne*, p. 38, item 52.
[50] *Bulletin de la Bibliothèque Royale*, V (1961), pp. 55-6. The third painting is reproduced on p. 56.
[51] Spencer, *Horloge*, p. 277.
[52] Spencer, *Ciboule*, p. 107. All three paintings are reproduced in pls. 11-13 of the article.
[53] *Ibid.*, p. 105.
[54] *Ibid.*, p. 106.
[55] *Ibid.*

She continues by listing a number of manuscripts in which, she believes, may be found the style seen in these three illuminations, and in particular, mentions the Salisbury Breviary. In a subsequent study of the artists who illustrated the Breviary, the work of this Chief Associate of the Bedford Master is designated 'Hand C': "The style of the third illuminator, C, is easy to recognize because he paints carelessly with strong mannerisms of form and colour ... the broad shape of ... heads, the blonde colour, the fussy folds of clothing and the carelessly rendered architecture betray his style. Unfortunately, this is the illuminator whose prolific production keeps the shop designs in use as late as 1465."[56]

Spencer's identification of the Chief Associate's work enabled Donal Byrne in 1974 to suggest that the same hand occurs among several that worked on a Book of Hours prepared *c.* 1443-1444 for the Admiral of France, Prigent de Coëtivy.[57]

In 1985 more light is thrown on the activities of the Chief Associate by James H. Marrow in an article announcing the discovery of the third fragment of a Book of Hours owned by a court official, Simon de Varie and dating from *c.* 1455.[58] Its other two fragments are in the Koninklijke Bibliotheek, The Hague. Among the hands represented in the volume as reconstituted are those of Fouquet, the Rolin Master and 'Hand A', working in the manner of the Chief Associate of the Bedford Master.[59]

[56] Spencer, "The Master of the Duke of Bedford: the Salisbury Breviary," *Burlington Magazine*, CVIII (1966), p. 612.

[57] Dublin, Chester Beatty Library, MS. West. 82; see D. Byrne, "The Hours of the Admiral Prigent de Coëtivy, "*Scriptorium*, XXVIII (1974), pp. 252-4.

[58] See J. H. Marrow, "Miniatures inédites de Jean Fouquet: les Heures de Simon de Varie," *Revue de l'Art*, LXVII (1985), pp. 7-32.

[59] *Ibid.*, p. 28.

CHAPTER THREE

Pictorial Programmes for the "Horloge de Sapience"

1. HEINRICH SEUSE, *alias* SUSO, AUTHOR OF THE HOROLOGIUM SAPIENTIAE

Heinrich Seuse was born in Constance *c.* 1300, his mother's name being Süsse, later spelt Seuse, and his father's Von Berg. The child so venerated his mother that when he was an adult, he adopted her name as his cognomen.[1] Its Latin form was Suso. By dispensation, Heinrich became a novice in the Dominican convent at Constance while only thirteen. Five years of teenage restlessness followed, before he began, *c.* 1318, to adhere strictly to the precepts of his Order. He adopted the practise of bodily mortifications and austerities to the neglect of preaching. He next attended the Studium Generale of the Order in Cologne and completed his formal studies. Returning to his home town, he was appointed to the important position of conventual reader, or *lector*. About 1339, because of a conflict with the burghers of Constance about the implementation of an interdict, the Dominicans went into exile outside the city walls. It was during this exile that Suso was elected prior and found a new orientation in pastoral care and preaching: "the lonely ascetic had become an itinerant preacher."[2] Physical hardships on the open road, jealousy among fellow Dominicans, human treachery, all visited him in one guise or another in his last years before his death at Ulm in 1366.

Like Vincent Ferrer after him, Suso's influence on his own generations was felt extensively in many parts of Europe. Most celebrated among his followers was Elsbeth Stagel of the Dominican convent of Töss, near Winterthur. Töss was a jewel in the Order's spiritual crown, being a centre for mysticism in the fourteenth century.[3] Suso visited it often and Elsbeth recorded for posterity her assessment of the man and his achievements.

Among Suso's important writings is the Latin *Horologium Sapientiae*, dedicated to the General of the Order, Hugo de Vaucemain (1333-1341), and composed *c.* 1334, principally for nuns.[4] Ac-

[1] For the brief biographical outline that follows, I have selected information from Jeanne Ancelet-Hustache, *Le Bienheureux Henri Suso*, Paris, 1946; from J. M. Clark, *The Great German Mystics Eckhardt, Tauler and Suso*, Oxford, 1949; and from Künzle, *Horologium*, pp. 1-6.

[2] Clark, *German Mystics*, p. 59.

[3] Cf. Jeanne Ancelet-Hustache, *La Vie mystique d'un monastère de dominicaines au moyen âge d'après la Chronique de Töss*, Paris, 1928.

[4] See Künzle, *Horologium*, pp. 19-27 for the date of composition. Suso is more famous for an earlier treatise on the same theme which he had composed in German, *Das Büchlein der ewigen Weisheit*, see K. Bihlmeyer, *Heinrich Seuse. Deutsche Schriften*, Stuttgart, 1907; *Henry Suso. Little Book of Eternal Wisdom and Little Book of Truth*, translated by J. M. Clark, London, 1953; an extensive bibliography is in J. A. Bizet, *Henri Suso et le déclin de la scolastique*, Paris, 1946.

cording to statements in the Prologue, the title recalls a celestial clock which Suso beheld in a vision. As a consequence, he divided the work into twenty-four chapters to symbolise the hours of the day. There are two Books, of eighteen and six chapters. At the beginning the reader meets Divine Sapientia,[5] who counsels her Disciple during the next four chapters to study and to imitate Christ's Passion. Sapientia then occupies the centre stage herself for four more chapters in which Suso places great emphasis on her potential role as a spiritual lover for the soul. Vivid descriptions of Hell and Paradise can be found in chapters 10, 11, 12 and 14. Christ and Sapientia both narrate Passion incidents in chapter 15, then the subject of the Virgin's insurmountable grief concludes Book I.

The atmosphere of Book II appears at times to be metaphysical. The Disciple passes through the schools of learning, evaluating the teaching as he goes. One cannot help feeling that the author's barbed comments on the self-interest and vanity of both students and teachers are based on his own youthful experiences in *studia* in Constance and Cologne. The Disciple next learns from Sapientia how to die and how to live with the Spirit. In chapter 4 he is reminded that Christ's Passion must be vigilantly recalled at every moment of life on earth, since at the Last Supper He gave mankind the Holy Sacrament of the Altar. In the remaining four chapters of this second Book the Disciple prepares for his spiritual union with Divine Sapientia. It is finally consummated with the recognition that he will enjoy everlasting bliss, for he has given himself totally to Christ.

If one were asked to characterise the structure of the *Horologium*, that is, the manner in which the mystical thought and human drama are interwoven, one would have great difficulty in finding an integrated approach. The principal element that Suso employs is the dialogue, not simply questions and answers on factual matters, but questions of principle, questions about the morality of proposed actions, dramatic interrogation about the cause and effect of human conduct. The answers are diffuse and circumlocutory, according as Sapientia thinks and speaks theologically or out of regard for the humanity of her pupil. Yet, neither the protagonist nor the antagonist in the disputation process have clear-cut and well-defined roles.

Sapientia is not merely the Wisdom of the Old Testament personified; she is also Christ, particularly during the Passion sequences. She remains feminine throughout the debates, to the point that Suso has no difficulty in allegorising her as Wisdom when the Disciple's soul takes her as its spouse. A leading Suso scholar has epitomised her transfigurations in these terms: "Eternal Wisdom is at first a woman, the personification of Wisdom, the spouse of the soul, as in the Wisdom Books of the Old Testament. Later Eternal Wisdom represents Christ."[6]

The Disciple's role is also diversified. At times he is the author, a Dominican friar succumbing to the world's vanities and temptations, and confessing his moral frailty to Sapientia. At others, he is a serious and dedicated student of Theology, trying to examine objectively the mysteries of the Passion. Towards the end of the treatise the Disciple assumes the guise of a Soul seeking its spiritual bride. He welcomes the epithet that Sapientia bestows on him in the epithalamium, Frater Amandus, the Brother needing Love.

Apart from the manifest debates between Sapientia and the Disciple, there are dialogues between the Disciple and his conscience. They occur either in the course of intimate and reflective meditation or when he is confronted with visions and imagery related to doctrines being expounded by Sapientia.

[5] A word of explanation is perhaps necessary for the use of the name Sapientia instead of Wisdom in the English renderings of textual extracts and in the discussion thereon. The form Sapientia recalls a triple association: the Latin name for the Book of Wisdom in the Vulgate, the Latin name for a personification of Wisdom or of the virtue Wisdom in the Middle Ages, and thirdly, the French form Sapience in the translation which the painter had in front of him. The form of the English word Wisdom is bereft of this rich etymological and homophonic heritage.

[6] Clark, *Little Book of Eternal Wisdom*, p. 13.

As for the narrative elements, they derive from several unembellished stories of the Old Testament, such as Susanna and the Elders, the Treachery of Haman, the Wrath and Death of Saul. Overshadowing everything and intruding at all points is the Passion narrative, from the Last Supper, through the incidents of the Garden of Gethsemane, the summary trials before Jewish and Roman leaders of Judea, to the *Via Crucis*, the Nailing to the Cross, the scene on Calvary, the Deposition and the Man of Sorrows.

The straightforward narration of Biblical events is augmented and heightened by descriptions of persons and places related by the Disciple when in an ecstatic or highly emotional state. The provocative elements or stimuli for the descriptive passages are visions. They enable Suso to suggest the reality of the person of Sapientia, as for example when the Disciple beholds her enthroned in the skies, or when he marries her at the end of the work. The vision is also a powerful agent for enhancing the terrors of Hell or the glory of the Court of Heaven. It furthermore facilitates the Disciple's comprehension of the presence of the Ram with the Iron Crown at the gates of the crumbling City of Religion.

2. THE PICTORIAL PROGRAMME IN MANUSCRIPTS OF THE FRENCH TRANSLATION *HORLOGE DE SAPIENCE*

The popularity of the *Horologium Sapientiae*, if the number of surviving copies is any guide, was extraordinary. A recent editor describes some 233 complete copies, mentions another 88 that are seemingly lost, and studies 150 extracts of varying lengths. Vernacular redactions were made in French, Netherlandish, German, Italian, English, Swedish, Danish, Polish and Czech.[7]

The principal French translation, known as the *Horloge de Sapience*, was made *c.* 1389 by a Franciscan of the Observant Convent in Neufchâteau (arr. des Vosges). The instigator of the translation was Demenge de Port, *licencié en droit civil et canon*, a son of Nicole de Port, knight and *seigneur en loi*. Nicole was a liegeman of Duke Henri de Bar and a trusty member of his Council. Maître Demenge, whose name perhaps represents a local form of the Latin *Dominicum*, would seem to have had St. Dominic as his patron saint.[8]

There is no critical edition of the French *Horloge*, and an exhaustive census of extant manuscripts is therefore not available for consultation. Spencer mentioned in 1963 that the illustrations of the sixteen French manuscripts known to her were destined to be the subject of a separate article in preparation.[9] Which sixteen she had in mind was not stated. Three years later Jeanne Ancelet-Hustache published a list of some fifty manuscripts with a brief statement about the number of paintings each contained. There was no description of the miniatures nor any attempt to reveal if the paintings illustrated the *Horloge* itself or both the *Horloge* and other texts within the same volume covers.[10] In a subsequent critical edition of the Latin *Horologium Sapientiae*, Brother Künzle O.P. describes fifteen complete or fragmentary French copies, but says nothing about illuminations.[11] Referring to copies of the *Horologium* and the *Horloge* in general, Spencer wrote that the

[7] The manuscripts of the *Horologium* are described by Künzle, *Horologium*, pp. 105-200 (extant ones), pp. 201-14 (lost ones), pp. 229-49 (extracts), pp. 250-76 (translations).

[8] The versified colophon containing the information about the translator and his instigator was published in part by P. Marot, "Note sur une traduction de l'Orloge de Sapience du Bienheureux Henri de Berg," *Bulletin mensuel de la Société d'archéologie lorraine et du Musée historique lorrain*, XXIII (1928), pp. 83-5. The year of the translation is given as 1399; this was corrected to 1389 by Jeanne Ancelet-Hustache, *Quelques indications* pp. 161-2, 167-9. See also Künzle, *Horologium*, pp. 251-2.

[9] Spencer, *Horloge*, p. 280, n. 15.

[10] Ancelet-Hustache, *Quelques indications*, pp. 161-70.

[11] Künzle, *Horologium*, pp. 250-5.

vividness of Suso's verbal imagery "would seem to have invited, if not compelled, the illumination of these manuscripts so that it is curious that the Bruxelles copy should be the only one that has come to light which has more than token illustration."[12]

It is beyond the scope of this study to discuss and describe every painting in the extant illuminated copies of the French *Horloge.*[13] It is of greater relevance to detail the nature and number of paintings in pre-1450 copies of the text, since a comparison between their art work and that of the Brussels volume will enable the reader to evaluate more critically the contribution of the Rolin Master to manuscript illustration. Only five illustrated books have been located that appear to date from a period before 1450.

The oldest dateable illuminated copy of the French *Horloge* is the one that Marie,[14] daughter of Jean duc de Berry, received in 1406 as a gift from her confessor, Simon de Courcy. At the commencement of the transcription the Luçon Master portrays a Dominican being shown the inner workings of a pedestal clock by Sapientia, whose hovering figure leans out of a cloud.[15]

In the same decade an anonymous illuminator painted a large representative frontispiece for another copy of the *Horloge.* Suso the Dominican is seated, like an Apostle, in the act of writing on a scroll, while God the Father looks down from a mandorla at the top of the picture. In the centre, a priest celebrates mass. The right-hand vertical section of the whole composition displays moments of the Last Judgment, with the damned disappearing into the Mouth of Hell and the chosen souls moving heavenwards assisted by angels.[16]

Another fifteenth-century *Horloge* opens with a full-page miniature portraying a Dominican friar, doubtless Suso, standing in a pulpit, preaching to an audience.[17]

In 1448 Jean d'Ardenay transcribed the text in Lille for Philippe le Bon and an unknown illustrator made an ink and wash drawing on fol. 6 by way of a frontispiece. It shows a centrally positioned pendulum clock with an alarm on top. To the viewer's left stands Sapientia in cowled habit, left hand on the casing, right hand holding a closed book. On the opposite side of the clock is a notable examining the mechanism.[18]

Another mid-fifteenth-century copy of the *Horloge* in two volumes was owned by Louis de Bruges, seigneur de la Gruthuyse. A Flemish artist of mediocre accomplishment painted on fol. 9 of the first manuscript a clock chamber dominated by a massive time-piece standing on the floor. A crowned and robed monarch, possibly Solomon, demonstrates the pulleys, chimes and inner

[12] Spencer, *Horloge*, pp. 279-80.

[13] I am very much indebted to Madame Anne-Françoise Labie, the Secretary of the Section Romane of the Institut de Recherche et d'Histoire des Textes, Paris, for supplementary information about extant manuscripts. Ancelet-Hustache, *Quelques indications*, p. 166 speaks of more than half of the identified manuscripts as being illustrated, but her own list gives the lie to this proportion. See also my paper "Pictorial Programmes in Manuscripts of the French Version of Suso's *Horologium Sapientiae*," *Archivum Fratrum Praedicatorum*, LVII (1987), pp. 31-43, 2 pls.

[14] A contemporary portrait of Marie de Berry is preserved in the Album of the herald Guillaume Revel, in Paris, Bibliothèque Nationale, MS. fr. 22297, fol. 17, reproduced in Champion, *Histoire poétique*, I, pl. X opposite p. 181.

[15] Paris, Bibliothèque Nationale, MS. fr. 926, fol. 113; see J. W. Bradley, *A Dictionary of Miniaturists, Illuminators, Calligraphers and Copyists*, 3 vols, London, 1887-1889, II, p. 175; and *Paris, Bibliothèque Nationale. Les Manuscrits à peintures en France du XIIIe au XVIe siècle*, Paris, 1955, p. 90, no. 185. The painting is reproduced in M. Meiss, *French Painting in the Time of Jean de Berry. The Limbourgs and their Contemporaries*, 2 vols, London, 1974, II, fig. 131 and discussed I, p. 35.

[16] The Hague, Koninklijke Bibliotheek, MS. 76.E.19, fol. 1, reproduced in Meiss, *Limbourgs*, II, fig. 130.

[17] London, British Library, MS. Additional 15288, fol. 1; cf. *Catalogue of Additions to the Manuscripts in the British Museum in the years MDCCCXLI-MDCCCXLV*, London, 1850. p. 123 for the year 1844.

[18] Brussels, Bibliothèque Royale, MS. 10981, fol. 6, reproduced in A. Chapuis, *De Horologiis in Arte*, Lausanne, 1954, p. 12, fig. 19; in *The Secular Spirit. Life and Art at the End of the Middle Ages*, Introduction by T. B. Husband and Jane Hayward, New York, 1975, p. 191. On the history of the manuscript, see G. Doutrepont, *La Littérature française à la cour des ducs de Bourgogne*, Paris, 1909, p. 208, and *La Miniature flamande. Le mécénat de Philippe le Bon*, Brussels, 1959, p. 80, no. 71.

wheels to a nobleman wearing the chain of the Golden Fleece, most likely Louis de Bruges, who is attended by a retinue. There is no Sapientia and no Dominican. But for a diminutive crucifix on top of the clock, one would say that the format has been secularised. At the opening of the second volume is another view of the same nobleman and his suite in court costume.[19]

It is clear that the total of the subjects collocated in pre-1450 illustrations of the *Horloge* rarely surpassed five or six.

3. THE PICTORIAL PROGRAMME OF THE *HORLOGE DE SAPIENCE* IN BRUSSELS, BIBLIOTHEQUE ROYALE, MS. IV 111

For the first time in the history of the textual illustration pattern, the Brussels copy contains at least one illumination per chapter, and some of the twenty-four chapters contain as many as three or four miniatures. A frontispiece is still a feature of the prologue. Miniatures may be complex in visual structure. If each segment of the polyscenes is counted as a pictorial unit in its own right, and if all the sections and monoscenes are added together, the sum total of the illustrations is one hundred and eight.

Few of the paintings have legends outside the frames, but in places the opening words of chapters that commence beneath miniatures could serve as a ready guide to pictorial content. Attempts by modern critics to describe the subjects have been of two kinds. The titles published in 1961 and 1969 by unidentified scholar librarians of the Bibliothèque Royale in Brussels matched each monoscene and nearly every segment of the large structures.[20] Spencer, on the other hand, selected titles that were sometimes vague, e.g. "Two exempla of Betrayal, f. 30v."; or interpretative, e.g. "Sapientia as Temperantia, f. 13v."; or misleading in iconographic emphasis, e.g. "Sapientia in the Tree, f. 20v." which was applied to a polyscene of four compartments.[21] Because of these inconsistencies, Spencer's titles have not been retained. A fresh attempt has been made to supply a title for each pictorial unit, based on the iconographic content and design.

The subjects of the pictorial units comprise cycles or groups of scenes whose thematic content is similar. The practice of illustrating manuscripts by cycles of pictures reflecting specific textual emphases became extensive in France in the thirteenth and fourteenth centuries owing to the popularity of stereotyped programmes for Latin Bibles, Psalters and Apocalypses.[22] As more texts were translated into French in the second half of the fourteenth century, the practice was extended to copies of their kind as well. Livy's *Histoire Romaine*, Boccaccio's *Cas des Nobles Hommes et Femmes* received profuse illustrations of the cyclic variety,[23] as did translations of the *Cité de Dieu*

[19] Paris, Bibliothèque Nationale, MSS. fr. 455-456. The painting of the Clock Chamber in the first volume on fol. 9 has been reproduced in Chapuis, *De Horologiis*, fig. 25 in colour, opposite p. 16; and in Meiss, *Limbourgs*, II, fig. 132.

[20] *Bulletin de la Bibliothèque Royale*, V (1961), pp. 48-58, and *Bruxelles, Bibliothèque Royale Albert Ier. Quinze années d'acquisitions*, Brussels, 1969, pp. 84-6.

[21] Spencer, *Horloge*, p. 281, n. 16.

[22] The programmes are discussed in detail in the following monographs and papers: S. Berger, *La Bible française au moyen âge*, Paris, 1884, pp. 282-91; F. Wormald, "Bible Illustrations in Medieval Manuscripts," in G. W. H. Lampe (ed.), *The Cambridge History of the Bible*, London, 1969, II, pp. 309-337, 524; S. Berger, "Les Manuels pour l'illustration du Psautier au XIIIe siècle," *Mémoires de la Société Nationale des Antiquaires de France*, LVII (1898), pp. 95-134; M. R. James, *The Apocalypse in Art*, London, 1931.

[23] M. Meiss, *French Painting in the Time of Jean de Berry. The Boucicaut Master*, London, 1968, pp. 50-7 discusses the pictorial programmes for these two texts. Across the Channel, Lydgate translated between 1412 and 1420 the *Troy Book* from Guido de Columnis's *Historia Destructionis Troiae* and it was not long before copies were receiving a standardised sequence of paintings; see Lesley Lawton, "The Illustration of Late Medieval Secular Texts with Special Reference to Lydgate's Troy Book," in *Manuscripts and Readers in Fifteenth-Century England* (ed. D. Pearsall), Cambridge, 1983, pp. 52-69.

of St. Augustine and the *Livre des Propriétés des Choses* of Bartholomaeus Anglicus.[24] The French translation of the *Horologium Sapientiae* was a late adherent to this iconographic trend.

One of the first things observed about the *Horloge*'s cyclic illustrations is the presence of several different groups, with sequences dislocated because of the author's erratic presentation of the subject matter. Suso had his own textual emphases which led to restatements of points or the suspension of thought to enable tangents to be followed.

In coping with the cyclic matter the artist employs two methods, integration or compartmentalisation. The *Susanna* painting with three separate sequences combined in the one picture plane represents the composition technique for the integrated type of narrative. The Sanhedrin series, on the other hand, is a collection of separate pictorial units, or compartments, presented sequentially, as Christ appears successively before Annas, Caiaphas, Pilate and Herod.

Prominent among the Rolin Master's cycles is the group of scenes concerned with Christ's corporeal existence on Earth. Standard cycles in Books of Hours range from the *Nativity* to the *Flight into Egypt*, and usually include events of the *Passion.*[25] The *Horloge* series has a broad sweep also: a *Nativity*, the *Virgin Suckling the Holy Child*, the *Temptation*, *Miracles*, a *Rejection at Nazareth*, the *Last Supper* with its echo in the *Elevation of the Host*, and the *Agony in the Garden.* Much has been omitted, in particular, scenes depicting Christ's early years, such as the *Presentation in the Temple*, the *Circumcision*, the *Flight into Egypt*, *Baptism*, *Entry into Jerusalem.*

The cycle for the Sanhedrin Trials may be thought to begin at the *Betrayal.* Although the Rolin Master executes a scene of Christ praying on the Mount of Olives, he chooses the moment before the arrival of the soldiers, the betrayal and the arrest. It is clearly an *Agony in the Garden.* The artist also includes the successive referral audiences with Annas and so on, concluding with Christ being presented by Pilate to the Jews as he utters the celebrated *Ecce Homo*! Thus, only the *Flagellation* has been suppressed from the series.

The sequel brings the beholder to the Passion Cycle or Stations of the Cross Cycle, in which some subjects are integrated in the same frame. Noteworthy omissions are Christ's meeting with His sorrowing Mother, the three Falls beneath the burden of the Cross, the spearing by Longinus (replaced by the Presentation of the Sop), and the Entombment.

The Resurrection theme could be said to be represented by two subjects only, the *Man of Sorrows* and *Christ Appearing to His Mother.* There are no views of the *Harrowing of Hell, Noli me tangere*, *Ascension* or *Pentecost.*

The next personage about whom a cycle of illustrations can be said to exist in the *Horloge* is Sapientia. There are three salient portrayals: *Sapientia in Majesty*, the frontispiece, *Sapientia in a Terebinth Tree* and Sapientia in the *School of Theology.* But like her pupil, she is ubiquitous in numerous paintings where she is seen to instruct, guide and to be ultimately united with him in spirit.

Her role as mentor is paralleled by the role of the Disciple as her devoted proselyte. Wherever he appears in Sapientia's cycle, it is to emphasise her spiritual training, its setbacks and ultimate triumph in the harmonious union of the two spirits.

Apart from this aspect of the Disciple, there is his mortal state and its interweaving with the Order of the Preaching Friars. The illustrations take us from the period of his *Youth in Dalliance* and *Carole* to the Postulant phase, subsequent admission, the listening to *legenda* passages while

[24] Cf. Sharon D. Smith, "New Themes for the *City of God* around 1400: the Illustrations of Raoul de Presle's Translation," *Scriptorium*, XXXVI (1982), pp. 70-8; and D. Byrne, "The Boucicaut Master and the Iconographical Tradition of the *Livre des Propriétés des Choses*," *Gazette des Beaux-Arts*, XCI (1978), pp. 153-9.

[25] Discussed in Margaret M. Manion, *The Wharncliffe Hours. A Study of a Fifteenth-Century Prayerbook*, Canberra, 1972, pp. 16-24, and Harthan, *Book of Hours*, p. 28.

eating in the refectory, the temptation, the studies in the quiet of his cell, the hardship of the shipwreck, a withdrawal from Sapientia, resumption of the ties, and final sublimation of his spirit with hers after a period of meditation. Thus, in places it is artificial to try and separate the Disciple from the *alter ego*, the Dominican.

A less structured cycle concerns mortal Man and his pilgrimage through the earthly world, and even here there is an allegorical overlap with Christ's life on earth. A pilgrim-like figure, personifying Christ at the Gates of the City of Religion in Ruins, is beset by Virtues and Vices which battle for domination of the body and soul. Life's Pilgrim later contemplates the Snares of the Devil, the Torments of Hell, and he suffers the dilemma of Tantalus. The frailty of the human body is recalled in a Death-Bed scene when gaunt Death lunges a spear at the moribund figure.

One finds also among the illustrations the usual *hors-série* ones, the isolated compositions such as *God the Father*, *Adam and Eve* and the *Court of Heaven*.

Although there are many Old Testament subjects represented in the pictorial programme, they are, strictly speaking, non-cyclic; they function as *exempla*. The *exemplum* had a long literary history before the High Middle Ages in France, according to Welter, an authority on moral tales. In Classical Literature Valerius Maximus popularised the technique of recounting anecdotes about historical personages and then drawing a moral. The Church Fathers continued the practice.[26] The exemplum was cherished by Dominican preachers whether they were on missionary *itinera* or delivering homilies closer to their convents.[27] Etienne de Bourbon (d. 1261) employed them extensively, as did the luminaries Humbert of Romans (d. 1277), Vincent of Beauvais (d. 1264), Martin the Pole (d. 1279) and the preacher Vincent Ferrier, whose links with the Dukes of Brittany were discussed in chapter I. Suso likewise has recourse to exempla from time to time in the *Horologium* in order to remind both the Disciple and readers about the need for moral rectitude in their lives.[28]

Exempla as art subjects were well known among French miniaturists of the fourteenth and fifteenth centuries, principally as a result of the need to illustrate texts containing literary moralistic tales of one nuance or other. Rosemond Tuve detailed their place in the decoration programmes for versions of the *Somme le Roi* and in the Belleville Breviary, a manuscript containing elements of the Dominican liturgy.[29] Sharon D. Smith has more recently studied the exempla in cyclic programmes of the *Cité de Dieu*.[30]

In her list of the titles of miniatures in the Brussels *Horloge*, Spencer supplies the names of many exempla, for example, the concept of Betrayal, David and the Jewish Chiefs.[31] One would also want to identify more precisely Jael and Sisera, Saul Threatening David, the Death of Saul, Esther's Banquet and Haman, Susanna, Absalom's altercations with his father, and the like. The only non-Biblical exemplum concerns Tantalus; it is derived from Ovid's *Metamorphoses*.

[26] The most informative reference works on the exemplum in the Middle Ages are J. T. Welter, *L'Exemplum dans la littérature religieuse et didactique du moyen âge*, Paris, 1927; F. C. Tubach, *Index Exemplorum. A Handbook of Medieval Religious Tales*, Helsinki, 1969; C. Brémond, J. Le Goff, J. C. Schmitt, *L'Exemplum*, Louvain, 1982.

[27] For the Dominicans in general, see Welter, *Exemplum*, pp. 137-43. For specific authors and preachers I mention, see also A. Lecoy de La Marche, *Anecdotes historiques, légendes et apologues tirés du Recueil inédit d'Etienne de Bourbon*, Paris, 1877; S. L. Forte, "A Cambridge Dominican Collection of Exempla in the Thirteenth Century," *Archivum Fratrum Praedicatorum*, XXVIII (1958), pp. 115-48; V. Almazan, "L'Exemplum chez Vincent Fierrier," *Romanische Forschungen*, LXXIX (1967), pp. 288-332.

[28] The Index in Künzle, *Horologium*, pp. 665-6 provides ready access to their placement in the narrative.

[29] Rosemond Tuve, "Notes on the Virtues and Vices," *Journal of the Warburg and Courtauld Institutes*, XXVII (1964), pp. 42-52 and 65-72. Further information about exempla in the Belleville Breviary is contained in Leroquais, *Bréviaires*, III, pp. 205-9; in Lilian M. C. Randall, "Exempla as a source of Gothic Marginal Illumination," *Art Bulletin*, XXXIX (1957), pp. 97-107; and in the paper by Lucy F. Sandler, "Jean Pucelle and the Lost Miniatures of the Belleville Breviary," *Art Bulletin*, LXVI (1984), pp. 73-96.

[30] Cf. Sharon D. Smith, *New Themes for the City of God*, pp. 74-5.

[31] Spencer, *Horloge*, p. 281, n. 16.

To appreciate truly the extent of the innovation this Brussels programme represents in the history of illustrations for the French *Horloge de Sapience*, one may compare it to the perfunctory iconography of pre-1450. First, the relationship between image and text has been greatly expanded, from one miniature comprising two to four subjects to thirty-six containing one hundred and eight. As a few of these are integrated minor cycles, there could be further subdivisions of the units which would in turn increase the total number of subjects placed in front of the beholder. Clearly, the visual statement was considered to be approaching that of the textual one in importance.

The second departure from earlier patterns is in the content and role of the frontispiece. Both have been completely reassessed. The secularising trend relating to the Clock and the undue emphasis it was receiving by its domination of the frontispiece has been halted. The true religious, didactic and mystical nature of Suso's work has been reasserted, redefined and restated in a powerful visual image of Sapientia. To be sure, the preaching element is retained, but relegated to a minor chord of grace.

Thirdly, the introduction for the first time of several cycles of illustrations allows one to rank the pictorial programme equal with the outstanding series of miniatures painted by the Virgil Master and the Boucicaut Master and even the Bedford Master.

The painter's interpretation of the subject matter has been made according to three procedures: he follows a 'standard' traditional format, e.g. the *Last Supper*; he modifies a traditional design, e.g. the *Seven Liberal Arts*; or he devises a completely fresh composition, as in the case of the *Ram with the Iron Crown*. "Fresh", of course, is understood to refer to a design not among the stock ones atelier craftsmen had at hand. The manner in which stock models have been employed to meet the textual imperative shows two further refinements of approach. For subjects that lie at the heart of Christian Belief, such as the Crucifixion, the Eucharist, Heaven and Hell scenes, hagiographical representations, the Rolin Master is traditional and totally orthodox, pictorially speaking. Modernisation occurs rarely in his traditional formats. One detects a Simon of Cyrene wearing the latest fifteenth-century male fashions in *Christ Carrying the Cross*. On the other hand, modern dress, armour and current architectural forms abound in several exempla, for instance, *Jael Slaying Sisera*, *Tantalus* and the *Carole*. Even the time-pieces of the *Clock Chamber* present contemporary shapes, dials and workings. The effect the modernising trend would have had on the viewer must have been considerable, for clothing in fashion on human figures increases their appeal to him and augments thereby the extent to which he can relate through the figures to the text proper.

Suso's work contains much that is biographical, but, most of all, it reflects Dominican outlook, experiences and concerns. Pre-1450 illustrations of the *Horloge* hardly acknowledged any of this inner pith. A Preaching Friar was portrayed in the prefatory miniature, but often one did not know if he represented the author or was an allusion to the Disciple. The Cyclic programme followed by the Rolin Master is revolutionary in this area. It transmits visually for the first time all these echoes of information about the Dominicans. The artist achieves this effect by portraying the Disciple, almost always garbed as a Dominican, in the role of a participant in the pictorial dramas, even in those of the highest moment, *Christ Carrying the Cross* and *Christ Nailed to the Cross*. Technically speaking, the Disciple is placed into the picture plane as an onlooker or a suppliant and so appears to be part of the action.

The Preaching Friars' concerns for religious and moral instruction of members of the community and for education in general are never far from the viewer's gaze because of the personification of Wisdom and knowledge by Sapientia whenever she is present in a painting. Maximum visual effect is achieved when half a folio is set aside for a representation of the School of Theology, after the imagined style of a Dominican *studium generale*. Its depiction is realistic in that a rich assortment of ecclesiastical figures, seculars and regulars, from all levels of the hierarchy, are in attendance.

The Rolin Master plays another but more silent role in the reader's understanding of the text. He subtly invites him to contemplate Nature. Suso mentions the countryside in two passages, once where the Disciple is counselled to contemplate the dumb beasts in the fields along with the Epicureans; another time, when youths are said to desport themselves with their friends. The artist has carefully included in his renderings of the episodes bright expanses of turf or meadows and fields. Yet, if we examine all the outdoor settings, and this means half of the units in the programme, we find they contain backgrounds which are the personal statements of the artist. He is consistent in his approach, which is to provide either distant hills with a river estuary on which ships move silently, or slopes and hills topped with ramparts, turrets and even a winding track. These naturalistic backdrops are always peaceful and dreamlike, inducive to contemplation, no matter how animated or violent the activities by human figures in the foreground or middle ground. Most noteworthy in this regard are *Christ Nailed to the Cross*, *Ram with the Iron Crown* and *Life's Pilgrim*. Such textual enrichments through illustration had not been achieved, even remotely, by any pre-1450 illuminator of the French *Horloge*.

One could have expected that several individual painters or a team of artists would have been assigned to handle a pictorial programme that was large in scope but not in size, as happened in the case of the Salisbury Breviary.[32] The fact that a single artist was entrusted with the extensive programme must surely speak highly about his reputation among fellow painters and clients in the Parisian manuscript trade of the period.

There remains still the question of the originator of the programme of illustrations. Was the Rolin Master, who filled with pictures the rectangles left blank by the scribes, the same person who chose the subjects to be illustrated and who decided their location relative to the textual matter? If he was, he chose traditional structures, but, compared with the meagre achievements by predecessors, he was totally innovative in the pictorial statements he made. Moreover, the placement of the paintings is closely linked to the position of textual passages on the folios. The artist must therefore have been literate and extremely knowledgeable of Suso's Latin text and of the French version. In other words, the master of the workshop was familiar with Suso's labyrinthine discourse, its mysticism, its allegory and theological teachings. Further, the Rolin Master, surely needed to be a Preaching Friar to be able to know when to register a pictorial emphasis for teaching, another for meditation, another for community life, and others for devotional practices, and so on. The hypothesis invites one to imagine a prospect of the Rolin Master at work in the Convent of Saint-Jacques-le-Grand in Paris, the Provincial Convent of France.[33] Apart from the fact that there is no documentation to confirm or negate the hypothesis, the identification of the artist with the originator of the programme runs up against more prejudicial doubts.

Why would a Dominican miniaturist be executing illuminations in a manuscript that was destined for a high-placed courtier such as Guillaume Gouffier? The so-called Dominican interiors in many pictures and the Cloisters with Fountain cannot be proven to reproduce the architectural structures of the Paris Convent, since these are not known. One or two small details in the Sanhedrin sequence suggest that the artist-cum-originator did not know the wording of the New Testament narrative as well as could be expected, if he were a Preaching Friar. The Latin phraseology on the banderoles for the Annas and Caiaphas segments is transposed.

These doubts and question marks that surround the contention that the Rolin Master was a Dominican illuminator who originated the programme, vanish if he is removed from consideration, as common sense now requires. The more plausible hypothesis is that the artist worked to a plan already prepared by someone else, who was scholarly, informed about Suso's work to the point

[32] Spencer, *Salisbury Breviary*, p. 608.

[33] See G. Rohault de Fleury, *Les Couvents de Saint Dominique au moyen âge*, 2 vols, Paris, 1903, II, unpaginated, see 'Paris' entry.

of intimate knowledge, and who was familiar with Dominican matters in general. In other words, we seek an author who was a Preaching Friar. His written programme is not preserved, it so happens, in the *Déclaration des hystoires* which prefaces the transcription of the French *Horloge* in the Brussels manuscript. The identity of its author and the nature of the *Déclaration* will be discussed in chapter IV. Suffice to say here that the *Déclaration* was composed after the pictures had been completed, not beforehand.

Written directions for French miniaturists to follow have survived in a few cases,[34] the most famous being Jean le Bègue's programme for a copy of Sallust, *c.* 1417.[35] Another written programme was employed by Maître François, *c.* 1470, for a copy of the *Cité de Dieu*, but it has not survived. Its existence is attested in a letter about the *egregius pictor* from Robert Gaguin, General of the Trinitarians to Charles de Gaucourt, Governor of Paris.[36]

The programme that guided the hand of the Rolin Master exists no longer either. All we have is the series of completed illustrations. The planner of the miniatures knew Suso's text and the Bible intimately and, it seems, in a Latin form rather than a French version. The Latin phrases in the scrolls correspond either to Suso's Latin words or to those of the Bible. French is hardly ever employed within the picture frame, and where it does occur, it has an identification role, not a narrative function. Another indication that the 'programmer' was versed in the Bible comes from the subject matter of numerous exempla. Whereas Suso may mention an Old Testament personage only by name, the picture that appears in response to the mention has detail in its statement that coincides with the written narrative in the Bible.

The same Dominican originator of the programme decided, it seems, that Sapientia must be rendered in a credible human form rather than continue to be represented as the disembodied spirit seen in early fifteenth-century illustrations of the *Horloge*. In planning her portraiture with doctoral robes for the School of Theology, he may well have had in mind the venerable first *studium generale* of the Preaching Friars, founded in Paris *c.* 1228.[37] Likewise, the designer of the pictorial programme was responsible for shifting the mundane clock out of the frontispiece, leaving the modified frontispiece to function ontologically in a more faithful representation of Suso's work: preaching and wisdom through scriptural and spiritual knowledge to perfect members of the Order of Preachers.

Then, there is the curious representation of a duke in Suso's narrative by the figure of a French king in the *Ram with the Iron Crown*. As will be shown in the full discussion of the iconography and format of this painting in chapter V, Suso narrates a sequence of incidents between a local duke and bishops in the regions of Constance. The French translation renders the text faithfully and *duc* is used consistently for the rank of the lay leader in the ecclesiastical dispute. The artist, however, portrays fleur-de-lis on a loggia over a crowned monarch receiving a deputation. It is most unlikely that the painter would have wilfully represented a duke as a king. He was most likely following written instructions of the programme which called for a French monarch. The Dominican originator of the instructions could have suggested a king be portrayed because he knew that a great Church Schism in living memory had been ended through the help and diplomacy of Charles VI in *c.* 1419, or because he was aware that his own king, Charles VII, had helped in 1448 to end the reign of the anti-pope Felix V and so conclude another Schism.[38]

[34] Two fourteenth-century descriptions are mentioned by Lucy F. Sandler, "Jean Pucelle and the Lost Miniatures of the Belleville Breviary," *Art Bulletin*, LXVI (1984), p. 73.

[35] Ed. J. Porcher, *Jean le Bègue. Les Histoires que l'on peut raisonnablement faire sur les livres de Salluste*, Paris, 1962.

[36] See Manion, *Wharncliffe Hours*, p. 8, and Plummer, *Last Flowering*, p. 65.

[37] See Mandonnet, *Order of Preachers*, p. 361.

[38] These historical details are discussed further in chapter IV.

It was, furthermore, the compiler of the programme of illustrations who was responsible for the new pictorial emphasis the text in the Brussels volume receives through the great cycles of the Stations of the Cross, the Life of Christ, the appropriateness of the Dominican way of life and thinking; the utter futility of the life led by the uninformed, the uneducated, the frivolous, the vain and the sinner.

The hypothesis that a Dominican was responsible for elaborating the illustrative programme fits neatly in with the role of executing it, entrusted to the Parisian artist, the Rolin Master. If the painter was like so many Parisian miniaturists who resided in the parish and quartier of Saint-Séverin, where the name Rue des Escrivains recalled the craft of the inhabitants, he had merely to ascend the nearby steep thoroughfare, the Grande Rue, also known as the Rue Saint-Jacques. At the top he reached the Porte Saint-Jacques in the perimeter walls of the *enceinte* built by Philip Augustus. Close by the Gate stood the Provincial Convent of the Order of Preachers.[39]

As a form of conclusion, we quote a comment by Spencer. She believed that the Brussels *Horloge* manuscript could be considered "a special order for a particular person or occasion, important enough to justify the collaboration of an imaginative and learned theologian with a vigorous and inventive illuminator able to sustain the quality of his work."[40]

[39] Cf. Rohault de Fleury, *Couvents*, II, unpaginated; see 'Paris' entry. The proximity of the Eglise Saint-Séverin to the Rue Saint-Jacques and the Rue des Escrivains can be clearly seen on a map of Paris from the early fourteenth century (redrawn according to modern topographical practices), reproduced in Françoise Baron, "Enlumineurs, peintres et sculpteurs parisiens des XIIIe et XIVe siècles d'après les rôles de la Taille," *Comité des travaux historiques et scientifiques. Bulletin archéologique*, n.s. IV (1968), pp. 37-121. fig. 3; on pp. 119-20 of the article are numerous references to enlumineurs resident in the Rue Erembourc de Brie in the same parish of Saint-Séverin in the fourteenth century.

[40] Spencer, *Horloge*, p. 292.

CHAPTER FOUR

The "Déclaration des Hystoires"[1]

When the reader or viewer opens the Brussels manuscript and begins to turn the folios, the first item that presents itself to view is a Table of Contents (fol. 2) for the texts transcribed in the volume. Moving on, one expects to arrive quickly in front of the first text, the *Horloge de Sapience*, but it is still not in sight. Instead, there are eighteen pages of a *Déclaration* or commentary on the paintings (fols. 3-11v.).

The *Déclaration* was not mentioned by the librarians of the Bibliothèque Royale in Brussels when they published their statements about the codex. Spencer,[2] on the other hand, did have something to say. She called it "a descriptive index to all the miniatures," and later expanded on her thoughts, saying "the miniatures are described or interpreted for us by the fifteenth-century writer of the *déclaration des hystoires*."[3]

She reproduces in footnotes and without comment the wording of the statements for the first two paintings, *Sapientia in Majesty* and the *Clock Chamber*.[4] She seems to suggest that the views about *Sapientia in a Terebinth Tree* are inadequate, at least for the associations of images she herself sees: "The fifteenth-century commentator of the *déclaration des hystoires* (f.3) merely describes the miniature, identifying her jar as containing sweet-smelling ointment, the twig as balsam, the book as her doctrine. Had he read further in Suso's text he might have added the Tree's own words..."[5] About *Sapientia asks the Disciple for his Heart*, Spencer records the fact that the "*déclaration* picks up the Latin from this scroll but adds no comment on the vision."[6]

As for the large composition of the *City of Religion in Ruins*, she writes, "The *déclaration* (ff. 5v.-6v.) explains that the saints who have left this, the city of religion, are standing at our left and that at the right, in five panels, are the victories of the vices over the virtues. Below, Dominicans crowd into a shop to sell church treasures. Even if we had no other evidence to indicate the persistence into the fifteenth century of these problems, the length and detail of the *déclaration* would suggest that attention was being called to matters of urgent importance."[7]

[1] The comment on the *Horloge* pictures is rubricated *Déclaration des hystoires*, while for the remaining three illustrations the term is in the plural *Déclarations des hystoires*, probably because three different texts are involved. In this chapter I employ the singular form *Déclaration*, irrespective of the group of paintings I discuss.

[2] Cf. *Bulletin de la Bibliothèque Royale*, V (1961), pp. 48-58, and *Bruxelles, Bibliothèque Royale Albert Ier. Quinze années d'acquisitions*, Brussels, 1969, pp. 84-6.

[3] Spencer, *Horloge*, pp. 277 and 281.

[4] *Ibid.*, p. 282, n. 18 and pp. 282-3, n. 19.

[5] *Ibid.*, p. 284.

[6] *Ibid.*, p. 285.

[7] *Ibid.*, p. 287.

Spencer further observes that when the painting of the *Ram with the Iron Crown* is discussed in the *Déclaration*, the reference to the time of the Schism and Charles VI is vague.[8] Commenting in her turn on the two *Hell* miniatures lost when fol. 57 was removed,[9] she fails to state that each one was a triptych. When in fact she reports the description given by the expositor, she omits the words of the first half of account of the first panel.[10]

She also touches upon the nature of the *Déclaration's* view of the last painting in the series: "In the last miniature (f. 127) the mystic marriage takes place, more explicit in the picture than in the text, for both the text and the *déclaration* virtually ignore what the painter is inspired to give us, the marriage of the Disciple to a member of the Trinity."[11]

In sum, Spencer did not raise several fundamental questions concerning the *Déclaration*, such as the manner adopted for viewing the formats, the exegetical methods employed by the expositor, the need for the work, its date, the identity of the author, and so on.

1. THE TERM "DÉCLARATION"

What then is the nature of a text that has in its title the word *Déclaration*? Examples of the term's usage recorded in standard reference dictionaries, such as those of Godefroy and Tobler-Lommatzsch,[12] reveal that by the fifteenth century the word had a conceptual field that included not merely the idea of 'statement' or 'something declared' but also the sense of 'explanation' or 'exposition'. This second meaning may well be due to the contemporary influence of the Latin word *declaratio*, circulating in scholastic and religious milieux with the meaning of 'exposition'.[13] The ambiguity contained in the title *Déclaration des hystoires* is mirrored in the two-fold perspective adopted by the author when discussing the paintings: he describes them or he expounds on the nature of their subject matter.

2. LOCATING THE PAINTINGS

In the *Déclaration* the remarks about each painting open with a serial number and a statement of the illustration's position in the French text. On four occasions the expositor associates a miniature on a recto side with another on the facing verso side or on the following verso side.[14]

In locating the illustration in the text, the author usually makes a brief statement about the textual matter. For example, on fol. 49v. is the triptych *Disciple Offering his Heart to Sapientia*, *Sapientia with Faith, Hope and Charity*, and the *Disciple Shutting out Sapientia*. The expositor includes the phrases "The seventeenth painting is in the eighth chapter at the place of the paragraph *O mon cuer*. In the chapter the author relates how Divine Sapientia visits the soul of her loyal friend..."[15]

Another aspect of the *Déclaration* is the fleeting mention of the artist, not by name, but through the use of the pronoun *he*, stated or implied: "And he depicts in this painting the religious form

[8] *Ibid.*, *Horloge*, p. 288. In n. 33 she prints the wording of the *Déclaration* for this miniature.

[9] See above, chapters I and II.

[10] Spencer, *Horloge*, p. 291, n. 38.

[11] *Ibid.*, p. 292.

[12] See F. Godefroy, *Dictionnaire de l'Ancienne Langue Française et de tous les Dialectes*, Paris, 1881-1892, IX, 282a; and A. Tobler and E. Lommatzsch, *Altfranzösisches Wörterbuch*, Berlin and Wiesbaden, 1925 +, I, 1253. I owe these references to Professor K. V. Sinclair.

[13] See Latham, *Medieval Latin Word-List*, p. 134.

[14] See above, chapter II, § 1.

[15] Cf.: La xvij^e hystoire est ou viij^e chapitre a l'endroit du parafe *O mon cuer*, ou quel chapitre il parle comment la Divine Sapience visite l'ame de son loyal amy (fol. 8).

of life..." (fol. 7); "First of all, he rehearses and depicts the story of David..." (fol. 9).[16]

Among the place assignations three may be considered important for the question of priority. Did the author write the *Déclaration* before the paintings were on the folios or after they were in place? His assignment of ordinal numbers that coincide with the pictorial reality of the Brussels *Horloge* series, and the quotation of the exact French words that open a paragraph below which or above which a painting is situated suggest that the pictures were present on the folios either in a completed form or at least in final sketch plan by the time he wrote the *Déclaration.* The fact that the expositor identifies segments within pictures confirms it.

3. READING THE MINIATURES[17]

The order in which the expositor reads the contents of the miniatures, or refers to the segments of the Rolin Master's complex formats merits a comment in itself. The pictorial structures are many and varied. There are monoscenes in one frame with unified narrative in time and place (*Youth in Dalliance*); polyscenes within one frame that have unified narrative in place, but not in time, as in the group of subjects on fol. 17: *Reception of a Postulant*, *Studying in the Cell*, *Lessons in the Refectory*, *Tree of Life.* The action in all four scenes takes place in a Dominican convent. Triptychs are a collection usually of separate scenes whose subjects appear unrelated and whose dramas occur in different places and in different time periods, for example, the *Disciple Offering his Soul to the Trinity*, *Life's Pilgrim* and the *Disciple's Contemplation.* There are also the polyptychs whose parts are in separate frames, each recording distinctive events in various places, as exemplified by the structure in nine compartments on fol. 23.

A progression for reading a polyscene with, say, four segments, two upper and two lower, may be to scan from left to right across the top, then come back to the left, and move again to the right. In the *Déclaration* the author departs from this pattern on several occasions, and one suspects that aesthetic considerations or chronological concerns may be contributing factors. A few examples will suffice.

The full-page illumination on fol. 17 to which reference has just been made is composed of four segments of almost equal dimensions, the upper two extending into a serrated, concave arch across the top of the whole painting. A burnished gold edge surround the structure while gold dividers separate the compartments from each other. The commentator's description emphasises the passage of time and alleges that the sequence chronicles the Disciple's progress in understanding Wisdom. The first section shows the *Reception of a Postulant* and the viewer is told that because the author of the text, namely Suso, was a Preaching Friar, the newly admitted brother will henceforth be seen in the habit of that Order. The expositor moves to the adjacent frame in the upper register and announces that the Disciple is in his cell studying the Bible and more particularly the Books of Solomon and of Wisdom, and learning to love what Sapientia inspires in him. The commentator goes next to the lower left and observes the refectory, where the Disciple hears the reader proclaiming more wisdom from the Bible. The last scene captures, we are told, the moment when Sapientia hands the Tree of Life to the Disciple. Thus, one could say that the temporal and metaphysical reading of the four compartments coincides with the logic of the pictorial programme.

[16] Cf.: Et met en ceste hystoire la forme de religion... (fol. 7); Premierement recite et figure l'istoire de David... (fol. 9v.).

[17] Elements of this discussion are contained in my paper "Reading Fifteenth-Century Miniatures: the Experience of the *Horloge de Sapience* in Brussels, Bibliothèque Royale, MS. IV. 111," *Scriptorium*, XL (1986), pp. 242-8, pls. 13-16.

The second chapter of the *Horloge* concerns the Disciple's search for knowledge; its iconographic projection is contained in the ninth miniature which presents episodes of Christ's life. The painting is folio size and divided into nine segments separated in most cases by burnished gold frames. The top border is arched and serrated. The structure is composed of three horizontal registers, each with three scenes whose subject matter is *God the Father*, *Man of Sorrows*, *Christ Crowned by God the Father*, *Adam and Eve*, *Miracles of Christ*, *Christ Rejected at Nazareth*, *Sapientia and Disciple*, *Nativity*, the *Temptation of Christ*. This reading follows the method of analysis already applied for the earlier polyscene, yet, it results in a chronological jumble and metaphysical dislocation. If the order of reading the painting's components employed by the writer of the *Déclaration* is followed closely, harmony is restored at both levels of interpretation.

His reading commences where Sapientia gives to the Disciple knowledge of the Earthly Paradise wherein are the parents of mankind, animals and plants, and the heavens on high; this segment should be studied before the one above it where God the Creator of the Universe says "...through knowledge of these creatures one comes to know God the Creator." The viewer is then directed to the lower sections depicting the Infancy of Christ and the Temptation by the Devil, then to move upwards to the centre right compartment. Here Christ preaches and performs miracles; a large twisted rock is the boundary for a scene where Jews threaten to hurl Him down the mountain. The reading concludes with a viewing of the Passion Instruments, and the Man of Sorrows, followed by a movement to the top right-hand corner of the whole structure where Christ in Glory is crowned by God the Father. The progression this time has been from lower left to upper left, downwards to lower centre, across to the right, then to centre right, onwards to upper centre and finally to the upper right. Such a path is very different from the downward progression of the fifth painting, for now both artist and author of the *Déclaration* emphasise an upward movement towards the splendid vision.

The simple burnished-gold frame and the serrated oval coping over the painting on fol. 33 differ little from previous ones. The subject matter, the *City of Religion in Ruins*, is as distinctive in format, however, as the miniatures we have been discussing. This time the formal divisions are three vertical panels; the left one has four compartments occupied by apostles, martyrs, confessors and female saints; the middle one offers the viewer a panorama of the City of Religion, below which is a vignette of a money-lender; the right-hand panel contains five segments, each portraying an elegantly dressed woman antagonised by a hag.

On this occasion the writer of the *Déclaration* commences his commentary in the upper centre panel depicting the City and proceeds to the saints. From the time when Christ was prince and ruler of the City it has fallen into decay through bad government. It is full of monsters and wild beasts disguised as humans, and Christ, now reappearing at the gates, meets the temptresses Wrath and Sloth. After drawing attention to crime in various quarters of the town, the commentator moves to the contest between the Virtues and Vices, represented by the noble women and the fractious harridans. In conclusion, the *Déclaration* directs our gaze to the lower centre where religious are selling church plate. Admittedly, until the centre panel has been examined and explained, the two lateral ones do not seem related to the focus of attention. Nevertheless, the visual path followed this time resembles a meander, with an upper centre focus, then one moves across to the left vertical which is studied with a downward glance. One crosses to the right side and descends through the five sections, before concluding the 'reading' in the lower centre. The Virtues and Vices face right, away from the City, in keeping with the movement of expulsion and flight. One is conscious of being a witness to an apotheosis of Saints, and yet of being present at a moral decline that has its nadir in the lower centre where Mammon is triumphant.

One disconcerting element in the *Déclaration* is the six occasions in illustrations VII, IX, XII, XXVIII, XXXIII, XXXV, when the words 'left' and 'right' do not appear to agree with the location of objects or figures in the miniatures or with the positioning of compartments that make up polyptychs. Could the commentator have been viewing the scenes in a mirror as he was composing his account? Such a procedure would certainly explain some of the placements, but the practice would have been cumbersome and somewhat eccentric.

Another explanation could lie in the reverse image theory. It can arise when a painter relocates the contents, say, of a right-hand panel of a sketched lay-out to a left-hand one in his own composition. If this were the case for the Brussels series of pictures, one would need to postulate that the expositor was using a sketch plan rather than the finished work. Yet, there can be little doubt here that he was studying the completed programme; he knows precisely, and says so, which compartments stand in physical isolation from others, even though they are part of a theme being treated in a larger contiguous illustration.

A third theory is that the author of the *Déclaration* was placing himself metaphysically in the role of a participant in the selected dramatic representations, and was thus able to look out at the viewer, in much the same manner as an actor on a stage. When looking about him, he finds left has become right and right is perceived as left.

A variation on this is to suppose that two persons were interpreting the pictures, one viewing from without the frame, the other's perspective being determined by the desire to present the action from within the pictorial confines.

Lastly, there is the appeal of a scribal intervention. In other words, one supposes that a copyist could have inadvertently replaced *dextre* by *senestre* and *senestre* by *dextre*. It is not unknown for medieval scribes to leave incoherent passages in works they transcribe. However that may be, until another copy of the *Déclaration* comes to light, it will be difficult to test this theory in paleographical terms.

One thing seems clear: reading the polyscenes and polyptychs of the Brussels *Horloge* is not always a matter of a straightforward progression from top left to lower right. Judging from two sample complex structures, one for episodes from the Life of Christ, the other for the City of Religion in Ruins, a fifteenth-century viewer could benefit from consulting the *Déclaration*. It assists him to comprehend the temporal, metaphorical and spiritual registers of the paintings and it orientates him in the ground plan set down for his edification.

4. DESCRIPTION

Colourless description or flamboyant exposition? Which form will the commentator employ after he has located the picture? *A priori* each miniature has the potential to serve as the subject of a description or of an exposition. An examination of all the wording in the *Déclaration* fails to reveal the reason for the author's choice of one or the other. There is, nonetheless, a correlation between the length of comment and the nature of what is said. The descriptive remarks are always the shorter of the two approaches. The extent of the criticism may also be related to the explicitness and popularity of the subject matter of the painting, that is, the degree to which the picture is self-explanatory on the one hand, and its theme is easily recognisable on the other.

The author's descriptive method is exemplified in the passage about *Christ Offered the Sop* (fol. 80), a painting filled with high drama, unresolved tension and spent emotions, and one that is perhaps among the finest of the Rolin Master's *Horloge* series. After locating the illustration in the narrative sequence of Suso, the commentator announces succinctly that a relevant part of the Passion is depicted by the artist. A similar procedure is adopted for the icon *Virgin Suckling the Holy Child*. Because Suso praises in particular the Virgin's cherished breasts and milk with which

Christ was nourished, "this painting shows the blessed matron" who uttered the encomium as reported in Luke's Gospel. In the light of this terse style for descriptions, the term needs to be understood as 'brief description'.

5. EXPOSITION

Expositions are more likely to occur when the commentator feels the need to be discursive about allegorised and metaphysical dimensions he encounters. One observes that the more unusual the elements are which are combined in a format, the more the writer of the *Déclaration* is prepared to amplify his comments. Three miniatures enable the viewer to appreciate the nature of this practice. In one, the commentator resorts to Biblical narrative; in a second, to allegory; in a third, to history, and recent history at that.

The component subjects of the triptych on fol. 30v. which portrays *Jael Welcoming Sisera*, *Jael Slaying Sisera*, *Samson and Delilah* have as their source of inspiration Suso's cryptic remark that an unidentified person, when asking for water, had received milk. The simple title supplied for all three pictures by the artist is *Ces figures signifient la vanité decevable du monde*, "These allegories represent the treacherous vanity of the world." The expositor is not content with such a perfunctory identification of the drama and offers his reader a more personal statement about the morality exhibited by Jael and Delilah. He names the Biblical sources and provides enough of its narrative for the observer or reader to orient himself in the drama evolving in the triptych. One notes that the writer of the *Déclaration* speaks of only two allegories, one concerning Jael and the other Delilah, even though there are three pictorial segments. Clearly, the first two scenes constitute two aspects of the one allegory, as far as the expositor is concerned. He offers an epitome of the abstract quality of the visual statement when he writes "it illustrates falseness and deception in the world." He then concludes his comment with his own moral for the hand of friendship that becomes the death-dealing instrument: anyone practising such deceit will receive everlasting death.

Turning to his explanation of the complex statement in the *City of Religion in Ruins*, one perceives that it is most elaborate for the actions of the characters; it is most perceptive in the identification of the personages; and it is most transcendant in the matter of allegories. The concepts he expresses are by and large not in Suso's text. To be sure, the Dominican mentions the idea that the Virtues were once in combat with the Vices, but does not develop it. He is more expansive about the ruin and decay that set in, once Obedience, Chastisement or Correction, Poverty, Chastity and Charity have been driven out of the City of Religion.

For its part, the pictorial narrative accompanying the text seems to be following earlier iconographic traditions and not Suso's words. The result is that select Vices are presented in conflict with select Virtues.

The expositor offers extensive descriptions of the figures' clothing and attitudes or stances; he explains the symbolism of their gestures and clarifies the allegories for his readers. Throughout the labyrinth of character delineations the commentator trails a thread of salvation: it is for the viewer or reader to grasp; it leads to a harsher awareness of a grim reality. Whatever Suso wrote and whatever the Rolin Master painted, the Virtues and Vices are battling ceaselessly to obtain control of every mortal. The verbs in the present tense serve to heighten the immediacy of the message. The swollen and leprous tongue of Mendacity causes a doleful lament to rise from the expositor's lips as he recounts instances of venality practised by persons in authority in the City of Religion.

The commentator refrains from declaring the source of his allegorical information; however, any fifteenth-century reader who was familiar with the literature about the Vices and Virtues would suspect that the *Déclaration*'s commentary could reflect a knowledge of the popular *Pèlerinage de la Vie Humaine*, composed by a Cistercian contemporary of Suso named Guillaume de Digulleville. This source is discussed in more detail in chapter V.

Whereas in general, the author of the *Déclaration* tends to present the pictorial narrative of abstract themes in a timeless eternity, there is one occasion where he believes the allegorical painting reflects historical events, to the point that he dates them. The illustration in question is the *Ram with the Iron Crown*, which, as it happens, follows immediately after the *City of Religion in Ruins*. The relationship between the two allegories is announced by Suso: a man of God in the ruined City sees the Ram in a vision. The expositor is the one who decides to interpret the episodic vision as an example of what happens to the Church when Vices hold sway and Virtues are cast out. The Church suffers villainous schisms. The Ram allegory is linked by the expositor to the Schism concluded in 1417 when Martin V was elected Supreme Pontiff.[18] In naming the period the author of the *Déclaration* intentionally sets aside the simple facts narrated by Suso in the Disciple's vision. Confirmation of this disregard for Suso occurs in the commentator's remark that the Ram represents the Pope, and the foxes symbolise self-seeking and ambitious religious and courtiers in the papal entourage. Thus, the modern reader or viewer cannot allow himself to be lulled into believing the pronouncement that all is clear in the painting.

One must further ask why the author of the *Déclaration* makes a favourable remark about Charles VI in striving to end the Schism. What was the appeal of such a remark to the viewer or potential viewer, that is, the purchaser of the manuscript? It would surely make good sense if the prospective owner of the volume was someone in the entourage of Charles VI's son, Charles VII. Since the coat of arms of the first and subsequent owners are not royal, the viewer was not a member of the blood line. The compliment paid to Charles VI would still be appreciated by a supporter of the reigning king, such as a mignon, an intimate or a high court official.

It is believed that these strands of ideas are well interwoven if the intent of the expositor is understood not in the context of the pre-1417 Schism, but in the context of the years 1448 and 1449 when the Schism was finally resolved.[19] The implication of the statement in the *Déclaration* is that the French king of the day should continue to pursue the diplomacy of his father when dealing with the Papacy. In fact, Charles VII was actively negotiating in 1448 for the rescission of his own 1438 Pragmatic Sanction and for the effacement of the dual papal image. Eugenius IV was deposed in 1439 by the Council of Basel but refused to recognise their mandate. The Council elected the widower, Amédée VIII, duc de Savoie; he took the name of Felix V and ruled from Lausanne.[20] Charles VII sent an important delegation to Eugenius IV in Rome in 1446, but nothing concrete was decided. On the death of Eugenius in 1448, the Cardinals elected Nicholas V whom the Council at Basel refused to acknowledge. The French king tried again to negotiate a solution. He sent to the new Pope in 1448 his most able prelates in the persons of the Archbishop of Rheims, Jean Juvenal des Ursins; the Bishop of Carcassonne, Jean d'Etampes; and the Chancellor of the University of Paris, Thomas de Courcelles; Jacques Coeur, the king's principal creditor, was also a member of the embassy. They arrived in July 1448. Two months later another delegation left Paris for Lausanne. The French diplomatic activity had its rewards in the following Spring: Felix V resigned on 7 April 1449 and the Schism ended two weeks later by official concordat.[21]

[18] Cf. N. Valois, *La France et le Grand Schisme d'Occident*, Paris, 1896-1902, 4 vols. and his *Le Pape et le Concile (1418-1450)*, Paris, 1909, 2 vols. On the election of Martin V, see the eye-witness accounts in *The Council of Constance, the Unification of the Church*, translated by Louise R. Loomis, ed. and annotated by J. H. Mundy and K. M. Woody, New York, 1961.

[19] Cf. J. Toussaint, *Les Relations diplomatiques de Philippe le Bon avec le Concile de Bâle 1431-1449*, Louvain, 1942; E. Delaruelle, E. R. Labande, P. Ourliac, *L'Eglise au temps du Grand Schisme et de la Crise Conciliaire 1378-1449*, Paris, 1962. For a survey of the pontificates of Eugenius IV, Felix V and Nicholas V, see M. Creighton, *A History of the Papacy*, London, 1899-1901, II, pp. 196-319 and III, pp. 3-110.

[20] Cf. Barraclough, *Medieval Papacy*, p. 184.

[21] Cf. Castelot and Decaux, *Histoire de la France*, III, p. 158.

6. PURPOSE

What can be said is the purpose of the *Déclaration*? To be sure, it is to explain and describe the paintings in the Brussels *Horloge*, but there still remains unanswered the question of need. Why should it be necessary to provide a guide to a series of pictures, when hundreds of texts in the Middle Ages that were illustrated as richly as the Brussels manuscript did not receive a viewer's guide? Specifically in the matter of the *Horloge* as a translation, why should it of all the copies be supplied with a *Déclaration*? The answer to this particular point must be sought in the number and pictorial method of presenting the formats. Firstly, it is the only manuscript of the *Horloge* to be illustrated with such a large number of paintings, one hundred and eight tableaux in all. Secondly, in the context of book illustration, there was a need to assist the viewer to orientate himself in the text, since only four of the total number of paintings have legends. Lastly, one could imagine a situation where the detail supplied in the narrative and exegetical comment is a function of the degree of intellectual and iconographic ignorance on the part of the *destinataire* of the manuscript.

7. AUTHOR

The didactic concerns of the expositor are a factor to be acknowledged when assessing his aims and objectives. He even strikes a preaching tone in several passages. The comments on the *Tree of Life*, for example, conclude on this note of hope: "Thus, by the fruit which the Tree of Life has borne, life everlasting is returned to us." The painting of *Saul's Death* is held to be a warning of the way "Our Lord, the true Judge, avenges the tribulations of His innocent servants. And by this illustration it is manifest that if tribulations or accusations are well founded, they are atoned for in this world. If they are unjust, vengeance falls on the heads of those who perpetrate them. Thus, in all things one must be patient." In his commentary for the painting of the *Battle for the Soul*, the author of the *Déclaration* asks rhetorically "Who is there who has lived in such a way that he has nothing to cleanse?"

What kind of person would need this enlightenment? It is most unlikely he was a member of the Church, whether a regular or a secular ecclesiastic. One may assume that such a person did not need to be told where to locate the story of Susanna or of Samson and Delilah. On the other hand, one could comprehend the usefulness of the "lessons" the paintings contain, if the future owner was a courtier, for he would most probably have been unfamiliar with the work of Suso, or with the Dominican's views on Christianity as practised in the previous century. And of course, these negative arguments *ex parte* add something positive to the portrait of the *Déclaration*'s author.

He was manifestly someone with the ability to consider paintings and their contents intelligently, to locate them in their chapter settings in a Dominican author's allegorical work on Wisdom, to comment in detail where necessary on Biblical passages which are narrated in the miniatures, to moralise on the subject matter and themes and to explain figures in terms of allegory wherever Suso had omitted to do so. Such refined and purposeful acts of intelligence point to a person who was knowledgeable in Latin, in the Bible and in the broad sweep of Christian theology. Moreover, this unidentified author of the *Déclaration* had most probably been educated in a *studium* since he explains two literary allusions from an educator's viewpoint. Suso was content to give merely names, but the expositor identifies more readily. He claims that the Epicureans "maintain that man's happiness comes from following his own pleasures and desires." Further on he glosses the name Tantalus in these terms: "The Disciple compares himself to Tantalus, who, according to the poets, was a man placed between a fountain and a tree; when he wished to drink, he could not lower himself; when he wished to eat, he could not reach." The authoritative poets the exegete had in mind must have included Vergil and Ovid.[22]

Further light is shed on the expositor's religious standing by random remarks in Dominican contexts. At one point in the *Horologium* Suso observed that the Disciple's custom of walking around the cloisters was in remembrance of the procession made by Christ to Calvary. The Rolin Master was inspired to paint a format showing *Cloisters and Fountain* (fol. 15v.). The author of the *Déclaration* in his turn was so enthusiastic about Suso's allusion to a current practice in the Order of Preachers, still observed in the fifteenth century, that he wrote "...this perambulation is represented in the present illustration which is structured like a cloister with the stretch of grass and fountain in the middle." But in the illustration there are no figures in the act of moving in a procession. He surely did not expect his reader and viewer to a see a procession that was not in front of his eyes. One deduces that the commentator's reference to the procession sprang from his own awareness that the cloisters of a Dominican convent were a venue still used for the Passion Procession.

The exposition for the *Reception of a Postulant* informs the reader that the youth enters into religion and that he belongs "to the Order of Preachers, because the writer of this Book belonged to the said Order. Henceforth, he will be wearing this habit." One wonders why this point is emphasised when it was quite clear that the artist had portrayed the friars receiving the postulant as Dominicans. Besides, the viewer or reader had learned from a study of the first painting and from the opening words of the *Horloge* that the writer of the work was a Preaching Friar.

When discussing the *Prior's Investigation* the expositor nearly lets his guard drop. He mentions the investigations of accusations that are conducted in Dominican convents and he employs phrasing that suggests he knew these occurrences from first-hand experience. He declares that the painter "depicts in this illustration the religious practice customarily observed daily or at intervals, by which, in order to preserve the discipline of the Order and to keep a brother from causing harm, one may accuse him with a view to correction and in *caritas*. And because no one has this virtue any more, it often happens that one accuses one's brother out of vengeance, bitterness or anger..." Suso, speaking through Sapientia, did not have occasion to make all these observations when he was describing a prior's role in litigious matters between friars.

Taken collectively, these disparate pieces of information suggest that the expositor was a Dominican. His nationality was most certainly French and his balanced remarks about Charles VI, the father of his own king, imply he was a royal supporter. Links between the Royal Family and the Dominican Order had been close since the second half of the thirteenth century. Many members of the Valois and Bourbon lines were interred during the thirteenth and fourteenth centuries in the chapel of St. Thomas Aquinas in the Church of the Provincial Convent of Saint-Jacques-le-Grand in Paris.[23]

One suspects also that the expositor was a confessor to a member of the Royal Household. The identification of all the Preaching Friars serving as confessors in the royal entourage about 1450 has not been made. However, a most interesting fact has emerged in a preliminary survey of the topic. Charles VII's confessor from 1432 to his death in July 1448 when aged 68, was the Bishop of Castres, Gérard Machet. He was not a regular. He was succeeded, at a date as yet undetermined, but presumably soon after 1448, by a Dominican from the convent at Evreux, one Robert Baygard.

[22] The statements in the works of both poets are discussed in articles by J. M. Steadman, "Tantalus and the Dead Sea Apples," *Journal of English and Germanic Philology*, LXIV (1965), pp. 35-40 and P. Ménard, "Je meurs de soif auprès de la fontaine. D'un mythe antique à une image lyrique," *Romania*, LXXXVII (1966), pp. 394-400.

[23] Cf. Chapotin, *Histoire dominicaine*, pp. 205-428.

Very little else is known about him.[24] Charles VII's second son, Charles de France, was served by a Dominican confessor, as was the friend of both Charles de France and Guillaume Gouffier, namely François II duc de Bretagne.[25]

8. DATE

The date of the composition of the *Déclaration* can only be conjectured. One fact emerged above, the work followed the completion of the pictures or of their final lay-outs in the manuscript. But this information is of little assistance, since the date of the Rolin Master's paintings in the Brussels *Horloge* cannot be accurately determined. On the other hand, if the role of Charles VI in having Martin V elected at the Council of Constance is understood as a moral for Charles VII and his ecclesiastic advisers to emulate in their own day, then the time for propagation of this attitude was in the period 1448-1449. It follows then that the French negotiations to oust Felix V were not complete by the time the *Déclaration* was composed. Therefore its date must antedate April 1449, the month of Felix's resignation.

[24] I am most grateful to Fr. Bernard Montagnes O.P. of the Convento S. Sabina (Aventino) in Rome for this information about the king's confessors.

[25] As discussed in chapter I of this study.

CHAPTER FIVE

Iconography and Text

I *Sapientia in Majesty* Fol. 13

Suso's *Horologium* opens with a prologue which is completely translated by the French Franciscan,[1] except that he preserves the author's first eleven words in the original Latin before giving his version of their sense: "Here begins the book called the Clock of Wisdom which was composed by Brother John of Swabia[2] of the German Nation [and] of the Order of Preachers. Solomon in his Book of Wisdom says in the first chapter *Sentite de domino in bonitate, et in simplicitate cordis querite illum.* Set your mind sincerely upon God, confirm yourself in His ways and His will, seek His presence with simple heart and pure mind... He shows Himself to those who have trust in Him, and who, after doing what they can and ought in good works, expect moreover the grace and help of God."[3]

Another relevant passage in the prologue is: "When the Holy Church was first founded, Divine Sapientia appeared in several ways and guises to those chosen for glory, who are in effect those chosen to be saved, and with her radiance she enlightened and directed their thoughts..."[4]

The *Déclaration*'s comment reads: "First of all, at the opening of the Book is Sapientia in the image and likeness of a woman, representing Jesus, Our Saviour, who is called the strength and wisdom of God the Father..." This identification of Sapientia as Christ is not in the prologue; moreover, the expositor paraphrases St. Paul, whose words can be seen on a banderole in the

[1] The reader will recall that the translator of Suso's *Horologium* into French was a Franciscan Friar of the Observant Convent in Neufchâteau (arr. des Vosges); see above, chapter III, § 2. The extracts of the *Horloge de Sapience* which are supplied in the notes are derived from one manuscript only: Brussels, Bibliothèque Royale, IV. 111.

[2] Several Latin manuscripts of the *Horologium* give Suso's first name as Johannes; Jean occurs in manuscripts of the French translation; see Ancelet-Hustache, *Quelques Indications*, pp. 161-2 and Künzle, *Horologium*, p. 10.

[3] (Fol. 13) Cy commence le livre appellé Horologe de Sapience, lequel fist frere Jehan de Souhaube de la nacion d'Almaigne [et] de l'ordre des Prescheurs, Salmon en son livre de Sapience ou premier chapitre dit: *Sentite de domino in bonitate, et in simplicitate cordis querite illum.* Sentez et entendez de Dieu en bonté, confermez vous en son ordonnance et a sa voulenté, querez sa presence en simplesse de cuer et en purté de pensee... Et il se monstre a ceulx qui ont fiance en lui [et] qui, aprés ce qu'ilz font leur povoir et leur devoir de bien faire, attendent au seurplus la grace et l'aide de Dieu.

[4] (Fol. 13) Au commencement de la fondacion de Saincte Eglise la Sapience divine en plusieurs guises et en plusieurs manieres se monstra et s'apparut a ceulx qui estoient esleuz a gloire qui sont des esleus pour estre sauvez, et de sa lumiere enlumina et endoctrina leurs pensees...

eighth illustration. One records, further, that the description of Sapientia in the *Déclaration* as being "...seated on a throne of Majesty like the true God..." is absent from the prologue as Suso composed it.

The commentator refers to the preacher and the persons of different stations who are exhorted "to set their mind sincerely upon God", but the religious kneeling to the left of Sapientia is not mentioned; likewise the figure in shining armour on the right-hand side of the painting. Spencer commented upon this.[5]

The *Déclaration* names the preacher as the "...Disciple, author and composer of this Book...", thus suggesting that the single figure is perceived as performing different roles. It portrays Suso the author, Suso the orator, Suso the member of the Order of Preachers. What is interesting is that the commentator lends an autobiographical dimension to the figure by suggesting it represents the Disciple, the follower of the teachings of Sapientia. In doing so, the author of the *Déclaration* fails to point out that the artist's portraiture of the Disciple is not the man in the pulpit but the pensive friar in the left foreground.

It would seem that the emphasis the fifteenth-century critic gives the icon of *Sapientia in Majesty* is a Pauline one. There is no hint that the portraiture alludes to the principal characters of the whole *Horloge de Sapience*.

A miniature that serves pictorially as a frontispiece both to a text and to a whole series of illustrations is really fulfilling two roles; it may epitomise the whole written work and recall specific elements of the prologue to the work. A closer examination of the themes in *Sapientia in Majesty* seems to confirm this view.

To the left of the centre oval four men[6] listen beneath a pulpit to a Dominican preacher uttering the words from Sapientia I, 1, here inscribed on a banderole *Sentite de domino in bonitate* "Set your mind sincerely upon God". The gesticulating orator may be intended to represent Suso himself. The kneeling figure at the left is balding, if not tonsured, and wears a blue gown; over his right shoulder is thrown a hat attached to a scarf of dark red hue. He has the appearance of a magnate or lord temporal. The man at the rear of the group is clad in a dark red gown and a small fur collar graces his neck; the belt, hat and scarf he sports are black. Next to him is a kneeling monk in grey habit with a white cord trailing on the ground from his waist-line. He could be a Franciscan, perhaps the translator of the *Horologium* into French. The last of the four figures is clearly a Dominican doubled over in a deep bow. The posture is one assumed by Sapientia's Disciple in other paintings of the Brussels *Horloge*.

The viewer recognises the orator as the symbol of the Preaching Friars' mission to take the word of the Gospels to all the corners of Christendom and beyond with a view to saving souls. Given the propinquity of Sapientia holding an open book, one must associate her with the exemplary zeal for instruction and knowledge that is also the characteristic of the Dominican way of life.[7]

At the lower right of the composition is a nobleman whose surcoat is emblazoned with the arms of Brittany. He wears a coronet. Both circlet and blason are overpaintings. Beneath the black and white ermine, one can discern another distinctive armorial bearing, that of the Gouffier family. The presence of these heraldic devices was discussed above in chapter I, § 4.

The seigneur kneels at a *prie-dieu* and faces towards Sapientia. Over the stool lies a rich, blue cloth with gold decoration; carefully positioned on the drape is an open book with fluttering pages.

[5] Spencer, *Horloge*, p. 282.

[6] The account given here of the four persons differs from that of Spencer, *Horloge*, p. 282. She saw "two laymen, a Franciscan and a Benedictine."

[7] See Mandonnet, *Order of Preachers*, p. 356 quotes the Constitutions: "Our Order was instituted principally for preaching and for the salvation of souls." He observed further (p. 357): "The Preachers consequently made study their chief occupation, which was the essential means, with preaching and teaching as the end."

Under the orant's knees one can see a bright red cushion, tasselled at the corners and with clasps of pearls.

The presence of a donor or owner, with or without armour, kneeling before an icon of his moral benefactor, patron or guardian is so traditional in medieval illuminations that when one encounters the imagery in fifteenth-century miniatures, one can rarely claim it is new, distinctive or unusual. Numerous Books of Hours contain portraitures of this kind.[8] But the knight in the *Horloge* frontispiece appears to see Sapientia in a vision. The ethereal atmosphere is conveyed by the position of the icon in the heavens and surrounded by seraphim as if they were the veil through which one passes to reach the presence of Sapientia.

The centre of focus in the miniature is the majestic figure of Sapientia seated on a Gothic throne. She is dressed in a mantle of blue and a blouse of crimson, with a robe held in place by means of a jewelled band. In her right hand is an open book and in her left hand rests an orb. From the yellow-coloured throne radiate gold lines, and, encapsulating all, is an ovoid representation of fluttering, bright-red seraphim. In both of the upper corners of the composition are three angels, arms crossed or hands put together in prayer.

Spencer's observations contain insights about the subject: "The frontispiece presents a vision of Sapientia as God enthroned... For this interpretation of Sapientia, the illuminator had, of course, not only his text but plenty of precedents from illuminated Bibles and *Bibles Moralisées* as well as representations of the Maiestas Dei in missals... The celestial mandorla, the throne and the inverted silver crescent beneath her feet imply also that Sapientia is linked here to the Virgin of the Immaculate Conception, die schöne Madonna, and the vision of the *Ara Coeli* as well as being the counterpart of the Maiestas Dei."[9] Additional observations about the *Ara Coeli* complement the statement: "The iconography of the *Ara Coeli* calls for a circle of gold about the sun in which appeared the Virgin holding the Child. The analogy here rests upon Suso's repeated references to Sapientia as the Virgin, and upon the interpretation of the Book as Christ, *Verba Dei.* The crescent at her feet, symbol of the Immaculate Conception, draws together these two visions..."[10]

It is difficult to lend credence to the claim of a link with the *Ara Coeli* theme. Meiss and Verdier, who have studied the subject, note two important features of it in French miniatures, the first is the presence of the Holy Child in the Virgin's arms and the second is the sun with lambent rays.[11] Neither is present in *Sapientia in Majesty.* There are rays emanating from behind the throne, but they are not lambent, that is, burning in the shape of tongues of fire.

A statement that is unfounded in Spencer's exposition is "...Suso's repeated references to Sapientia as the Virgin..." It has no basis. An examination of the *Horologium* reveals that the author refers to her epithets in one passage, to the *Stabat Mater* theme in another, and to her name several times in three pages devoted to the need to recite the *Ave Maria* and the *Pater Noster.*[12] Suso also devotes one chapter to the Virgin out of the twenty-four in the *Horologium.* It is the sixteenth and it bears the title *Commendatio singularis Beatae Virginis et de dolore eius inaestimabili quem habuit in passione filii.*[13] A similar scrutiny of all the mentions of Sapientia reveals no reference to her as the Virgin.[14] There are no grounds therefore for considering the iconography of the Ara Coeli, the Immaculate Conception, die schöne Madonna as relevant to the miniature in question.

On the other hand, Spencer's remark about the presence of the *Maiestas Dei* (or *Domini*) theme in the painting merits further attention. The simplified elements of the iconographical imagery of

[8] Cf. Harthan, *Book of Hours*, p. 26.

[9] Spencer, *Horloge*, pp. 281-2.

[10] *Ibid.*, p. 282, n. 18.

[11] See Meiss, *Limbourgs*, Text vol., pp. 139-41; P. Verdier, "A Medallion of the *Ara Coeli* and the Netherlandish Enamels of the Fifteenth Century," *Journal of the Walters Art Gallery*, XXIV (1961), pp. 8-37; Verdier's paper was mentioned by Spencer, *Horloge*, p. 282, n. 18.

[12] Künzle, *Horologium*, pp. 463; 499-500; 601-3, respectively.

[13] *Ibid.*, pp. 506-18.

[14] *Ibid.*, pp. 683-4 ('Register').

the theme are a central oval or lozenge within a rectangular frame, each of the four corners being occupied by one of the figures of the Tetramorph. Angels may also be present. Within the oval or lozenge is a represention of God in Majesty seated on a throne in a frontal pose. The icon may represent either God the Father or God the Son, often it is difficult to distinguish the identity of the subject.[15] When the figure holds in its left hand a crucigerous orb, the symbol of sovereignty, it is usually considered to represent the Almighty; if an open book is held in lieu of the orb, the figure is the Logos or Second Person of the Trinity.[16]

God the Son often resembled God the Father in early Christian art forms, especially when portrayed in Majesty, as the studies of Van der Meer have shown for the Apocalypse typology, and Christa Ihm for absidal mozaics in the fourth to eighth centuries.[17] In fifteenth-century French miniatures, particularly those in Missals, one meets God the Father in Majesty, he may then display other distinctive traits, such as a conical tiara and a long beard.[18]

The viewer will notice that Sapientia in the *Horloge* painting is nimbed and seated on a throne like a Maiestas figure and she holds an orb. In this guise she recalls the iconography of God the Father. However, she supports an open book, the symbol of the *Verba Dei*. She thus represents the Logos or Christ in Majesty.

At a textual level, the deified image of Sapientia prefigures several personalities and character roles in Suso's narrative. A leading Suso scholar has given us this profile of Sapientia in the *Horologium Sapientiae*: "Eternal Wisdom is at first a woman, the personification of Wisdom, the spouse of the soul, as in the Wisdom Books of the Old Testament. Later, Eternal Wisdom represents Christ."[19]

II *Clock Chamber* — Fol. 13v.

Suso had adapted his own German text, *Das Büchlein von der ewigen Weisheit*, or *Little Book of Eternal Wisdom*, into Latin and given it a different sounding title, the *Horologium Sapientiae*. He chose the vocable *Horologium* because of its rich association of ideas[20] between twenty-four hours and the twenty-four chapters of the text, between a clock's awakening function by chimes and bells and the text's aim to awaken and call the pious to their devotions, and between a clock's set path of numerical progression to record the passing of time and his treatise's objectives which included presenting his readers with a set plan of contemplation, meditation and ultimate spiritual union with Sapientia and Christ as the chapters are read.

There are two particularly relevant textual passages, one of a moral nature, the other allegorical and centred on the image of a clock: "And yet, Divine Sapientia is always on the alert to save all human beings, by wishing to improve the life of the said Elect of God, and is ever ready to remove and destroy all vices from their conscience. In this book she desires greatly above all to rekindle the dead sparks, to re-enthuse those who have grown old, to arouse sinners, to summon and urge

[15] See L. Réau, *Iconographie de l'art chrétien*, 6 vols, Paris, 1955-1959. *Ancien Testament*, p. 10.

[16] *Ibid.*, p. 9. The typology of the Logos from earliest times in Byzantine and Western Art is discussed in Gertrud Schiller, *Iconography of Christian Art*, 2 vols, Greenwich, Conn. 1971-1972, I, pp. 4-12.

[17] F. van der Meer, *Maiestas Domini: théophanies de l'Apocalypse dans l'art chrétien*, Paris, 1938, pp. 223 ff.; Christa Ihm, *Die Programme der christlichen Apsismalerei vom 4. Jahrhundert bis zur Mitte des 8. Jahrhunderts*, Wiesbaden, 1960. I should like to thank Professor Margaret Manion for suggesting I should consult these two publications.

[18] Réau, *Iconographie, Ancien Testament*, pp. 8-9.

[19] Clark, *Little Book of Eternal Wisdom*, p. 13. A key passage in Suso's chapter I reads: *Verum licet quaelibet persona per se accepta sit sapientia, et omnes personae simul una aeterna sapientia, tamen quia sapientia attribuitur Filio, et quia ratione suae generationis sibi convenit, ideo dilectum Patris Filium sub eodem significato consuetudinaliter accepit...* cf. Künzle, *Horologium*, p. 381.

[20] See Künzle, *Horologium*, pp. 57-8.

the slackers back to their devotions, and to awaken the negligent somnolents to the study of virtuous works."[21]

The relationship of the objectives of Sapientia to their attainment is suggested in a short passage of the Prologue: "...and it pleased our sweet and merciful Saviour Jesus Christ to show in a spiritual vision the manner and merit of this book to its author when He showed him a clock of beauteous and noble line whose wheels were remarkable and whose bells were ringing out melodiously."[22]

The commentator of the *Déclaration* stresses the aims and achievements of Sapientia when she is in contact with frail mortals. Following Suso closely, he points out that to represent this, the work is called *Horloge de Sapience.* He then turns for a brief moment to the illustration in which he sees and states the obvious: Sapientia's regulatory control of horological mechanisms. He reproduces the words on the facing of a time-piece located on Sapientia's left: *Ante secula qui Deus et tempora / homo factus est in Maria*[23] "God, who precedes time and the ages, became incarnate in Mary." This message proclaimed by the bells is said by the expositor to be appropriate for Our Lord Jesus Christ. The role ascribed to the Disciple is contemplative. The commentator's concerns are primarily with the religiosity of the picture. His viewer or reader may interpret how they like the pictorial presentation of the workings of the numerous time-pieces. Even the secular clothing of Sapientia does not arouse comment, nor the fact that, to all appearances, a woman is knowledgeable about clocks and their parts.

If the fifteenth-century expositor was unexpansive in his appreciation of the iconography, modern scholars have had much to say about it. In the course of her discussion of the subject matter Spencer remarked: "Suso says that Christ, i.e. Sapientia, acting as Temperantia, *lui monstra ung horloge ou ologe...*" and she adds the remaining words of the French sentence, the one which was our second 'source' quotation above. At face value, Spencer appears to be attributing to Suso the identification of Christ as Sapientia and her role as Temperantia. Suso's narrative is quite unambiguous; for him it is Sapientia who shows the Disciple the clock.[24] However, the French states at this point that Christ reveals the clock in a vision. It would seem that the text of the French version caused Spencer to proclaim Christ as the instructor.

The notion that Sapientia acts as Temperantia in the painting is plausible. According to one iconographical tradition, which has been the subject of much academic interest and research, the virtue Temperantia was regularly portrayed in late fourteenth and fifteenth-century illuminations executed in manuscripts of Christine de Pisan's *Epître d'Othéa*[25] and of tracts about the Virtues.[26] Her symbol or attribute is a clock, usually a portable one. She demonstrates the mechanism of her clock as a model for regulating human conduct.

[21] (Fol. 13v.) Et pourtant la divine Sapience qui tousjours est songneuse du sauvement de toute humaine creature, en desirant d'amender la vie des diz esleus et en voulant oster et aneantir tous vices de leurs consciences, veult et entent en ce livre principalment ralumer les estains, renflammer les (*MS.* les les) refroidiez, les pecheurs resmouvoir, les mal devots a devocion rappeller et promouvoir, et les endormis par negligence a l'estude des euvres vertueuses esveiller.

[22] (Fol. 14) ... a pleu a la douceur et pitié de Nostre Sauveur Jhesucrist de monstrer par vision espirituelle la façon et la valeur de cest livre a celui qui le composa quant il lui monstra ung horloge ou ologe de tres belle et de tres noble forme dont les roes estoient excellentes et les cloches doulcement sonnans...

[23] I have not found the source of the inscription. The lettering contains the abbreviation *m'a*, which I have read as *Maria*, in line with the reading adopted by the expositor himself, by Spencer, *Horloge*, pp. 282-3, n. 19, and by Künzle, *Horologium*, p. 62. On the other hand the abbreviation *m'a* may be read at times as *materia*, cf. A. Cappelli, *Dizionario di abbreviature latine ed italiane*, Milan, 6th ed., 1961, p. 209. Michel, *Horloge*, p. 294 preferred *materia.*

[24] Künzle, *Horologium*, p. 364.

[25] Cf. Meiss, *Limbourgs*, figs. 127, 128, 129.

[26] Numerous examples are mentioned in the papers of Charity C. Willard, "Christine de Pisan's Clock of Temperance," *L'Esprit Créateur*, II (1962), pp. 149-54; Rosemond Tuve, "Notes on the Virtues and Vices — Part I," *Journal of the Warburg and Courtauld Institutes*, XXVI (1963), pp. 264-303; Lynn White, Jr. "The Iconography of Temperantia and the Virtuousness of Technology," in T. K. Rabb and J. E. Seigel (eds.) *Action and Conviction in Early Modern Europe. Essays in Memory of E. H. Harbison*, Princeton, 1969, pp. 197-219.

Another minor textual problem concerns the reference to wheels in the same French passage. In Suso's Latin, there are no wheels, merely roses:

> Unde et praesens opusculum in visione quadam sub cuiusdam horologii pulcherrimi rosis speciossimis decorati et cymbalorum bene sonantium...[27]

Spencer's view[28] that *rosis* in the Latin is a later corruption of an original *rotis* was criticised by Künzle[29] who pointed out that of the two hundred and thirty-three codices of the Latin text he had examined, only four from the fifteenth century contained the reading *rotis*; he also made the point that from the choice of words employed by Suso, it was ornamentation on the clock not its workings that was uppermost in the Dominican's mind.

Suso mentions one clock with bells sounding and decorated with roses; he says nothing about the size of the time-piece, its location, whether it was standing on a floor or was portable, whether it was on a building or inside a chamber. The image that is presented to the viewer includes these textual elements, but it offers much more. The clock is in a splendid chamber with other horological pieces. Spencer's remark about the display of scientific devices rings true, it is "worthy of a collection of any fifteenth-century prince."[30]

As the viewer's eye contemplates the interior of the Clock Chamber, a large free-standing structure is prominent on the left: a weight-driven mechanical clock of some grandeur.[31] The upper half of the casing is open on three sides, allowing a view of the wheels and cogs that drive the time-piece; a dial screens the fourth side. One presumes the weights are suspended in the lower half of the structure which is enclosed by wooden panelling. What, however, is of interest is a thin rod which emerges from the conglomeration of cog-wheels near Sapientia's head and connects to a bar fixed parallel to the barrel ceiling of the room. From this bar projects a further rod which extends out of the picture plane of the miniature, traversing the text of the folio and ending in the upper margin where it is joined to a bell.[32] The movement of the rods, similar to the oscillation of a piston, would then ring the clapper of the bell. One may deduce that the mechanism of the clock was placed in a position accessible for the 'governor'[33] of the device to regulate easily the idiosyncracies of these still less-than-reliable machines.

In the Brussels miniature the dial is marked with two sets of Roman numerals I to XII, which of course conforms to a twenty-four hour reckoning. The single indicator is in the shape of a human arm, slightly crooked and with the index finger as a pointer. There are twenty-four rays in the sunburst. The clock may be among the last to display a twenty-four hour dial, for in numerous miniatures dating from after 1450 twelve hours only are registered. They obviously represent a new trend in marking the number of hours.[34]

[27] Künzle, *Horologium*, pp. 364-5.

[28] Spencer, *Horloge*, p. 282, n. 19.

[29] Cf. Künzle, *Horologium*, p. 61, n. 1.

[30] Spencer, *Horloge*, p. 283.

[31] Cf. F. A. B. Ward, *Time Measurement. Science Museum Descriptive Catalogue of the Collection*, London, 1966, p. 35.

[32] Cf. H. Michel, "L'*Horloge de Sapience* et l'histoire de l'horlogerie," *Physis*, II (1960), p. 294.

[33] Cf. C. M. Cipolla, *Clocks and Culture 1300-1700*, London, 1967, p. 41, n. 6.

[34] In recent years historians of horology have drawn attention to many representations of such clocks: Froissart's *Chronique*, London, British Library, MS. Harley 4379, fol. 3, reproduced in F. A. B. Ward, "A 12-Hour Public Clock Dial of the 15th Century in Paris," *Antiquarian Horology*, 8 Dec. 1973, pp. 495-6. In a further article entitled "15th Century Clocks and Clock Dials," *Antiquarian Horology*, 9 June 1976, pp. 762-67 Ward comments on and reproduces additional models of time-pieces from Wavrin's *Chroniques d'Angleterre*, London, British Library, MS. Royal 14.E. IV, fol. 16; from Gerson, *Oeuvres*, Brussels, Bibliothèque Royale, MS. 9305-6, fol. 76 (an earlier reproduction occurred in *La Miniature flamande*, pl. 45); from Wauquelin's translation of the *Chroniques de Hainaut*, by Jacques de Guise, Brussels, Bibliothèque Royale, MS. 9243, fol. 1 (earlier reproduced in *La Librairie de Bourgogne... Cinquante Miniatures*, Brussels, 1970, no. 20); and from the *Histoire de Charles Martel*, Brussels, Bibliothèque Royale, MS. 6, fol. 9 (earlier reproduced in *La Librairie de Bourgogne*, no. 36 and on the front cover of *Antiquarian Horology*, 8 Dec. 1973). S. E. Holbrook, "Clock Dials on a fifteenth-century Country Estate, and other Burgundian Clocks," *Proceedings of the Antiquarian Horological*

One would like to think that this particular model in the *Horloge* illustration was a copy of an existing one which the Parisian artist had seen and sketched. Unfortunately, descriptions or plans of clocks that do survive are not sufficiently detailed to permit an identification of Sapientia's one with contemporary royal, ducal, episcopal or conventual models. About 1430 Philippe le Bon had a Gothic style clock constructed,[35] but doubts have recently been raised about its authenticity.[36] In the summer of 1459 Charles VII purchased five clocks from a Paris maker and four had chiming mechanisms.[37]

Two among many church time-pieces can be usefully mentioned at this juncture. Spencer expressed the view that the *Clock Chamber* miniature "may well be visual testimony to the activities of such men as Jean Fusoris, canon of Reims and Notre-Dame, who constructed astrolabes and clocks, notably the great clock for the cathedral of Bourges..."[38] There was also a celebrated clock in the Cathedral at Autun where Cardinal Jean Rolin II was Bishop. The name of its *gubernator* in the period between 1440 and 1458 was Mathieu de Poisson (Matheus de Pisce).[39] The Cardinal himself owned three models of clocks, but in the period after the execution of the Brussels *Horloge* manuscript. They are recorded in the inventory of possessions found in his Paris *hotel* at the time of his death in 1483.[40]

The next object in the Clock Chamber that catches the eye is the very large astrolabe hanging from a hook on the lower portion of the clock's panelling. According to the eminent horologist, Henri Michel, it is exaggerated in size so that the details of the astronomical markings may be observed.[41] An almost identical instrument may be seen in the Science History Museum at Oxford (item 21) which Gunther dates *c.* 1400.[42] Various scientific characteristics lead Gunther to believe that the Oxford instrument is Italian in style. Michel agrees that an Italian origin for the astrolabe in the Brussels picture is quite possible, but feels compelled to mention that ten instruments of this kind are recorded in a 1379 inventory of Charles V. In all probability a French School of instrument making became established in the second half of the fourteenth century.[43] Be that as it may, Michel notes that Italians with a highly sophisticated awareness of scientific knowledge were attached to princely courts in the reigns of Charles V and VI. The name of one such scientifically skilled person that comes to mind is, of course, Thomas de Pisan, father of the celebrated Christine, who settled in Paris in 1368, and who was, among other accomplishments, an astrologer.[44]

To Sapientia's left is another prominent tall wooden structure with open sides, and consisting of three disparate sections: drive wheels activating levers connected to hammers aimed at five

Society, VII, 10 (1978), pp. 819-26 cites the Master of Margaret of York or a workshop associate who included eight representations of clocks in the illuminations of the *Profits Ruraux* by Pierre de Crescens, New York, Pierpont Morgan Library, MS. Morgan 232. Holbrook draws attention to further clock models in another copy of the same text, Paris, Bibliothèque de l'Arsenal, MS. 5064. His article reproduces a fragment of a panel painting at the Musée National de Versailles and known as the *Duke of Burgundy's Hunting Party* in which a twelve-hour clock graces a gable of the wooden hunting lodge.

[35] Cf. M. de Leber, *Notice sur l'horloge gothique construite vers 1430 par Philippe III, dit le Bon, duc de Bourgogne*, Vienna, 1877.

[36] Cf. E. Poulle, *Un Constructeur d'instruments astronomiques au XVe siècle, Jean Fusoris*, Paris, 1963, pp. 39-40, pls. X-XII, and Michel, *Horloge*, p. 297.

[37] Cf. M. G. A. Vale, *Charles VII*, Berkeley, 1974, p. 185.

[38] Spencer, *Horloge*, p. 283, n. 20.

[39] See A. de Charmasse, "L'Horlogerie et une famille d'horlogers à Autun et à Genève aux seizième et dix-septième siècles," *Mémoires de la Société Eduenne*, n.s. XVI (1888), p. 180.

[40] "Deux horloges dorees de fin or, garnies de leurs contre-poids, prisees ensemble xii l." A third clock is said to be "a façon de singe, en un tabernacle de bois, prisee lx s.p." A receipt for the first two time-pieces indicates that they had been acquired by a brother, Guillaume Rolin (1406-1476), long before the Cardinal's death. See Charmasse, *Inventaire de livres liturgiques*, pp. 301-2.

[41] Michel, *Horloge*, p. 294.

[42] Cf. R. T. Gunther, *The Astrolabes of the World*, Oxford, 1932, II, p. 317, no. 168; see also Poulle, *Un Constructeur d'instruments*, pp. 95-124.

[43] Michel, *Horloge*, p. 294.

[44] M. J. Pinet, *Christine de Pisan 1364-1430. Etude Biographique et Littéraire*, Paris, 1927, p. 7.

saucer-shaped bells, over which is a beautifully proportioned Gothic cupola. On the side of the frame is an inscription which has already been noted when discussing the *Déclaration*'s comment about it. Michel holds the view that this machine was not designed *après nature*, but more likely it has been inspired by diagrams in a *document plus ancien*, for "la position inexplicable d'une roue oblique résulte évidemment d'une erreur du dessinateur."[45] At first glance one may surmise that the instrument is not a time-keeping device in the modern sense of the word but rather an alarm. Michel takes this idea one step further; it is, he contends, a carillon of five bells which are attached to the circular support below the cupola and struck by hammers joined to slim rods. The horologist observes that the carillon may have been, or was intended to be, an *horloge auxiliaire*; a crankhandle would have been the means of setting it in motion.[46]

On the far right of the Brussels miniature are divers portable instruments for measuring time. The table where they lie or hang carries in fact a sundial or *cadran portatif horizontal*; a sundial with compass or *cadran portatif équatorial*; a flat table clock or *horloge de table*; a shepherd's calendar or *cadran de berger* and a Profatius quadrant, an instrument for measuring the height of the sun in the sky.[47] Early fifteenth-century paintings of such portable devices are preserved among the works of the Limbourg Brothers in Paris, Bibliothèque Nationale, MS.fr. 166, fol. 5v. Meiss writes "This manuscript is, we believe, the Bible for which Philippe le Hardi paid Paul and Jean de Limbourg from February 1402 until January 1404. The two brothers completed the first three gatherings of the manuscript, fols. 1-24v..."[48]

Portable sundials are recorded by Michel in 1451 or slightly earlier.[49] One such instrument belongs to the Ferdinandeum Museum in Innsbruck and is very similar in design to the one pictured in the Brussels *Horloge* with a finial or gnomon attached to a plateau which folds across the surface of the dial to allow it to be transported conveniently.[50] Michel cites E. Zinner's observation that Pierre de Maricourt mentioned in a document dated 1269 that time can be measured by the sun's azimuth.[51] However, the first reference to a "style-axe" device with a gnomon inclined towards the celestial pole occurred, according to Zinner, in a manuscript reliably dated 1431.[52]

In the Brussels painting of the *Clock Chamber* appears another instrument that is a variation on these models, it is a portable sundial with compass, enclosed in a round metal box. When the "style-axe" of the dial and the circumscribing ring of the compass are folded flat, both are protected by a circular, hinged lid. Michel believes that a similarly constructed instrument is described in the same 1431 manuscript cited by Zinner,[53] but stresses the point that it is described, not illustrated. Only the Brussels *Horloge* miniature does that. More importantly for horological research, the *Clock Chamber* instrument antedates by at least some thirty years the oldest extant model, dated 1479, now in the Chicago Planetarium Museum.[54]

As for the representation of the *horloge de table* in the Brussels *Horloge*, Michel observes that "il a certainement été dessiné d'après nature."[55] This flat table clock lacks a dial and hour-hand, and is also without a case. The shape of the frame holding the machinery is octagonal, which prompts Michel to wonder "...s'il s'agit d'une horloge tambour ronde ou octogonale, attendu qu'on n'a jamais rencontré celles-ci avant le XVIe siècle..."[56] What is, however, of prime importance with this piece of mechanism in the miniature is the high visibility of a coiled spring with

[45] Michel, *Horloge*, p. 294.
[46] Michel, *Horloge*, p. 295.
[47] *Ibid.*
[48] Meiss, *Limbourgs*, p. 342 (Text vol.) and fig. 289.
[49] Michel, *Horloge*, p. 296.
[50] Cf. S. Guye and H. Michel, *Time and Space Measuring Instruments from the 15th to the 19th Century*, London, 1971, pp. 234 and 237.
[51] Michel, *Horloge*, p. 296.
[52] Cf. E. Zinner, *Die ältesten Räderuhren*, Bamberg, 1939, p. 74.
[53] Michel, *Horloge*, p. 296.
[54] Mensing Collection, item no. 279, formerly in the Heilbronner Collection.
[55] Michel, *Horloge*, p. 297.
[56] *Ibid.*

a *fusée* drive encased in a barrel. This invention was still quite new in the middle of the fifteenth century; Ward claims the Brussels picture is "the earliest known illustration of this important mechanism..."[57] For dating purposes of the first fusée-driven clock, Michel cites the celebrated clock of Philippe le Bon, already mentioned in this discussion, and which specialists consider to have been made between 1429 and 1435. Almost identical mechanical parts are visible in a clock that was included in the portrait of the chronicler Jean Lefèvre, seigneur de Saint-Remy, known as Golden Fleece King of Arms; it is dated *c.* 1450 and painted in the style of Rogier van der Weyden.[58]

The Profatius quadrant in the Brussels *Clock Chamber* hangs to the right of the flat table clock and is probably intended to represent an ivory original. Profatius was a Jewish astronomer who lived at Montpellier during the last decade of the thirteenth century.[59] The instrument prototypes date from Antiquity and through mathematical calculations assist in measuring the height of the sun in the sky. Guye explains the technical points in a lucid but succinct fashion: "The quadrant ... is formed by two straight edges at right angles to each other with a limb between in the form of a quarter circle. At the apex of the angle a plumb-line is attached. When the instrument is held vertically with one of its edges pointing at the sun, the plumb-line shows the angle of the sun to the vertical on the circular edge. It is then a simple matter to graduate the line in the opposite way so as to make the line indicate the angular height..."[60]

In conclusion, one may say that the *Clock Chamber* announces to the viewer at least two, possibly three, dimensions of visual interpretation. It is a splendid evocation of precision instruments for recording the passage of time in the mid-fifteenth century. They are clearly the best the artisans and craftsmen could make. The major clock and its dial declare loudly their prime function when attention to devotions is required. But the concentration of the Disciple on the words of his mentor reminds the beholder that instruction has commenced. The teacher is Sapientia demanding moderation of conduct, contemplation of time's fleet passage, and meditation on the Cardinal Virtue, Temperantia, whom she personifies.

III (1) Ecce Homo Fol. 15v.
(2) *Christ Carrying the Cross*
(3) *Christ Addressing the Daughters of Jerusalem*
(4) *Cloisters with Fountain*
(5) *The Disciple Lamenting before the Crucified Christ*

In the prologue of the *Horologium* Suso relates an anecdote[61] about the Disciple's spiritual progress: "Once upon a time it happened that the Disciple, about whom this book is a record, used to go in procession after Matins around the cloisters or through the church of his convent in honour and memory of the most piteous walk made by Our Saviour Jesus Christ when He was led from Jerusalem to Calvary. And the Disciple was accustomed to make such a perambulation each night after Matins. And when on this particular night he had completed his said procession, he knelt before the representation of the crucifix, and with a most grievous heart he began to lament and

[57] Ward, *Time Measurement*, p. 35.

[58] The painting is in the Koninklijk Museum voor Schone Kunsten at Antwerp and was included in the 1962 Order of the Golden Fleece Exhibition, see *Bruges, Musée Groeninge. La Toison d'Or. Cinq siècles d'art et d'histoire*, Bruges, 1962, p. 94, no. 12. The clock is discussed in H. A. Lloyd, *Some Outstanding Clocks over Seven Hundred Years 1250-1950*, London, 1958, p. 26, pl. 25,

[59] Michel, *Horloge*, p. 295.

[60] Cf. Guye and Michel, *Time and Space*, 234.

[61] Suso is thought to be echoing here a similar story told about a Metz Dominican by Gérard de Frachet, O.P. in his *Vitae Fratrum Ordinis Praedicatorum*, see Künzle, *Horologium*, p. 369, n.

complain that his heart was not so fervent, so ardent or so piously atuned to the blessed Passion of Jesus Christ as it should be.''[62]

The statements by the writer of the *Déclaration* concerning the cloister with its central fountain, and the Crucifixion sited in the church, appear to be based on those in the French translation of Suso's text. The comments in the *Déclaration* are ambiguous, however, and give cause for confusion to the reader: ''the ... Disciple was wont ... to perambulate ... around the cloisters...'' and further, ''...this perambulation is represented in the present illustration which is structured like a cloister...'' One observes that the commentator does not say the Disciple's perambulation is represented by the depiction of the cloister, and the reader may well be inclined to look for a figure in the cloister. The *Déclaration* announces that the painting exists in order to depict the Disciple's remembrance of Christ's procession of suffering. Yet, the *Ecce Homo* and *Via Crucis* themes are not mentioned specifically in Suso's work at this point. There is an allusion to the walk from Jerusalem to Calvary. So clearly, the expositor's eye is following the pictorial segments when he comments on the Passion narrative. Curiously, he explains the detail of the praetorium's emptiness, and in doing so, he introduces an element of realism by direct speech. However, the moment of the Flagellation and the Crowning with Thorns are not depicted in the illustration. There is an echo of both of these themes in a later segment called the *Buffeting of Christ* on fol. 25.

The expositor's account of the next three compartments is descriptive. The last one is understood correctly, that is, to be unreal in its pictorial dimension: the Crucifixion is shown *as if* in a church. The author of the *Déclaration* concludes where he began, with reference to the perambulation.

The modern eye would perceive two or possibly three themes skilfully interwoven in the textual and pictorial fabrics; these are the Stations of the Cross, the Dominican support for their close observance, and the moral issue of humanity's lament for the wrong reasons about Christ's suffering.

The predominant imagery that strikes the viewer is a selection of the Stations of the Cross. This representative programme of the *Horloge* differs from the full fifteenth-century cycle,[63] which, for instance, Rogier van der Weyden used on the central panel of the Miraflores Altarpiece, *post* 1437 (?).[64] The Flemish artist placed a Lamentation group in the foreground, and the other six scenes of the Stations of the Cross are depicted in the stone archivolt painted in a *trompe l'oeil* fashion.

In the *Horloge* painting, by contrast, Christ leaves Pilate's praetorium and the waiting Jews are addressed with words from John, XIX, 14: *Ecce rex vester* ''Behold your king.'' Next, He is seen carrying the Cross with Simon assisting as He nears the city gate. Tradition has it that this was the point that the Jews are thought to have ordered Simon to carry the Cross, fearing Christ was not strong enough to reach Golgotha before sunset.[65] The beholder then sees the leading horsemen blow trumpets to announce the departure of the procession from the city's boundary. The progress is visually interrupted when Christ chides the Daughters of Jerusalem with the celebrated words from Luke, XXIII, 28: *Filie Iherusalem nolite flere super me* ''Daughters of Jerusalem, weep not for me.''

The episode of the women in sorrow who lament loudly as Christ passes receives iconographic emphasis because of its position in the foreground. Here on the left, Christ is seen looking over

[62] (Fol. 16) Advint une fois que le disciple dont ce livre fait memoire faisoit aprés matines une procession entour le cloistre ou par l'eglise de son couvent en l'onneur et souvenance de celle tres piteuse procession que Nostre Sauveur Jhesucrist fist quant on le mena de Jerusalem en Calvaire. Et avoit a coustume le dit disciple a faire chascune nuit aprés matines une telle procession. Et quant il ot fait celle nuit sa dicte procession, il se mist a genoulz devant l'ymage du crucifix et en grant douleur de cuer se mist a guermenter et a complaindre de ce que son cuer n'estoit pas si fervent ne si ardant ne si devot a la benoite passion de Jhesucrist comme il deust.

[63] The reader will recall that at the time the miniature was executed, the Stations of the Cross comprised only seven positions, cf. Schiller, *Iconography*, II, p. 82.

[64] The altarpiece is extensively discussed and illustrated in E. Panofsky, *Early Netherlandish Painting*, see in particular, I, pp. 259-62 and II, pl. 182, fig. 320.

[65] Cf. Schiller, *Iconography*, II, p. 78.

His shoulder across the picture plane to the right foreground where the matrons grieve outwardly. Christ's rebuke to them is well known and the iconography is a concise visual representation of the motif of misplaced sorrow, which will presently be given additional attention and exposure in another segment of the miniature.

Any lay person viewing the painting in the fifteenth century would surely recognise the devotional images which prompt him to experience the same torment as he follows Christ on the *via dolorosa*. But it should not be forgotten that the textual imperative is a Dominican one, one by which Suso describes the actions of a friar of the Order. The religious in his perambulation of the conventual cloister follows mentally, step by step, the Stations of the Cross. Such a practice among the Preaching Friars finds confirmation from another contemporary source. As Spencer observed, "A Dominican Missal with additions for use in Paris in the fourteenth century confirms this as a practice in the Order which parallels the directions for the processions through Paris and other cities on Palm Sunday, eventually to become part of the French Passion plays and of the liturgy of the Stations of the Cross."[66]

A centrally placed, darkened and lonely cloister, with a fountain, reminds the viewer of the venue for the Preaching Friar's perambulation. The shape of the fountain gives the impression that it is functional rather than mystical or symbolic. It has no crucifix rising above the basin which receives the Holy Blood from the five wounds of Christ. Moreover, St. Mary Magdalene and St. Mary the Egyptian are not in attendance, one on each side of the fountain. The object is not therefore a Fountain of Life, of the type that was often represented in fifteenth-century art.[67]

The final segment of the miniature has a twofold function, it acts as the climax of the Stations of the Cross motif and as the conclusion of the anecdote about the Dominican Friar who, after his perambulation of the cloister, kneels lamenting before the crucifix. As he kneels, he sees a small mound with bones present, on which the Cross has been positioned; from Christ's wounds, the blood continues to stream. Bearded Jewish elders mingle with Roman soldiers. The phrase which the centurion of the guard is said by Matthew, XXVII, 54 to have uttered just after Christ expired: *Vere filius Dei erat iste* "Truly, this man was the Son of God" has been included on a scroll to announce to the viewer the point reached in the Final Agony.

The suppliant posture of the friar suggests he is wrestling with his conscience; he has not been very fervent in his devotion to the Passion as he could be; the inner grief and torment causes his lament. But the Passion he sees presents a devotional image expressive of the physical pain of the journey to Golgotha, of the wounds inflicted at the Crucifixion, and of the total humiliation from mortals on all sides. The image of the Crucified Christ in the miniature in question also interprets His death and what Schiller might call "its redemptive effect."[68] Already the gaze of the friar is directed upwards with a gesture of receptiveness, linking his doubts with the victory of the dead Christ hanging before him. The message of the image is thus one of sorrow, of victory and of redemption for the mortal who studies and comprehends this Great Memorial of Sacrifice.

One further observation is pertinent for the imagery. It is the only painting in the *Horloge* series to depict a Dominican praying at the foot of a crucifix. The link with St. Dominic himself would

[66] Spencer, *Horloge*, p. 280, n. 10 cites a Dominican Missal, Paris, Bibliothèque Nationale, MS. lat. 8884, fol. 93v. and refers her readers to Leroquais, *Sacramentaires*, I, xix and II, 104, n. 287. The practice was still very much alive in the French capital in the eighteenth century, according to an anonymous pamphlet *La Solide Dévotion de la Passion de Nostre Seigneur Jésus Christ en l'Eglise des Jacobins de Saint-Jacques*, Paris, 1782. I know of the existence of this work only through the bibliography supplied by Rohault de Fleury, *Couvents*, II, unpaginated; see 'Paris' entry. Chapotin, *Histoire Dominicaine*, p. 217 prints a plan of the medieval Convent of Saint-Jacques, showing the cloisters along one side of the church, and the refectory near the main entrance door of the church.

[67] The theme and its liturgical connections are discussed in E. Mâle, *L'Art religieux à la fin du moyen âge en France*, 6th ed., Paris, 1969, pp. 108-117.

[68] Schiller, *Iconography*, II, p. 116.

have been recognised by any friar contemplating the kneeling figure. Although the typology of the themes representing the saint's life does not include a specific motif of his devotion to the Crucified Christ,[69] it seems it was a well known icon for him and his followers. In a year post 1442, Fra Angelico had painted an almost identical subject on the cloister wall of the Convento S. Marco in Florence.[70] The viewer sees St. Dominic, with halo and wearing traditional garb of the Order he founded, kneeling at the foot of a crucifix, which stands on a small mound. He is positioned three-quarters facing the beholder, his hands clasping the upright of the Cross. The Holy Blood streams down the Rood from an image of the Dead Christ, hanging limp and spent. As in the *Horloge* miniature, the detail of the identification label INRI appears above His bowed head.

IVA *Youth in Dalliance* — Fol. 16v.

"There was once a youth who, in the prime of life, gave himself over to dalliance and worldly vanities spending his time in frivolous, mundane pursuits, and of his own free will strayed from the path of salvation on to the path of error and ruin. He was brought back by Divine Sapientia, who wondrously filled his heart with Light..."[71]

The author of the *Déclaration* notes that this small miniature has been separated from the larger illustration concerning the Disciple's conversion. From his one sentence description of the *Youth in Dalliance*, one could say that he is paraphrasing the words of the text rather than relating the action of the picture. Because of this, it seems that the expositor is not desirous of drawing attention to the types of "vanities and sensuous pleasures of the world." Moreover, he alone employs the term *voluptez* and thereby raises the spectre of sensuality, more usually associated with adult amorous pursuits than with the pleasures of youth.

The artist in his turn depicts joys of the inncocent variety when he portrays the Youth surrounded by objects and pets that symbolise his pleasure-seeking, such as toys with which to gratify his whims; musical instruments to answer to his lyricism; a hawk and a hound to indulge his hunting instincts and fantasies. Even the setting is idyllic, a kind of *hortus conclusus*[72] where man's hand has exercised control over Nature by providing well-trimmed shrubs, a stream zigzagging artificially, and an arched pergola along two sides of the enclosure.

Nevertheless, this is an empty paradise,[73] all action and movement, even the hound's spirited leap, are as if suspended in time and space by the instantaneous flash of Enlightenment. The passage of Sapientia's transcendant wisdom to the Youth's heart is suggested by rays emanating from her person, seen in a mandorla, from where she radiates wondrous light and holds up her open book for the young man to behold. She leads him back from the pleasure garden of mankind towards spiritual salvation.[74] Like Christ in the Garden of Gethsemane, the Youth confronts his destiny in a garden.

[69] Réau, *Iconographie, Saints*, I, pp. 391-8 does not mention such a format in the typological themes for St. Dominic.

[70] J. Pope-Hennessy, *Fra Angelico*, 2nd ed. London, 1974, pl. 66.

[71] (Fol. 17v.) Jadis fut ung jeune homme qui, en la fleur de son aage, se donna fort a la joliveté et a la vanité du monde et occupa son tempz en diverses plaisances vaines et mondaines et de sa propre voulenté se eslongna de la voye de salut et se transporta ou chemin d'erreur et de perdicion duquel la divine Sapience le retraÿ et par merveilleuse maniere lui enlumina son cuer...

[72] The most informative monograph I have consulted on the subject is by Dagmar Thoss, *Studien zum locus amoenus im Mittelalter*, Stuttgart, 1972.

[73] Man's distrust of bountiful and beauteous Nature in her landscapes and-gardens was highlighted by K. Clark, *Landscape into Art*, London, 1949, pp. 1-15.

[74] An overview of the symbolism attached to gardens is supplied by Marilyn Stokstad, in M. Stokstad and J. Stannard, *Gardens of the Middle Ages*, Lawrence, 1983, pp. 19-24.

IVB (1) *Reception of a Postulant* Fol. 17
(2) *Studying in the Cell*
(3) *Lessons in the Refectory*
(4) *Tree of Life*

Having briefly presented the disoriented Youth to his readers at the opening of chapter I, Suso immediately passes to incidents in the novitiate. "Now it happened at the time the said Youth was still a minor and a novice and not skilled in divine knowledge that be began to read the Bible. Among the words he found some which made him realise that he did not well understand how to love Sapientia. Then he passed over all the other phrases and writings in the Bible and applied his thinking with pleasure to this teaching, and he was wont to write down on a small scroll the phrases he had found about how one should love Sapientia. This novice was often present during the reading of lessons at meal-times, for he was devout..." A sentence or two later, Suso elaborates: "Often he heard readings and accounts of praise of the said Divine Sapientia, and on one particular occasion he heard the reader say that Sapientia was worth more than all the riches of the world, and nothing one can wish for or long for can be compared to her. She holds in her right hand longevity or length of days, and in her left she holds honour and riches. She is the staff of life to all who grasp her, and he who holds her fast is blessed."[75]

The expositor in the *Déclaration* treats each of the four segments in turn, after remarking that the group presents phases of the novitiate. He makes a point of identifying the Order in which the Youth is received as the one to which the author of the text belonged. When describing the Disciple's zeal in writing down select phrases from the Bible after hearing them quoted in the refectory, the commentator goes further than the textual source. He expresses his belief that the novice read specifically the Books of Solomon and Wisdom. He is at pains to repeat the exact Biblical wording about Longevity and Riches, but he does not state that Suso obtained them from the Book of Proverbs, wherein the Old Testament Sage personifies Sapientia.[76]

Similarly, the Staff of Life quotation is reproduced with a vague source indicated through the words "another saying". The author of the *Déclaration* proclaims that the Cross Sapientia passes to the Disciple is representative of the Tree of Life which stood in Paradise. Everlasting life belonged to those who ate its fruit. Suso's opinion about equivalence is confined to the view that Sapientia or Wisdom is the Staff of Life. The association of a tree in Paradise with the Holy Rood's fruit being able to bestow Everlasting Life originates in the mind of the commentator. His concluding remark is not suggested by the painting or by Suso's work. However, as we shall see below, it was not untoward for a fifteenth-century theologian to be equating the imagery of the Old Testament's Staff of Life with the New Testament's Tree of Life.

When comparing the polyptych with Suso's narrative, a modern critic may observe that the Dominican does not describe the ceremony of profession or of the taking of vows, yet, the *Reception of a Postulant* is the subject of the first tableau. It is clearly a logical transition for the hero

[75] (Fol. 17v.) Advint pour le temps que le dit jouvencel estoit encore enfant et novice et non pas habile a la congnoissance divine. Ung jour commença a lire en la Bible, et entre les autres parolles lui en vindrent aucunes au devant, lesquelles il n'entendoit mie bien d'amer Sapience. Lors trespassa l'enfant toutes les autres parolles et escriptures de la Bible et appliqua son engin et print son plaisir en celle doctrine, et fist qu'il escrivi en une cedule les parolles qu'il avoit trouvees comment on doit aimer Sapience. Cest enfant estoit souvent a la leçon que on lisoit a table a heure de mengier, car il estoit religieux... Plusieurs fois il oioit lire et raconter la louenge d'icelle divine Sapience, et par especial une fois il entendi le lisant qui disoit: "Mieulx vault Sapience que toutes les richesses du monde et chose que on puisse souhaidier ne desirier ne se peut comparer a elle. Elle tient en sa main dextre la longuesce ou la longueur des jours, et en senestre main tient gloire et richesses. Ceulx qui la prennent sont bien apuiez et soustenus de l'Arbre de vie..."

[76] Proverbs VIII, 11 *Melior est enim sapientia cunctis pretiosissimis et omne desiderabile ei non potest comparari*, and III, 16 *Longitudo dierum in dextera eius et in sinistra illius divitiae et gloria*, and III, 18 *Lignum vitae est hiis qui apprehenderint eam, et qui tenuerit eam beatus.*

of the narrative, the Disciple, from his period of Dalliance, as suggested in the previous painting, to studies in his cell, which follow. It is not only the Disciple's life we behold, the scene recalls quite forcefully the time when Suso himself was a postulant. Thus, both setting and subject are totally Dominican and biographical.

Studying in the Cell illustrates the text's references to concepts of Sapientia that were so new and obscure to the novice that he felt the need to copy them out when in his cell. The Word, however, is more impressive if it is personified. Sapientia appears to the novice, her right hand raised in a sign of peace, and the Bible held open to emphasise open access to the Word of God.

The refectory is the venue for the *lector* who quotes from Proverbs. The text's abstract notion of longevity is interpreted as a sun with lambent rays, while the riches are transformed into two bags of coins, brooches and a necklace. Sapientia holds the sun, but merely touches the other prominently displayed objects. The novice is flooded by rays of Enlightenment from Sapientia and wealth is no longer a metaphor but a tangible reality.

In the fourth segment, some of the Latin words from Proverbs serve as a legend, *Lignum vite est hiis qui apprehenderint eam et qui tenuerit eam beatus.* The subject of the verb *est* in the context is Sapientia; thus, "she is the staff of life to all who grasp her." But Sapientia has already been personified in a nearby segment as a radiant woman holding an open book; this time the imagery is reinforced by her reappearance holding a Tau Cross and her commanding stance in front of the novice.

Since the Old Testament words *lignum vitae* in the *Horologium* are translated as *arbre de vie* by the French Franciscan, and since the term is rendered pictorially as a cross and not as a staff, one may deduce that something effected an evolution in the symbolism surrounding *lignum vitae* in the intervening centuries after the writing of the Book of Proverbs. On the language side, the word *lignum* acquired a referential zone touching Calvary among medieval Christian writers. Attested in *c.* 1125 is *Lignum Domini*, and in *c.* 1220 *Lignum Dominicum*, both meaning the Cross on which Christ died.[77] This verbal usage reflects shifts in the symbolism being attached to the words *lignum vitae*. G. B. Ladner expressed the point admirably in 1979 in his paper on the meaning of symbolism. He wrote: "Thus, the *Lignum vitae* of Paradise was the prototype or prefiguration of the Cross, the *Lignum vitae* of the new dispensation."[78]

The symbolism of the Holy Rood was enriched by the treatise *Lignum Vitae*, composed *c.* 1259 by the Franciscan, St. Bonaventura. His work on the Cross of the Passion had a wide circulation and was often translated into vernacular tongues in Europe. The popularity and influence of the work in Franciscan circles might easily explain the anonymous translator's substitution in the Brussels *Horloge* of *arbre de vie* "tree of life" for the *lignum vitae* of the Latin original. As the spiritual influence of St. Bonaventura grew, the iconographical representation of the motif of the Tree of Life assumed a new shape. To quote Lucy M. Sandler: "The typical representation of the Tree of Life shows the crucified Christ on a tree with branches, leaves and fruit, and sometimes the tree is set in the Garden of Paradise. Such trees bear twelve fruits, each standing for a phase of the infancy, public life, Passion, or glorification of Jesus Christ, and each event is commented upon by one of the twelve Prophets of the Old Testament."[79]

[77] See R. E. Latham, *Revised Medieval Latin Word-List*, London, 1965, p. 277, s.v. *lignum*.

[78] G. B. Ladner, "Medieval and Modern Understanding of Symbolism: a Comparison," *Speculum*, LIV (1979), p. 236.

[79] Lucy F. Sandler, "Jean Pucelle and the Lost Miniatures of the Belleville Breviary," *Art Bulletin*, LXVI (1984), p. 81. She reproduces on pp. 82-5 several formats derived ultimately, but not without contamination, from the Bonaventuran one, and she refers her readers to R. Ligtenberg, "Het lignum vitae van den H. Bonaventura in de ikonografie der veertiende Eeuw,"*Het Gildeboek*, XI (1928), pp. 15-40. Equally informative on the format of the Cross of St. Bonaventura is Schiller, *Iconography*, II, p. 134. The *Lignum Vitae* is included in the standard Quarracchi edition, *S. Bonaventurae Opera Omnia*, Ad Claras Aquas, 1898, VIII, pp. 68-87.

Turning back to the visual statement in the *Horloge* miniature, one quickly realises that Sapientia does not hold a Bonaventuran Cross of leaves and flowers, but a miniature Cross of Golgotha. The shape is most symbolic in the context and refers not only to Christ's Passion which the novice is being exhorted to study while he transcribes passages from Proverbs, but also to Christ's Passion which she will relive with him in the days and months and years to come, as she leads him to a greater awareness of the mystery and true meaning of the Crucifixion. The simple Tau Cross held in front of the novice thus prefigures the moving tableaux on a later folio in which both Sapientia and the Disciple view in person on Calvary representations of *Christ Nailed to the Cross* and *Christ Offered the Sop* (fol. 27v.). One recalls the discussion of the large composition on fol. 15v. of which one segment present the *Disciple Lamenting before the Crucified Christ*. This miniature had illustrated sections of the prologue to Suso's work and as such presented an overview of the whole *Horologium*. With the *Tree of Life* painting the viewer has gone back in time, as it were, to the early period of the novitiate.

V A (1) *Temptations of the Flesh* Fol. 18
(2) *Repentant in his Cell*

A point has by now been reached in the Disciple's progress towards Enlightenment and Wisdom that he is burning to have Sapientia as his bride. "It happened that the Devil, who is the cause of all evil, became envious of the Disciple's sound purpose and advantage, so he induced evil thoughts in him, saying: Hey, there! foolish one, do you want to love and marry the creature you do not know and have never seen? ... This bride whom you speak of could be loved, if she did not forbid wordly and carnal pleasures and delights. I would advise you to love her, if she allowed you to live hedonistically and drink and eat your fill..."[80]

The *Déclaration*'s wording focuses on the theme of carnality, derived from the textual passages, and in particular, attractions to the sight are emphasised. The other four senses are not channels employed by the Devil on this occasion to tempt the Disciple. The repentance phase is next mentioned by the *Déclaration*, but here the expositor records the audible register of culpability and remorse, as well as the ritualistic gestures and posture adopted by the Disciple in the painting.

Today we may read more deeply into the manifestation of the Devil's wiles, for the source of inspiration for the imagery is manifestly a guileful apparition of wordly and carnal pleasures. Towards the front of the Disciple's cell appears a white-clothed trestle table, sumptuously set with a feast of bread, meats and wine. To its rear, there is a plain, four-legged table, placed at right angles to the other and displaying on its undressed surface two full money-bags whose contents spill out. A grinning female devil acts as an *entremetteuse* and lures the Disciple with three young women who pose seductively behind the drapes of a diaphanous curtain on the right.

The theme of carnal appetite and seduction is, in the annals of manuscript illuminations, not unique to the Brussels *Horloge*. Early in the fifteenth century, *c.* 1402-1404, Paul and Jean de Limbourg executed several vignettes on the same subject in a *Bible Moralisée*. One portrays an amorous monk and a woman (fol. 7), another a monk and a youth (fol. 7); in a third miniature a monk and his female companion make love (fol. 18v.), while a fourth tableau presents a monk and temptress (fol. 24).[81] These last two scenes are the most salient in the group because of the drama being

[80] (Fol. 18v.) Advint que l'ennemi qui est cause de tout mal ot grant envie du bon propos et du prouffit d'icelui, et lui mist au devant aucunes mauvaises pensees en disant: "Hé, hé, fol! Fol, veulx tu amer la chose et espouser celle que tu ne (*MS.* ne ne) congnois et ne veis oncques?... Celle espouse dont tu parles seroit a amer, s'elle ne defendoit les plaisances et delices mondaines et corporelles. Je conseilleroie bien que tu l'amasses, se elle te souffroit vivre joieusement et boire et mengier largement..."

[81] Paris, Bibliothèque Nationale, MS. fr. 166, fols. 7 (two scenes), 18v., 24; reproduced in Meiss, *Limbourgs*, figs. 292, 315, 324, respectively.

enacted. On fol. 18v. two winged demons, like fiendish armorial supporters, encapsulate the seated friar and the buxom wench, his doxy. In the vignette on fol. 24, a lady, more dignified and elegantly dressed than her counterpart in the other illustration is held firmly in the grasp of two demons, one of whom tugs forcefully on the hood of the monk kneeling, back to his antagonists, at the portal of a chapel or church.

When engaged on illustrations of the life of St. Jerome in the *Belles Heures* of Jean de Berry, the illuminator, Jean de Limbourg, again represented the temptation theme. In this instance, Jerome kneels at a church portico and turns his head, at a devil's invitation, to contemplate two dancing girls "Who kindled the fires of lust," as the Latin legend expresses it.[82]

The spirit of these miniatures, but not their design, recalls in particular the moralised theme in the *Horloge*'s *Temptations of the Flesh* where the viewer is required to decide for himself whether the Disciple resists the carnal allurements offered to him. There is an emotional and spiritual hiatus between the *Temptation* miniature and the penitence illumination in the adjoining frame. The friar's reaction is to retire to his cell and to kneel, bending low, in an act of contrition.[83] Sapientia's radiance shines down on him as bright as ever.

VB *Epicureans* Fol. 18v.

Sapientia mentions, quoting Ecclesiasticus XL, 20, that wine and music soothe the heart, but the love of Sapientia exceeds both.[84] An inner conflict takes place now between the Disciple's mind and his heart. The Disciple speaks: "And since it is thus that my heart is restless night and day, I have advised it to give up everything and abandon such a love (for Sapientia) and go off and cohabit with the Epicureans who live according to the desires and pleasures of the flesh." His heart replies: "See here now if you think that the fat rams and ewes are blissful and content which graze on that luxuriant and rich pasture in the green meadows? And they speak by action, not words... The dumb beasts are fat, lively and well fed, while you through your love become thin and your body wastes away each day because you do not find what you seek, and what you find does not suffice ..."[85]

The *Déclaration*'s author contends that the illustration presents what the Disciple saw when again he was roused by the Devil: he formed a mental picture of the Epicureans whose doctrine requires man to yield to desires and pleasures. The commentator mentions the scene wherein humans lead the good life in an inn, while dumb animals nearby lead the good life in their terms, which means they are unconcerned about the hereafter. The remarks in the *Déclaration* are discursive when compared with Suso's text, even if they remain true to the broad subject matter of the painting.

The artist presents the text's inner conflict of eye and heart, or body and soul, as an episodic tug-of-war between the Devil and the Disciple. As we shall see presently, the painter's source for

[82] New York, The Cloisters, MS. 54.1.1, reproduced in Meiss and Beatson, *Belles Heures*, fol. 186.

[83] Spencer, *Horloge*, p. 286 suggests the Disciple retires "to prostrate himself in the Dominican *venia*..." Suso does not employ the term *venia* at this point in his narrative; he had used it in the Prologue (cf. Künzle, *Horologium*, p. 369, n.) but the Franciscan translator did not allude to it or render it into French.

[84] Cf. Künzle, *Horologium*, p. 377.

[85] (Fol. 19v.) "Et puis que ainsi est que mon corage nuit et jour est sans repoz, je lui ay conseillié qu'il se cesse du tout et qu'il se departe de telle amour et s'en aille vivre avec les Epycuriens qui vivent selon le desir et le plaisir de la chair.—Veez cy maintenant se tu y penses que lé gras moutons et les brebis sont bieneureuses qui sont par ces prez verdoians en celle pasture belle, plaine et habondant, et dient non mie de parolle maiz de fait... Or sont les bestes mues grasses, reveleuses et bien nourries, et tu par ton amour amaigris et aneantis ton corpz par chascun jour, car tu ne treuves pas (*MS.* par) ce que tu quiers et ce que tu treuves ne te souffist pas."

this interpretation was the French translation, not Suso's Latin *Horologium.* In the front of an inn, on the left of the tableau, a devilish creature harangues the Disciple, while to the right, the Disciple is seen again, but now gazing at the beasts in the fields. Satanic guile and deceit are carried over from the preceding painting of the *Temptations of the Flesh.* Indeed, we do not see the Disciple engaging in animated conversation with his fickle heart as Suso states he does, but rather he debates with the wildly gesticulating devil who seems to represent its fallibilities. It is to this demon that Spencer refers when writing that another devil confronts the Disciple "with a substantial two story inn, crowded with customers..."[86] Having identified them as the Epicureans and cited Ecclesiastes XL,20, instead of Ecclesiasticus, as the source for the reference to wine and music, Spencer states that Suso puts into the mouth of the devil the sentence "I have advised... pleasures of the flesh," which was reproduced at the opening of this section. Spencer also believes the devil talks to the Disciple about the scene of peaceful husbandry, at the right of the tableau, and even reproduces his utterances from the French translation. It is, however, necessary to clarify two points about the quotations. In the first place, the French wording is attributed to Suso by her; and secondly, Suso does not put any speeches into the mouth of the devil. Spencer did not appear to appreciate the fact that the French translator had freely introduced the illusive devil. The passage reads in translation "It happened on one occasion that the Devil advanced strong reasons and clear and subtle arguments, saying that he should give up God and love the world..."[87]

On the other hand, Spencer was quite correct in describing the two-storied inn as filled with customers; she saw "...some twenty laymen and friars happily eating and drinking in the warmth of two fire-places."[88] In portraying the Epicureans thus, the gastronomic side of the pleasures of the flesh are emphasised. Framed under the open arch of the inn, two well dressed men sit feasting; the elder's clothes are blue and his figure is portly; the younger one wears pink, his angular jaw contrasts markedly with his companion's round face. Both sport elegant hats. Opposite them are seated two valets or attendants who serve food to their masters. Another servant stands to the right of the older person and guides his right arm and hand to his mouth so that he may ingest more food into his plump form. The other figures at table are less elegantly clad and do not seem to be of the same social station; they could be members of the two gentlemen's entourage. A food-stall nearby is tended by a *bonne femme*,[89] busily threading wool on her distaff. On a low table in front of her are partridge-like birds, packed together on a linen-covered basket. Another basket contains what appears to be pig-trotters; behind her, on a shelf, are three more baskets heaped with eggs and fruit. A heart-shaped pastry or pie completes the range of viands.

This single tableau is open to several interpretations. One may read it literally as a group of episodes through which the Disciple, now out of the convent and in the world, moves in turn. He inspects a tavern where humans are gathered to eat and drink, he passes by a stall, he gazes at animals in the countryside. One may further observe that both animals and Epicureans eat without a thought for the hereafter. Thirdly, the carousing of the wealthy revellers in the tavern contrasts with the decorum of the simple *bonne femme.* Perhaps most important of all, there is the moral depravity of the secular life, as opposed to the spiritual rectitude of the regulars, symbolised by the Disciple in the habit of the Preaching Friars.

[86] Spencer, *Horloge*, p. 286.

[87] (Fol. 19) ...advint une foiz que l'ennemi lui mist au devant raisons fortes et apparans et soubtilz argumens en disdant qu'il devoit Dieu relenquir et amer le monde... Suso's text states: *Quadam vice, cum subtilissimis argumentationibus ad amorem mundi et Dei desertionem urgeretur, cor eius amoris impatiens semet ipsum coram Deo effundebat...* (cf. Künzle, *Horologium*, p. 377).

[88] Spencer, *Horloge*, p. 286.

[89] I cannot agree that she is a fish-wife, as Spencer contends, *Ibid.*

VI (1) *Sapientia in a Terebinth Tree* Fol. 20v.
(2) *Sapientia asks the Disciple for his Heart*
(3) *Lessons in the Refectory*
(4) *St. Paul Addressing the Disciple*

On the preceding fol. 20, it is recorded that the Disciple heard a different lesson in the refectory from an earlier one. The translator actually retains the Latin of Suso before he renders it into French: "*Ego quasi therebintus extendi ramos meos et ramy* (sic) *mei honoris et gracie*;[90] *quasi libanus non incisus vaporavi habitationem meam et quasi balsamum non mixtum odor meus.*[91] *Qui me invenerit inveniet vitam et hauriet salutem a Domino.*[92] Sapientia said: "I have spread out my branches as does the tree called the terebinth. All my limbs and branches are laden with honour." (The terebinth is a tree, wondrous in flowers, in leaves, in fruit and whenever distilled. For it weeps and sheds resin like the pine-tree.) And Sapientia continued: "I am like the forest that is filled with trees and timber; my scent is that of the balsam when unadulterated. Whoso discovers me, discovers everlasting life and shall have salvation in the abode of all goodness, that is Jesus Christ." When the Disciple had heard Sapientia speak, he was overwhelmed by her love, and said with great fervour: "It is true! One could not find a truer thing!"[93]

The Disciple continues his monologue with a few more observations. "Would it now please God that I could behold Sapientia once of my own accord! O! everlasting Lord, how beautiful, how good and how singular is this lady about whom I have heard so many good things.' It happened that he had such a desire to see her that his heart and body burned. And at that moment a figure was shown to him that was of wondrous beauty and very special in form which portrayed the being and person of Sapientia. The Disciple saw in a cloud the august figure, seated on an ivory throne; she wore striking garments that were of supreme beauty. In that position the lady appeared to be a pleasing, fine and beautiful bride, resplendent in golden raiments and crowned with colours. Her crown was everlasting life and her vestments were consolatory bliss. Her speech was honeyed and courteous and her embrace was gentle and true intercourse."[94]

After several more reflexions about her changing presence in the sky and the wonderment of the Disciple, the action continues. "Sapientia leaned forward affably in his direction; she had an inviting look in her large eyes; the divine and spiritual evocation of her was pleasing, rubescent and

[90] Ecclesiasticus, XXIV, 22. Spencer, *Horloge*, p. 284 erroneously attributes the terebinth tree comparison to Wisdom, VIII, 1.

[91] Ecclesiasticus, XXIV, 21.

[92] Proverbs, VIII, 35.

[93] (Fol. 20) Disoit Sapience: "J'ay espandu mes rainceaulx si comme fait l'arbre nommé therebin. Tous mes membres et mes branches sont chargiez d'onneur." (fol. 20v.) (Le therebint est ung arbre vertueux en fleur, en fueilles, en fruit et en distillacion. Car il pleure et jette poix comme le sapin). Et dit: "Je suis semblable a la forest plaine et peuplee de boys; mon odeur est odeur de baulme qui n'est point meslee d'autre liqueur. Qui me trouvera, il trouvera vie pardurable et aura son sauvement en pays de tout bien, c'est Jhesucrist." Quant le disciple ot ouÿ Sapience parler, il fu seurpris de son amour et par grant ardeur d'esperit disoit: "C'est vray! On ne pourroit trouver chose plus veritable." It is interesting to see the liberty the translator has taken with the three Latin quotations. The end of the extract from Proverbs is more usually rendered "...shall discover life and shall obtain favour of the Lord." The order of the two passages from Ecclesiasticus is the reverse of the accepted one. The words of the *quasi libanus* excerpt, generally understood to mean "Like the cedar of Lebanon, I fill my abode with incense," have not been accurately conveyed at all in the French phraseology "I am like the forest that is filled with trees and timber."

[94] (Fol. 21) "Or pleust a Dieu que je la peusse veoir une fois a mon gré! Ha! Dieu pardurable, comme belle, comme bonne et comme singuliere est celle dame de qui j'ay tant ouÿ dire de biens." Advint qu'il avoit tel desir de la veoir que tout le cuer et le corps lui ardoit. Et a celle heure (*MS.* heure heure) une figure lui fut monstree de merveilleuse beaulté, de tres especial façon, qui representoit l'estre et la personne de Sapience, et vit celle precieuse ymage en une nue assise en un throne d'yvoire, vestue de vestemens excellens et de souveraine beaulté. En celui estat apparoit celle dame, espouse plaisant, belle et clere, et reluisant en vestemens dorez, et de couleurs couronnee. Sa couronne estoit vie pardurable et sa vesteure estoit beneurté consolable. Son parler estoit fondé en doulceur et en courtoisie et son acoler estoit doulce et loyale compaignie.

wondrous for all people to behold; from her beatified gaze came a ray of dominant light that flooded the whole area with its splendour. She greeted the Disciple: 'Dear son, give me your heart!' As soon as he heard her, he was struck with love and by the sweet request made of him by the lady. He fell in a swoon at her feet and offered her grateful thanks for the solace she had given him.''[95]

The last segment of the polyptych has been inspired by quotations from St. Paul. ''It came to pass that the Disciple heard read out (in the refectory) a teaching of St. Paul which ran: 'Jesus Christ is the power of God and the wisdom of God; in Him lies hidden all God's treasury of wisdom and knowledge.''[96]

The author of the *Déclaration*, reading from both text and picture, practices two arts in his account of these four paintings, he extends the linearity of the narrative about the Disciple's relationship with Sapientia, and he explains why the action takes place. In the first segment the commentator sees that Sapientia is seated in a terebinth tree and says so, whereas Suso merely states that she spreds her branches like the tree in question. The properties of the tree are next described in the *Déclaration* and an interpretation is supplied to the reader ''all this means that her teaching is gentle, fragrant, full of grace...''

When he turns to the adjacent compartment, the commentator makes a sweeping generalisation about Sapientia's request to the Disciple to hand over his heart. In the vision splendid reported by Suso, she greets her pupil with the words from Proverbs, XXIII, 26 *Prebe, fili, cor tuum michi* ''My son, give me your heart'', the Latin words being reproduced in a scroll by the artist. The expositor declares the request is made by Sapientia ''...to anyone who loves her...'' Moreover, the apparition, which is carefully documented by Suso and imitated by the artist, is not mentioned by the writer of the *Déclaration.* He seems to have been more intent on explaining the spiritual love between the Disciple and Sapientia than its visual statement.

Further, one may observe that in response to the order of the pictorial narrative, the expositor refers to the ''fine sayings'' about God's Wisdom heard by the Disciple in the refectory, and then mentions that the occasion for the exclamatory *Verum est, verum est* is recorded in the third compartment of the illustration.

When it comes to the last segment dominated as it is by St. Paul's portraiture, the commentator offers a remark that harks back to his rejoinder about the frontispiece when he was viewing *Sapientia in Majesty.* By juxtaposing Pauline quotations to the effect that Christ is ''... the wisdom of God'' and the ''treasury of wisdom'' with the idea that the Disciple wishes to learn the true identity of his mentor Sapientia, the expositor seems to imply that Sapientia is Christ.

Twentieth-century art historians in their turn can provide a wider focus for three of the visual statements. The transformation of the themes into symbols or realistic narrative has been conducted with a certain liberty of emphasis. Suso mentions the refectory first, but this motif has been relegated to the third position by the artist, to allow for a more striking image to occupy the first compartment. Moreover, Suso links the exclamation *Verum est* with the terebinth image, but the painter separates these two and transfers the realisation of truth by the Disciple to the *Lessons in the Refectory.*

Spencer adduces several symbols in connection with the first tableau where we see Sapientia, garbed in her customary manner for the *Horloge* series, seated in a tree. ''Is the terebinth tree really

[95] (Fol. 21) Adonc s'enclina humblement vers lui, et d'un attrayant regard d'uns yeulx grans, et de une divine et espirituelle remembrance plaisant et vermeille et merveilleuse a regarder a toutes gens, et de son tres reverend regard yssoit un ray de souveraine lumiere resplendissant et toute la place enluminant, et salua le disciple en disant: ''Beau filz, donne moy ton cuer.'' Si tost que le disciple l'ouÿ, il fu feru d'amours et de la doulce requeste que la dame lui avoit faitte qu'il cheï tout pasmé a ses piez et lui rendi graces et mercis du confort qu'elle lui avoit fait.

[96] (Fol. 21v.) Advint adonc qu'il ouÿ lire un enseignement de saint Pol qui disoit: ''Jhesucrist est la vertu de Dieu et la sapience de Dieu ou quel sont encloz et contenus tous les tresors de la science et sapience de Dieu.''

an oak? What was in the mind of a painter who put a white-robed Sapientia into its branches? Did he refer to Abraham's oak of Mamre, said to prefigure the Cross? or to Seth's vision of the Christ Child, the Child in the Tree signifying Christ upon the Cross? Or was he perhaps more familiar with Guillaume Deguulleville's (*sic*) description of *la dame blanche, Virginité*, who sits in the green apple tree to guard the one apple, white as snow ...[97] If this is so, then may we not assume that the dead stump is the Dry Tree, whose apple Adam ate, whose life Justice asks Virginité to restore by transferring to it the apple, Christ's crucifixion, by which Adam's sin may be redeemed? The decay at the base of the terebinth tree may indeed allude to the state of the Church which was of such concern to Suso."[98]

On the evidence one could support the symbolism Spencer evokes for Virginity derived ultimately from the poetic imagery of Digulleville's *Pèlerinage de l'Ame.* Sapientia is no longer herself speaking the wisdom of Ecclesiasticus and Proverbs, she has been endowed with the allegorical role of a Virgin in a Tree, a role that is represented later in the century by Nicolas Froment when he portrays Mary with the Holy Child seated in a Burning Rose-Bush.[99]

The passage Suso quotes from Proverbs VIII, 35 *Qui me invenerit inveniet vitam et hauriet salutem a Domino* also occurs in the *Missale Romanum*, and significantly for our discussion, in the Feasts of the Nativity of Mary and the Immaculate Conception, being part of the lesson from the *Liber Sapientiae.* With reference to the Missal, E. Harris wrote: "Thus the liturgy suggests not only the connection between the Burning Bush and the Virginity of Mary but also the relation of this symbolism to the main feasts of the Virgin and to the idea of Salvation which it implies ('whoso findeth me, findeth life and shall obtain favour of the Lord')."[100]

Further, the *Horloge* painting suggests yet another Digulleville theme, as Spencer saw: the dead stump placed beside the flourishing terebinth recalls the symbolism of the Dry Tree whose apple Adam had eaten long ago.[101] Numerous legends of the Cross exist[102] which have been conveniently summarised by Schiller, and in one of them the withered tree prefigures the Cross.[103] In Paradise, on a mission for Adam,[104] Seth beholds a dead tree which, the Archangel Michael informs him, had withered after the Fall. According to a later variant of the legend as told by Mandeville, the Dry Tree stood in the Earthly Paradise and lost its leaves when Christ died. It will bear foliage and fruit when the Holy Land is won again for Christianity.[105]

The upper right miniature is an almost complete pictorial rendering of the Disciple's vision, even to the detail of Sapientia's radiance flooding the terrain with light. Spencer noted that the ray of light of the text has been changed "into a scroll for her words, making it seem to come from her left eye."[106] The source for the light in Sapientia's portraiture, however, is the traditional one: a *rayonnement* of beams from behind her figure. The burning seraphim near the throne suggest she

[97] I omit here four lines quoted from Digulleville's *Pèlerinage de l'Ame.*

[98] Spencer, *Horloge*, pp. 284-5.

[99] See E. Harris, "*Mary in the Burning Bush.* Nicolas Froment's Triptych at Aix-en-Provence," *Journal of the Warburg and Courtauld Institutes*, I (1937-1938), pp. 281-6.

[100] *Ibid.*, p. 282.

[101] Spencer, *Horloge*, p. 285.

[102] The bibliography for legends of the Cross is extensive because the textual versions are many and varied; they include the *Legenda Aurea*, travel accounts, and vernacular poems in several languages. The most informative discussion about the legends I have consulted are by R. Morris, *Legends of the Holy Rood, Symbols of the Passion and Cross Poems*, London, 1871; W. Meyer, *Die Geschichte des Kreuzholzes von Christus*, Munich, 1881; F. Kampers, *Mittelalterliche Sagen vom Paradies und von dem Holze des Kreuzes Christi*, Cologne, 1897; W. N. Seymour, *The Cross in Tradition, History and Art*, New York, 1898.

[103] Schiller, *Iconography*, II, pp. 12-13.

[104] A cycle of pictures of the legend of Seth can be viewed in a Book of Hours prepared by the Cleves Master in the 1440's, see J. Plummer, *The Hours of Catherine of Cleves*, London, 1966, pls. 79-86 in the section for the Friday Hours of the Compassion of God. In the surviving miniatures, a withered tree as such is not depicted.

[105] Cited in Meiss, *Boucicaut*, p. 45. An image of the Barren Tree near Hebron is preserved in a format from the Boucicaut workshop, reproduced by Meiss in fig. 94.

[106] Spencer, *Horloge*, p. 285.

is truly celestial, a figuration of the Godhead. The narrative moment of the picture corresponds to that which precedes the Disciple's swoon.

The dominating scrolls in the last segment provide the means to an understanding of the meeting between St. Paul and the Disciple. The reference to power and wisdom is from I Cor. I, 24: *Christum Dei virtutem et Dei sapientiam*, and the remainder of the passage is taken from Coloss. II, 2-3: *Christi Iesu in quo sunt omnes thesauri sapientiae et scientiae absconditi.*[107] The scroll next to St. Paul in the tableau combines both Latin phrases. The wording in the Disciple's banderole is *Que est ista tam r... forma pre filiis hominum*, where the letters preceding *forma* are obscured by a twist in the scroll. The second half of the statement may be derived from Ps. XLIV, 3 *Speciosus forma prae filiis hominum* "You surpass all mankind in beauty." The venue is the cell, not the refectory, for it is there that the Disciple reflects on the passages he hears mentioned at meal-time. As he ponders, he sees St. Paul standing before him. The portrayal of the Apostle is intended to convey a positive force and presence in his life, evocative of the wisdom and power of God. For those viewers who are familiar with the life of St. Dominic, the *Horloge* painting presents an echo of St. Paul's visit to the saint to deliver the Book of Gospels, symbol of the doctrine to be taught to the world by the Order of Preachers.[108]

VII (1) *God the Father* Fol. 23
(2) *Man of Sorrows*
(3) *Christ Crowned by God the Father*
(4) *Adam and Eve*
(5) *Miracles of Christ*
(6) *Christ Rejected at Nazareth*
(7) *Sapientia and the Disciple*
(8) *Nativity*
(9) *The Temptation of Christ*

The heading of the second Chapter of the *Horloge* "How one can come to know the Divinity through the Passion of Jesus Christ; the form God took by the said Passion of Jesus Christ" sets the tone for the context. The Disciple observes to Sapientia, "...since you require me to love none other than yourself and you wish to be my only loved one, cause me to have clear knowledge of you so that I can better carry out what you command and teach." Sapientia answers, "When God made all the things that are his creations, such as the heavens and the earth and all else, he so ordered and prescribed for Nature so that one creature depends naturally on the other..."[109] There follows a short section about dumb animals and mortal humans and then Sapientia returns to the point, "And, however, if you intend to come to the path and sure knowledge of me, that is, of the Divinity, you must begin right at the bottom on earth, just like someone wishing to climb a ladder starts at the rung closest to the ground, then goes to the next highest, and thereafter from one rung to the next. Thus, the first rung to mount, in order to know and have knowledge of the Divinity is the Humanity of Jesus Christ. It is the creature on earth which above all and most properly represents the Divinity. After this step you will climb to the second rung, that is, Humanity's

[107] *Ibid.*, she gives the source of the quotations in a confused manner as "(I Cor. i. 24; II: 8)."

[108] Cf. Réau, *Iconographie, Saints*, I, p. 395.

[109] (Fol. 23v.) "Et puis que tu me requiers que je n'aime autre amie que toy, et que tu veulz estre ma seule amie, fay donc tant que j'aye clere congnoissance de toy afin que je puisse mieulx acomplir ce que tu me commandes et enseignes." Sapience: "Quant Dieu fist toutes les choses qui sont cree[e]s, si comme ciel et terre, et les autres choses il donna une telle loy et une telle ordonnance a nature que une creature despend naturelment de l'autre..."

suffering; then, from suffering to patience; thence to the miracles that occurred at the Passion...''[110]

In writing his account of this large illustration the author of the *Déclaration* has drawn on evidence from both text and paintings. His eye scans the left column from bottom to top, and he declares their import to be knowledge of God the Creator through knowledge of His creations. The commentator then commences to 'read' the paintings of the second and third columns by pairs, starting at the bottom and concluding at the top. As mentioned above in chapter IV, § 3, the description of the illustration contains the terms 'left' and 'right' employed in a manner that does not correspond to a modern viewer's position.

The theological thread is lengthened by the expositor in arguments which explain the presence of the segments depicting the Nativity, the Temptation, the Miracles, and the Jewish Rejection of Christ at Nazareth. These illustrate aspects of Christ's Humanity as suggested in brief by Suso's narrative.

The expositor reiterates the logical process whereby knowledge of one aspect of the Divine is the *sine qua non* to access to a new level of cognition about the Divine. Suffering is seen to be epitomised by the Passion, Entombment and Resurrection events which are all pictorially represented in a single compartment. The commentator does not feel the need to interpret for his reader or viewer and declare that the Instruments of the Passion symbolise Christ's Passion, that the Entombment is recalled by the sarcophagus, and that the Resurrection has taken place, since the Risen Christ sits on the edge of the sarcophagus.

The explanation supplied for the last upper segment, *Christ Crowned by God the Father*, conforms to the traditional portraiture and the positioning of the sitters.

To summarise the comments in the *Déclaration*, one would observe that their author perceived the nine compartments in terms of the three ways laid down by Sapientia for the Disciple to follow if he were to gain an intimate knowledge of the Divine.

If we now examine the purely pictorial achievement in less theological terms, we view a full-page illustration whose nine compartments are not arranged as a twentieth-century observer might expect, that is, for reading across one register at a time after commencing at the top left, a progression followed for the list of paintings at the beginning of this section. In fact, we need to depart from the lower left corner and arrive at the upper right segment.

The scene where Sapientia and the Disciple are in discussion has a conventual interior, probably a *parloir*, as the venue. He seeks more knowledge of God, Christ and Divine Sapientia from the figure who personifies all three: *Da michi te cognoscere* "Cause me to have knowledge of you" (represented in a scroll).

If the theme of the Creation were to receive a standard treatment in the *Horloge* at this point, the beholder would be a witness to representative scenes of God's activities on each of the seven days, in a manner similar to the cycles that occur in copies of *Le Livre des Propriétés des Choses* which is Jean Corbechon's translation of a work by Bartholomaeus Anglicus.[111] But the single pic-

[110] (Fol. 23v.) "Et pourtant, se tu as entencion de venir a la voye et certaine congnoissance de moy, c'est a dire de la divinité, il te couvient commencier au plus bas vers la terre si comme fait celui qui veult monter une eschielle commence a monter le premier eschielon plus pres de la terre, et puis le second en hault et tousjours aprés de l'un a l'autre. Semblablement, le premier degré pour monster a congnoistre et avoir congnoissance de la divinité, c'est l'umanité de Jhesucrist; c'est la creature en terre qui plus proprement et premierement represente la divinité. Aprés cestui degré monteras tu au second, c'est a la souffrance de l'umanité; de la souffrance a la pacience; puis, aux miracles qui apparurent a la mort." The metaphor of the ladder in the French translation is not in Suso's *Horologium*, where the path to knowledge of the Divinity is the *via regia*, see Künzle, *Horologium*, p. 384.

[111] A typical representation of the theme can be seen in Paris, Bibliothèque Nationale, MS. fr. 9141, fol. 9; reproduced by Meiss, *Boucicaut*, fig. 447. A fuller discussion of cyclic programmes executed by French miniaturists is contained in D. Byrne, "The Boucicaut Master and the Iconographical Tradition of the *Livre des Propriétés des Choses*," *Gazette des Beaux-Arts*, XCI (1978), pp. 149-64.

torial unit in the *Horloge* may be said to be an epitome of the work of the Creator.[112] The separation of Light and Darkness and the creation of the angels are symbolically recalled by the light above the horizon fading to a darkened starry firmament, and by the seraphim burning brightly in the empyrean. The separation of water and sky on the Second Day is marked by the horizon line. The toil of the Third Day included the separation of Water and Earth and the creation of plants. The estuary scene with a clear demarcation between the water mass and the land mass denoted by the line of curtain-wall between bastion and turrets recalls God's work. The trees, shrubs and grass betoken the presence of plant-life.

The Sun and the Moon in the heavens signify the completion of the Fourth Day. The Seventh Day of Rest was the time for God's contemplation of His Creation. The iconography conveys this state of mind. The Almighty is seated on a rainbow whose arc symbolically embraces the scene below; His left hand covers the *orbis mundi*, signifying His domination and power over the world; His right hand blesses His Creation. On His nimbed head the Creator wears the four-arched crown of imperial power. Peace and calm is everywhere, suggested by the ships idling at anchor or with furled sails. Nature is still, and God the Creator is omnipresent. In spite of this analysis of the symbolic content of the painting, the immediate impression is not one of the Creation, rather it is of the omnipotence of the Almighty. This consideration has determined the retention of the name *God the Father* for the iconography.

Missing from the motifs represented in the upper-left segment is a figure that could be said to represent the animal kingdom and the human race, the creation of which occupied the Fifth and Sixth Days. The compartment below *God the Father* certainly contains both subjects and perhaps should be read in conjunction with it. Yet the creatures are no longer in their first moments of innocence. The scene depicts, instead, the Fall of Man. All the traditional elements of the iconography of this theme are present.[113] In an earthly paradise frequented by two weirdly formed animals, moving among well spaced trees that grow in a beautiful landscape, stands the Tree of the Knowledge of Good and Evil. Around its trunk is entwined the Tempting Serpent with a humanoid head, grinning maliciously at Eve. The act has already taken place. Eve, one hand and fig-leaf over her pudenda, the other near her breasts, moves spiritedly towards Adam standing on the other side of the Tree. He is slightly turned away and covers his genitals with a hand.

The iconography of Christ's Humanity in the tableau has its beginning in a traditional *Nativity* scene, depicted in the centre of the lower register.[114] The humble, balding Joseph tends a fire, while behind him the Virgin is seated on a comfortable bed in a stable. She is seen in the act of suckling the Holy Child, who is helpless and inert in swaddling clothes and cradled in her arm. A heavenly light sends down its rays on the Child, perhaps suggesting His origins in Divine Light. Just as that descending light seems to herald a white dawn behind the low hills of the landscape, so the Infant will lighten the darkness of the universe. The star is an allusion to the words of Revelation XXII, 16: "I am the root and offspring of David, and the bright and morning star."[115] The tenderness of the Virgin, the solicitude of Joseph and the innocence of the Child combine to heighten the spiritual expression of the image. The painting constitutes an important statement on the first period of Christ's Incarnation in the Earthly World.

[112] The Work of the Creator is conveniently summarised in the repertoire of Réau, *Iconographie, Ancien Testament*, pp. 68-76, which I follow here.

[113] *Ibid.*, pp. 86-7.

[114] The theme of the *Nativity* underwent many emphases in Western Art from the ninth to the mid-fifteenth century. A survey of attitudes is supplied by Réau, *Iconographie, Nouveau Testament*, pp. 213-224 and Schiller, *Iconography*, I, pp. 66-76. For the icons in manuscript painting, see Rosy Schilling, "The *Nativity* and *Adoration of the Christ Child* in French Miniatures of the Early Fifteenth Century, *Connoisseur*, CXXX (1952), pp. 167-9 and 221.

[115] The analogy is made by Schiller, *Iconography*, I, p. 76.

Another aspect of His Humanity is illustrated by one of the three Temptations, as recorded in Matthew, IV, 3-4. Schiller observed: "In the Late Middle Ages it became customary to present the first Temptation as the main image and to depict the others as small subsidiary scenes in the background."[116] The visual representation in the *Horloge* concentrates on the first theme. A winged, dragon-like tempter utters the Latin words carried in a scroll over its head: *Si filius Dei es, dic ut lapides isti panes fiant* "If Thou art the Son of God, tell these stones to become loaves of bread." Another banderole carries Christ's rejoinder: *Non in solo pane vivit homo, sed in omni verbo, quod procedit ex ore Dei* "Man cannot live by bread alone but by every word that God utters." Both quotations are derived from Matthew, IV, 3-4. Behind Christ is an ornate chapel, and, nearby, a mound and trees. The chapel may be thought to suggest the presence of a hermit, but Christ's garb does not confirm the illusion. He wears the tunic and robe of a missionary preacher. The natural surroundings are clearly not intended as a wilderness, but rather as signs of sequestration and isolation. It is the spiritual isolation that is stressed; this in turn makes Christ's challenge to the Devil and triumph over his wiles all the more incontestable.

Christ's Mission on Earth continues to unfold before the viewer as he beholds Christ in the open air, preaching from a pulpit. The listeners appear to be persons He has cured. One is a woman who may be Simon's mother-in-law gripped with fever;[117] another figure is a paralytic on the point of taking up his bed and walking;[118] a man in the foreground raises a withered arm;[119] a grateful and unshrouded Lazarus sits up in an open-air grave, far from Bethany.[120] The walled city on the horizon is bathed in light, evoking an image of the Holy City in the Earthly World to which Christ will soon return.

The selection of the Miracles appears to have been undertaken in a haphazard manner, but there is a common theme for all. Christ's supernatural powers are not visually suggested, only his powers of healing.[121] The presence of Lazarus nevertheless is not merely to remind the viewer that Lazarus died of a sickness and was later revived to continue life as a fit and healthy man. As Schiller remarked, "More often than any other resuscitation, the Raising of Lazarus has been represented at all periods from the early third century onwards and has never lost its important position in art."[122] In the *Horloge* painting, however, Lazarus appears visually as yet another miracle of the group. The symbolism of his resurrection is much more important than the icon seems to convey. It could be seen as a manifestation of Christ's ability to overcome death. Schiller further observes that "Even in the Early Christian period the Raising of Lazarus was regarded as a prefiguration of the Resurrection of Christ and the resurrection of the dead at the Last Judgment."

A portent for the future is registered on the folio in the adjoining pictorial unit. Christ, still garbed as a preacher, is rejected by a group of Jews at Nazareth and appears to be on the point of vanishing into a natural geological and also spiritual abyss. But Luke IV, 30 offers a different outcome for the confrontation. Across the lower edge of the painting runs the legend *Ihesus autem transiens per medium illorum, ibat* "Jesus, however, walked through their midst, and went off."

One of the most salient and pathetic features of Christ's Humanity was suffering, as Sapientia reminds the Disciple in the textual passage at the head of this discussion. The theme is the reason for the inclusion of a *Man of Sorrows* on the folio. The representation is marked for its portrayal

[116] *Ibid.*, I, p. 145. Visual renderings of the theme are comparatively rare, according to Schiller, *Ibid.*, and Réau, *Iconographie, Nouveau Testament*, pp. 304-9.

[117] Cf. Luke, IV, 38-9.

[118] Luke, V, 17-26.

[119] Luke, VI, 6-11.

[120] John, XI, 43.

[121] The typology of these miracles is discussed by Schiller, *Iconography*, I, pp. 169-81. They often occur individually in cycles of the Life of Christ.

[122] Schiller, *Iconography*, I, p. 181. One may also consult E. Mâle, "La Resurrection de Lazare dans l'Art," *Revue des Arts*, 1951, pp. 44-52, but he has little to say about fifteenth-century French miniatures.

of tenderness, conveyed by the gestures of the succouring angel,[123] centrally placed, who supports Christ's body as He sits on the edge of the sarcophagus. Two motifs, however, cause the viewer to remember the physical hardships and torture undergone by Christ's whole physical being. The first is His gesture of placing the left thumb into the gaping wound in His side. The other is echoed by an array of the Instruments of the Passion, traditionally part of the Man of Sorrows theme,[124] and here serving as a solemn backdrop.

The top right-hand segment of the whole cycle of pictures is occupied by the crowning glory of the Disciple's knowledge of the Divinity. The apotheosis is clearly a pictorial triumph and an unusual embellishment to Suso's text in which no specific mention is made of Christ's Coronation by God the Father. The upward movement of the viewer's gaze is abetted by the figure of King David looking heavenward at a richly tapestried, high-backed throne where Christ sits on the right of the Almighty. David recites the words of his Psalm CIX, 1: *Dixit Dominus Domino meo, sede a dextris meis* "God said to my Lord, sit on my right." God the Father passes to Christ the Sword of Justice, symbol of His power to judge, and prepares to place on Christ's head His own four-arched crown of omnipotence. Angels are seen behind the group, intoning *Deus, iudicium tuum* and the words of Psalm VIII, 6: *Gloria et honore coronasti eum* "Thou hast crowned Him with glory and honour."

VIII *Sapientia and the Disciple behold Christ Carrying the Cross* Fol. 24

The discussion between Sapientia and the Disciple about the beauty of the true love of Christ prompts him to observe: "It is not the case with the Crucified Christ. For He was so disfigured and rendered obscene through the torments of the Cross and the harrowings of Death that He carries no sign of beauty on His body for which one should love Him, to my way of thinking."[125] Sapientia draws three morals: lovers pay no attention to rose thorns, provided they succeed in plucking the flower; wise people do not prefer an attractive-looking chest to a dilapidated one; wives, however beautiful they appear to paramours, can in the end be poison. The sentence relating to the second parable is missing from the French translation in the Brussels *Horloge*, fol. 24. Suso's words in Künzle, *Horologium*, p. 385 read: *Nec vere sapientes scrinia pulchra et deaurata quae tamen in se vilia continent, capsulis vilissimis praeferunt, quae pretiosissima ornamenta inclusa abscondunt* "Nor indeed do the wise prefer beautiful gilded chests which contain filth within, rather, they prefer the ugliest ones which hold, unrevealed, objects of great beauty."

The *Déclaration* offers a rare instance of the author's attempt to provide a complete *raison d'être* for a tableau. The argument advanced paraphrases the textual statement, and omits the parable of the wives and paramours: the painting, after all, contains no trace of it. Next, the expositor concentrates on the principal subject: Christ carrying the Cross on His own without soldiers and an unruly populace surrounding Him.

The painter's response to the textual statement of Suso is both simple and complex. He does not depict the third parable at all. As for the lovers plucking the rose amid its thorns, it is merely suggested. The rose and thorns are held delicately by Sapientia; one is required to comprehend the symbolism implied and to recognise the Disciple as the lover.

The second parable is the most graphically presented. Since information about it is not in the

[123] The importance of the ministrations of angels in the icon is discussed by Schiller, *Iconography*, II, pp. 215-19.

[124] *Ibid.*, II, pp. 207-10.

[125] (Fol. 24) Or n'est pas ainsi de Jhesus le crucifié, car il fu si desfiguré et enlaydis par tourmens de la Croix et de la mort qu'i n'a en son corpz signe de beaulté par quoy, a mon adviz, on le doye amer.

French text, the artist must have derived it from the Latin of Suso. A worm-eaten chest filled with plates and goblets lies at Sapientia's feet, while, along the rear wall of the loggia, can be seen another chest draped with a cloth of gold. One can just glimpse the corpse and rotting flesh inside.

It is difficult to comprehend why the first textual passage has been visually interpreted in the manner it has, namely as a version of Christ Carrying the Cross. One may well have expected a Man of Sorrows, which has the capacity to emphasise more pointedly the "torments of the Cross and the harrowings of Death", as Suso puts it. Perhaps it was felt that a subject of that nature, being somewhat stereotyped and providing an exposition of the Instruments of the Passion, would eclipse any visual impact of the secondary theme concerning the parables.

The beholder may, furthermore, be forgiven for thinking that the miniature captures the moment when Christ loads the Cross on to His shoulders, and in the presence of mockers, sets out from the praetorium on the way to Calvary. But, on reflexion, one realises that the departure is a symbolic movement in this instance. The selected posture of a figure that is already pitiful and hapless, with a brow bround by thorns, and with marks of the Crucifixion on hands, feet, and right side, is not really betokening a Station of the Cross. The figure reflects more profound symbolism: however unsavoury or odious appearances are, they hide a wondrous inner beauty waiting to be uncovered. The imagery of Christ's sorrow merges with the less important motifs of the attractive and repellent chests. In one small illumination the viewer finds the hidden truth of the inner beauty of Christ's Love in spite of His repulsive outward form.

IX (1) *Agony in the Garden* Fol. 25
(2) *Christ before Annas*
(3) *Christ before Caiaphas*
(4) *Christ before Pilate*
(5) *Christ before Herod*
(6) *Buffeting of Christ*
(7) *David's Soldiers at the Well*
(8) *Soldiers bring Water to David*
(9) *Sapientia and the Disciple*

As the Disciple listens, he hears from Christ's lips the nature of the great moments of the Passion. "I went off to the Mount of Olives with eleven disciples and I prayed to my Father at length, greatly troubled in heart and body..."[126] Seized by soldiers and dragged before Annas, Christ continues, "They spat in my face, they poked my eyes, they broke my neck with their blows, and just like mad and infuriated creatures they rose against me. In the morning at the house of Caiaphas they stated and decreed that I was worthy of death because of a truth I had uttered. She who had conceived me, borne me, and given birth to me was so overcome with pity for me that her heart was deeply wounded; she wept copiously all the time without comfort or relief; she was cut to the quick in seeing me in such shame and with such grievous bodily harm. From there I was taken before Pilate, accused and condemned... Next, Herod, out of mockery and derision, clad me in white robes and each mocked me as they would a fool. They lashed and beat my body, and crowned me with a circlet of sharp thorns which pierced my brain. They sullied and besmirched my face with blood and filth."[127]

[126] (Fol. 25v.) Je m'en alay en la montaigne d'Olivet avec xj disciples et priay a mon pere longuement en grant angoisse de cuer et de corpz...

[127] (Fol. 25v.) "Ilz me crachoient ou visage et me estoupoient les yeulx et me rompoient le col par force de horions; et si comme foles gens et enragiez ilz se eslevoient

The Disciple is so despairing and disturbed by the account of events that he begins to reproach Sapientia for the tragedy and sorrow it entails. She reassures him that suffering is the path to Glory, and she draws an analogy from the annals of King David. "The servant should not take his ease when he sees his lord fighting valiantly. Reflect and call to mind the three courageous and resolute knights of David. If you wish to be one of them, you must be bold and abandon fear. You must be courageous and patient and carry yourself with honour in all circumstances, in prosperity and in adversity. Thus, you will be able to draw water at the Bethlehem Well and offer it to your sovereign lord. That is to say that by this path you will present yourself to God in order to suffer and endure all things and for His love; and you will obey His will completely. This parable is written in the Second Book of Kings in the twenty-third chapter."[128]

Suso's text speaks of Christ going to the Mount of Olives with eleven disciples. *The Déclaration*'s statement is based on the miniature where only three of them are present at the moment of Christ's prayer to God the Father. The expositor offers their names, even though Suso is silent on the subject of their identity. From here on the commentator takes each of the other eight segments in turn and paraphrases or summarises the trials before members of the Sanhedrin. He also follows Suso's wording for the explanation of the parable of David's knights, but he adds details that could only have come from the Biblical story itself. When describing the last compartment which depicts Sapientia and the Disciple, he makes a logical assumption: Sapientia is concluding her doctrine by drawing the analogy of David's soldiers and the Disciple's suffering for Christ. In reality, of course, the conclusion is relative to the pictorial series, not to Sapientia's total teaching efforts. Finally, one observes that the terms 'right' and 'left' are used where the modern eye would expect 'left' and 'right', as was discussed in chapter IV, § 3.

The nine-part miniature is positioned at the very opening of the third chapter which commences on the verso side of the folio. As on fol. 15v., the arrangement for viewing is three registers, each of three scenes. Six compartments present incidents of the Passion narrative from the time of Christ's entry into the Garden of Gethsemane to the Buffeting. They are chronologically in step with the textual account.

The themes of suffering and isolation are movingly captured by the iconography in each scene. Saints Peter, James and John seem oblivious to Christ's moment of truth as He prays, distant from them, just as the three are alienated from the main body of the Disciples. The olive grove is symbolically circumvented by a wicker fence and is thus set apart from the sloping hills. God the Father in the heavens observes the Cup of Sorrow placed on a pinnacle of rock and listens to His Son say the words recorded on the scroll *Mi si fieri pater, transeat a me calix iste* "My Father, if it could happen that this Cup pass me by." The wording is close to Matthew, XXVI, 39: *Pater mi, si possibile est, transeat a me calix iste.*

contre moy. Le matin en l'ostel de Caÿphe ilz dirent et jugierent que je estoie digne de mort pour une verité que j'avoye ditte. Celle qui m'avoit conceu, porté et enfanté avoit telle compassion de moy qu'elle estoit amerement grevee jusques au cuer, et tousjours habondamment plouroit sans remede et sans confort et lui perçoit le cuer quant elle me veoit a telle honte et a si griefz tourmens. De la je fus mené devant Pylate, accusé et condempné... Aprés, Herode me vestit d'une cotte blanche par moquerie et derrision, et chascun se moquoit de moy si comme d'un fol. Mon corpz batirent et flagellerent; mon chief d'un chappeau d'espines poignans couronnerent et me percerent jusques au cervel. De sang et d'ordure ma face me honnirent et broullerent."

[128] (Fol. 26) "Car le varlet ne doit mie prendre ses aises quant il voyt son maistre combatre si puisamment. Considere et te souviengne des trois fors et vaillans chevaliers de David. Se tu desires a estre en leur compaignie, il te couvient doncques prendre hardiesse et laissier paour. Il te couvient estre courageux et pacient et toy honnourablement maintenir en tous estas, en prosperité et en adversité. Ainsi pourras tu puiser de l'eaue de la cisterne de Bethleem et la presenter au souverain roy. C'est a dire que par ceste voye tu te presenteras a Dieu pour tout souffrir et endurer et pour son amour, et obeiras a sa voulenté du tout." Ceste figure est escripte ou second livre des Roys, ou xxiij^{e} (*MS.* xxiiij^{e}) chapitre.

The supreme isolation of Christ is carried into the Sanhedrin trial scenes. He is positioned centrally and motionless. Because He is surrounded on three sides each time by guards, accusers and tormentors, all animated, gesticulating wildly and hurling abuse or mockery, the beholder perceives all the more readily the noble character of Christ's Humanity, His natural dignity. Yet, His corporeal frame weakens as He appears before Annas, Caiaphas, Pilate and Herod. In *Christ before Caiaphas* the sorrowing Virgin is placed prominently on the viewer's left, where, in conformity with Suso's narrative, she weeps copiously.

Banderoles remind the viewer of high points in the unfolding events. Whoever transcribed the statements on the scrolls in the scenes of *Christ before Annas* and *Christ before Caiaphas* seems to have been confused about the speakers. One should therefore read the utterance of Annas from the Caiaphas scroll: *Expedit ut unus moriatur homo pro populo* "It is in the interest (of the Jews) that one man die for the whole people," which may be compared favourably with John, XVIII, 14: *Quia expedit unum hominem mori pro populo*. By the same token, the words recorded in the Annas tableau belong to the Caiaphas one: *Quid me cedis? Ego palam locutus sum mundo. Sic respondes pontifici?* "Why do you strike me? I have spoken openly to the world. Is that the way to answer the High Priest?" The three phrases have been copied out of order and represent John, XVIII, 23, 20, 22.

In *Christ before Pilate* the Roman governor's question is carried in a scroll: *Tu es rex iudeorum?* "Are you King of the Jews?" Christ replies: *Rex sum ego* "I am King." A kneeling courtier pulls apart the corners of his mouth in mockery as he shouts *Ave, rex iudeorum!* "Hail, King of the Jews!" The three quotations are from John, XVIII, 33 and 37, and XIX, 3. The last of these belongs, strictly speaking, to the Flagellation of Christ in the Gospel narrative; it is an episode which the Rolin Master does not represent pictorially. In the *Buffeting of Christ* the phrases one reads in the banderole are: *Tolle! Tolle! Crucifige eum!* "Away with Him! Away with Him! Crucify Him!" These words were shouted at the conclusion of the Buffeting, according to John, XIX, 15.

The allegory of courage in adversity is dramatised by King David's soldiers, when they won through and brought water to the monarch, as is related in II Samuel, XXIII, 13-17. These two scenes are drawn to the attention of the Disciple by Sapientia as she points to her right in the direction of the visual statements. The Disciple too must submit himself to suffering out of love and devotion to Christ, his Lord and Saviour.

X (1) *Christ Nailed to the Cross* (2) *Christ Offered the Sop* — Fol. 27v.

Christ describes to the Disciple His sufferings at the Crucifixion: "When I had been led to the site on Calvary outside the city, they hanged me on the Cross between two thieves, only so that my death would be more shameful in the eyes of the people. When I was crucified and so grievously and painfully beaten and surrounded, my eyes, which had been fair, clear and sparkling, lost their brightness and were blinded. My ears heard much blasphemy and mockery. My nostrils smelled powerful stench. My mouth received a mixture of bile and bitter wine. They struck me, and wounded me so much that their blows broke all my wretched flesh; I was wounded in such a way that both from the harshness of the punishment they meted out to me and from the nails driven into my hands and feet, great gushes of blood poured down my body."[129]

[129] (Fol. 28) "Quant je fus amené au lieu de Calvaire hors de la cité, ilz me pendirent en la Croix entre deux larrons pourtant seulement que ma mort feust plus honteuse devant le peuple. Quant je fus crucifié et si douloureusement et angoisseusement feru et avironné, mes yeulx, qui estoient beaulx, clers, reluisans, perdirent leur clarté et furent

Were a reader of the expositor's words in the *Déclaration* to lack recourse to the visual interpretation of the drama, he may well conjure up imagery far different from the broad sweep of vistas and rolling hills depicted in this illumination. The comments are in general correct, but interpretative colour is wanting. The expositor mentions the narrative nexus existing between this painting and the preceding format, the *Agony in the Garden*, but his remarks are a summary of comments made at that point. The commentator's interest in both the textual and visual content of the illustration is slight; he merely speaks of "...the Crucifixion of Our Lord Jesus Christ on an ignoble site and between two malefactors...".

To a modern viewer the iconography inspires much more comment. It will be noticed, for example, that Suso's narrative describes in Christ's own words several of the Stations of the Cross. As some of these had already received pictorial dimensions elsewhere in the *Horloge* series, the drama or dramas selected for enactment needed to be presented for the first time. The resultant visual narrative is a panoramic sweep that embraces simultaneously the Holy City and the Heights of Calvary, and presents with an original lay-out two major Passion themes in which Christ is the focus of pity and suffering, and a minor one, minor only because He is not present, the procession on the *via dolorosa*.

From a distant Jerusalem in the left background, a procession winds its way, and two thieves struggle with their crosses on a stony path. A squadron of horsemen in the rear try and hurry them along. Christ has already reached the slopes of Golgotha, has been stripped of His purple cloak and laid out on the Cross. A man is seen in the act of nailing His right hand, another pulls His arm at full stretch with a rope. As the eye moves instinctively to the right to survey the other major theme, it meets a sequel to the Nailing,[130] the Cross is elevated to an upright position while workmen secure its base in the ground; Christ's mouth is about to receive the Sop; His blood already streams down the Rood.

Since Suso's narrative presents Christ's words as told by Sapientia to the Disciple, both figures are included within the picture plane. As their dialogue is the pretext for the two salient themes, the figures' position on a nearby knoll, symbolises on the one hand, their distance in time from the enactment, but on the other, their spiritual participation in every move and gesture that Christ makes, in each torment and act of humiliation He suffers. The horror-stricken faces of the two onlookers register their sensibility. The imagery serves as a graphic exposition and elaboration of the symbolism of the simple cross that Sapientia had earlier shown to the novice in his cell, calling it then, without explanation, the Tree of Life.

As Schiller noted when discussing the forms in art assumed by the preparations for the Crucifixion, the Nailing is a motif with a long history in Byzantine illumination, and occurs in Western art in Italy and North of the Alps, particularly in Passion altars. Reference to French manuscript illumination is not made, but one observes that the theme popularly functions as a "subsidiary motif to the image of the Crucifixion."[131]

aveuglez. Mes oreilles oÿrent moult de villenies et de derrisions. Mes narines odorerent de grans punaisies. Ma bouche receut ung buvrage de fiel amer et de vin aigre. Ilz me ferirent. Ilz me navrerent en tant que par force de coupz ilz derompirent toute ma povre chair, et fu navré en telle maniere que tant par la durté de la discipline que ilz firent sur moy, tant pour les plaies des clous qui me furent fichiez es piez [et] es mains, les undes de sang couroient par tout mon corps."

[130] Schiller, *Iconography*, II, p. 86.

[131] *Ibid.*

XI (1) *Jael Welcoming Sisera* Fol. 30v.
(2) *Jael Slaying Sisera*
(3) *Samson and Delilah*

Copied beneath the triptych is a long doleful passage, as the Soul pours forth its bitterness on the subject of deception: "O vile and obscene world, who with your wickedness and fallacies deceives everyone. He who serves you is cursed... O wordly pleasure, you showed me an attractive face of flesh and blood, a smooth bosom, a pair of laughing eyes. And in doing so, you wounded and pierced my poor, grieving heart. I was seeking water and you gave me milk. I was asking for a drink and you satisfied me with poison and venom..."[132] One observes that Suso does not record the names of the persons in the story he alludes to.

After stating the theme of the triptych and, incidentally, that of the whole of chapter IV of Suso's work, in terms of the falseness and deception of the world, the author of the *Déclaration* turns to the visual statement about Jael and Sisera. He retells it at length, using the Biblical narrative and locating it in the Book of Judges. However, when he reaches the last compartment of the triptych he merely announces the subject matter of the exemplum and indicates the Biblical source.

The words from Judges, IV, 18: *Intra, domine mi; intra, ne timeas* "Enter, my lord, enter; fear nought" remind the viewer of the scroll in the first panel of the triptych that Jael is welcoming Sisera. The legend across the bottom of the picture *Aquam petenti lac dedit* "She gave milk to him who sought water" recalls the next incident, cf. Judges, IV, 19. Jael is presented as the personification of deception. On the one hand, she is alluringly dressed in the latest court fashion of the mid-fifteenth-century. The modelling of the times allows the bosom's dimensions to be boldly displayed. On the other hand, her eyes are full of laughter and ardour.

The allegory of poison and venom is reserved for treatment in the second panel where the hostess betrays Sisera's presence to his enemy, Barak. He arrives on the scene just as she drives a tent-peg into the skull of her slumbering guest. Her greeting to him is contained in a scroll: *Veni, ostendam tibi virum quem queris* "Come, I shall show you the man you seek" (Judges, IV, 22). The pictorial statement is startingly macabre, which serves to amplify the villainy and to reinforce the moral appeal. Speaking of this particular exemplum, Réau observes "Dans l'art misogyne du Moyen Age, c'est un exemple classique des ruses de femmes. On l'assimile à Phyllis chevauchant Aristote, à Dalila coupant les cheveux de Samson."[133] Certainly, it seems that the association is attested even in the group of pictures on this folio of the Brussels *Horloge*.

The last section of the triptych presents Delilah's duplicity during her liaison with Samson.[134] The source of the anecdote is the same Book of Judges, which the visual imagery follows closely. The deception occurs in the open air, she cuts the great judge's hair while he sleeps on her lap and she utters the prearranged words of alarm: *Philisti(i)m super te, Samson* "The Philistines are upon you, Samson" here copied on a banderole from Judges, XVI, 20. In the miniature the action is rendered more theatrical by the enemy soldiers' attempt to seize their victim.

[132] (Fol. 30v.) O mondaine vanité, o mondaine plaisance, tu m'as monstré ung viaire charnel et attrayant, une gorge polie, uns yeulx rians, et par ainsi faire tu as navré et percié mon povre dolent cuer. Je demandoye l'eaue et tu m'as donné le lait. Je demandoie a boire et tu m'as enyvré de poisons et de venin...

[133] Réau, *Iconographie, Ancien Testament*, p. 328.

[134] Réau, *Iconographie, Ancien Testament*, p. 236.

XII (1) *Apostles*[135] Fol. 33
(2) *Martyrs*
(3) *Confessors*
(4) *Female Saints*
(5) *City of Religion in Ruins*
(6) *Money-Changer*
(7) *Charity and Envy*
(8) *Chastity and Lechery*
(9) *Obedience and Sloth*
(10) *Chastisement and Pride*
(11) *Poverty and Avarice*

The fifth chapter of the *Horloge* is very elegiac in tone. Using allegory, Suso bewails the moral decay of Religion. The Disciple had a vision "and in it he saw a poor pilgrim at the gate of the City of Religion and he held a staff in his hand and he seemed wretched and indigent. Near him was the site of a city which had once been fortified and apparently enclosed with walls and encompassed by moats, towers, gates, locks and other strong means of protection. But with age the city was all destroyed and had gone to ruin. It happened that through the shortcomings of the inhabitants, a part of the city had been captured by enemies, while another quarter, through neglect, had fallen down. One section remained intact, but it was leaning noticeably and was ready for collapse in places, and its wall was rent from top to bottom. In this city great wild beasts looking like humans appeared to the pilgrim and when he sought their help, they basely refused. Even so, some took pity on him and wanted to comfort and aid him, but others prevented them..."[136]

Sapientia now attempts to explain the allegory of the vision to the Disciple. "Here is a wondrous vision that is full of much sorrow. Since you wish to know its meaning, I shall tell you. I am the pilgrim you saw and I was formerly Prince and Lord of the ruined city which was shown to you. The city thus wasted and in ruins is the Holy Church and Mother of Christian Religion which contained from the very beginning persons who were pious, fervent, religious and filled with faith and charity, loyalty and harmony, who with one heart and one will served God and led a holy and devout life; they had renounced the world in heart and deed and had centred their heart in God alone. They were for ever in holy meditation and set a good example for their neighbours and each other through their good works. They did not disturb or harm anyone, but led a pure and simple life, crucifying and tormenting their bodies mercilessly. They regularly undertook devotional exercises and founded divers religious institutions through their exemplary practices. They were filled and endowed with the good virtues and they fought effectively against the vices..."[137]

[135] The order of numbering of the segments would normally be across the registers, starting at the top. Here exceptionally, the titles are arranged in accordance with the vertical functioning of the panels of the whole composition.

[136] (Fol. 33v.) Car en sa vision lui apparut un povre pelerin qui estoit a son huis et tenoit ung baston en sa main et sembloit estre povre et indigent. Et pres de lui apparoit la place de une cité, laquelle avoit esté autresfoiz fermee et close de murs, se sembloit, et avironnee de fossez, de tours, de portes, de serrures et d'autres fortes garnisons. Mais par force de viellesce elle estoit toute destruite et alee a ruyne. Et est advenu que par deffaulte des habitans une partie de celle cité avoit esté prinse de ses ennemis et l'autre partie par deffaulte de secours avoit esté cheute a terre. Une partie estoit demouree entiere, mais elle s'enclinoit fort et se disposoit a ruyne deça et dela. Et estoit le mur fendu des le hault jusques au fondement. En celle cité lui apparurent bestes moult sauvaiges en figure d'omme. Le pelerin demandoit ayde de celles bestes, mais elles lui refuserent villainement. Toutesfois aucunes d'elles avoient pitié du pellerin et le vouloient conforter et aydier, mais les autres les empeschoient...

[137] (Fol. 33v.) Sapience: "Vecy vision merveilleuse et de grant pitié plaine, et puis que tu la desires, je t'en diray la signifiance. Je suis le pelerin que tu as veu, qui autresfois fus prince et seigneur de la cité destruite qui t'a esté monstree, et celle cité ainsi gastee et ruyneuse, c'est Saincte Eglise et la mere de religion crestienne en laquelle furent au commencement personnes devotes, ferventes et religieuses, plaines de foy et de charité, de loyaulté et de concorde, qui

Sapientia's pronouncements about corruption in the City of Religion contain a section on the Virtues of Obedience and Correction or Chastisement. "We see with our own eyes that some of the young, and several of the aged and the professed in religion have become emboldened and so used to rebellion that they rise against their leaders to the point that chastisement is neither spoken of nor carried out. They are seen arming themselves with malice, ruse and new sophistries and bringing legal counsel to defend themselves, to cover their wickedness and to put themselves beyond chastisement. And if some are, on occasion, punished by the prelate, they go away and perform worse than before. Of no worth are the fences of good order, the moats of correction, the bastions of the constitution or the rule of religion, because men such as those transgress and exceed all boundaries. By their presumption and disobedience they break down and undermine the solid structures of peace, harmony, purity, prayer, humility and obedience; and often by their disputations and wrath they put body and soul in peril. Certain causes and signs of this ruin and destruction have been evident and are still manifest. How long do you think the edifice can last now that you see the foundations, the blocks of stone and the pillars are being, and will soon all be removed; in other words, when humble obedience will be cast out, and each will do what he likes, and no prelate or anyone else will dare to say: Why do you act thus? Alas and alack! where is the obedience of St. Paul the Simple, of St. Maurus, of St. James, of St. Benedict and of the other good, perfect or recognised holy men and disciples of God, who in bygone days were so truly obedient that they all desired keenly to do and carry out everything commanded of them; they were not even able to disobey those who asked them to do things they could not do."[138]

These remarks lead into a section on the blatant flouting of poverty by religious. After Poverty, Sapientia speaks of the Virtues of Chastity and Charity. "Of Chastity, who is the honour and beauty of the Virtues, there is no need to say very much." Further on, she continues: "What shall I say about Charity which is the mortar and cement of the whole spiritual edifice? I shall say that the structure's oblivion is nigh, it has been wasted and destroyed completely, right down to the foundations."[139]

d'un cuer et d'une voulenté servoient a Dieu et menoient saincte vie et religieuse, lesquieulx avoient renuncié au monde de cuer et de fait et avoient mis leur cuer en Dieu seulement; qui tousjours estoient (fol. 34) ententis et occupez en saintes meditacions et par leurs bonnes euvres edifioient eulx et leur proesme. Ilz ne baretoient nullui ne bleçoient mais deduisoient leur vie en purté et en simplesse, et crucifioient et tourmentoient leurs corpz sans pitié. Ceulx yci se excercitoient et acustumoient en devocion et par leurs bonnes (*sic*) exemples fondoient diverses religions. Iceulx estoient garnis et remplys de bonnes vertus et se combatoient puissamment contre les pechiez..."

[138] (Fol. 34) "Nous veons a l'oeil que aucuns enfans et plusieurs bien aagiez et profés en saincte religion sont venus a celle ha[r]diesse et aussi acoustumee rebellion que ilz se lievent contre leurs souverains en tant que correction n'y a point d'audience ne de lieu. On les voit maintenant que ilz se arment de malices, de cautelles et de nouvelles subtillitez, et admainent a la fois allegacions de drois pour eulx defendre, pour couvrir leur mauvaistié et pour eulx mettre hors de correction. Et se il advient que le prelat en punisse aucuns, ilz en ont telle desplaisance que ilz en font pis aprés que devant. La ne vault la haye de bonne ordonnance ne les fossez de correction ne la fermeté de constitucion ne la regle de religion, car ilz passent et tressaillent tout; par leur presumption et inobedience rompent et defont les bons edifices de paix, de concorde, de purté, d'oroison, d'umilité et d'obedience. Et souvent advient que par leurs debas et courroux ilz se mettent en perilz d'ames et de corpz. De ceste ruyne et destruction on a veu et voit on encores certains signes et raisons apparans. Combien cuides tu que cest edifice doye durer puis que tu voys que le fondement, la pierre quarree et les pillers en sont et seront tost ostez; c'est assavoir quant humble obedience en sera hors gettee et que chascun fera sa voulenté et ne sera homme, prelat ne autre qui ose dire: Pour quoy fais tu cecy? Las! las! ou est obedience de saint Pol le Simple, de saint Mor, de saint Jacques, de saint Benoist et des autres bons, parfaiz et approuvez religieux et disciples de Dieu, qui ou tempz passé estoient si vrays obeissans que ilz ardoient tous de faire et d'accomplir tout ce que on leur commandoit; mesmes ilz ne povoient desobeir a ceulx qui leur commandoient les choses qu'ilz ne povoient faire." One observes that whereas the French translation supplies the names of four saints, Suso only mentions St. Paul the Simple, cf. Künzle, *Horologium*, p. 406: *...ubi est oboedientia Pauli simplicis et quorundam discipulorum probatissimorum.*

[139] (Fol. 34v.) "De chasteté qui est l'onneur des vertuz et la beaulté ne fault il mie faire grant parlement... (Fol. 35) De charité qui est le mortier et le ciment de tout l'edifice espirituel, qu'en diray je? Je diray que la fin en est venue; elle est effacee, aneantie jusques au fondement."

Beneath the painting of the *City of Religion in Ruins* is yet another tableau which is not strictly related to the archaic pattern of Virtues-Vices, though one could imagine it standing alone in any collection of illustrations as an exemplum in its own right. Furthermore, it could be said to "restate" iconographically the pith of Sapientia's remarks: "There are those alms-gatherers who go through the world, wearing the habit and in the guise of poverty, seeking alms. More often than not, it's not out of necessity, but in order to maintain their already considerable estate and to amass treasure and to transgress intemperately the vow of poverty. But indeed, in order to provide for themselves covertly for times to come, they convert their money into precious objects, namely, books, plate, church ornaments, reliquaries and other movables. In this way, when money and provision for the common good are lacking, they can take counsel of the god of Ekron, as if there were no God in Israel; in other words, when spending is over, they can have recourse to the movables and to the jewellery, and turn all into cash. So what is given piously and dedicated to God and to the Church is converted into food and drink and common living."[140]

As we read the *Déclaration*'s statement and view this illustration we are reminded of an earlier point made in chapter IV, § 3, that the employment of 'right' and 'left' is not always in line with our own placement of the subjects. More importantly, it is necessary to acknowledge that, with seven columns of textual explanation devoted to the pictorial statement, the expositor wishes to place emphasis on the visual and spiritual importance of the polyptych. One may also wish to note that he perceives the ruinous state of the Church and Religion as a contemporaneous situation, and not as a vision beheld by the Disciple and explained by Sapientia. The groups of holy men and women in the left vertical panel are called "apostles, martyrs, confessors, virgins and women saints". Perhaps the traditional attributes for each person, supplied by the miniaturist, were thought by the expositor to be adequate for identification purposes.

The physical ruin of the City is graphically presented by the artist, it is true, but it is the moral decadence that the commentator dilates on, in imitation of Suso. Christ as the pilgrim at the gate is identified in conformity with the text, as are the humanoid monsters on the walls. Yet, Sapientia's interpretation of the pilgrim as herself, "I was formerly Prince and Lord of the ruined city..." is ignored by the expositor. Instead, he launches into a long account of the Vices and Virtues.

The figure of the pilgrim wishing to enter the City of Religion seems to have aroused in the commentator's mind a parallel he knew to be contained in a widely read dream text of the previous century, the *Pèlerinage de la vie humaine* which the Cistercian monk, Guillaume de Digulleville, first composed *c.* 1331 while at the Abbey of Chaalis. Indeed, the writer of the *Déclaration* can be shown to derive the garb, accoutrements and gestures of the Vices and Virtues from those of the same allegorical characters encountered by the pilgrim in the *Pèlerinage.* Of the two recensions of the poem, it is the first that has been employed.[141] The modern reader, or even a fifteenth-century one for that matter, is not informed about the textual sources by the expositor.

[140] (Fol. 34v.) "Ainsi font telz questeurs, car ilz vont par le monde en habit et en semblance de povreté pour aumosnes demander, et ce n'est mie par necessité souventes fois, mais est pour leur grant estat tenir et pour tresor amasser et pour le veu de povreté excessivement trespasser. Mais quoy? pour eulx pourveoir plus couvertement pour le tempz advenir ilz convertissent leurs pecunes en joyaulx, c'est assavoir, en livres, en vaiselle, en paremens d'eglise, en reliquiaires et en autres meubles, afin que, quant la monnoye et la provision du commun fauldra, ilz puissent prendre conseil a Dieu a Charon, si comme s'il ne fust point de Dieu en Israel, c'est a dire que, quant la despense fauldra, ilz puissent recourir aux meubles et aux joyaulx et faire du tout argent. Et ainsi, ce qui estoit saintement donné et dedié a Dieu et a l'Eglise est converti en boire et en mengier et en vivre commun." Suso's words about the god at Ekron: *quasi Deum non sit in Israel, deum Accaron... consulere possint* (cf. Künzle, *Horologium*, p. 407) allude to the passage in IV Kings, I, 3: *Numquid non est Deus in Israel, ut eatis ad consulendum Beelzebub deum Accaron*?

[141] I should like to thank Professor K.V. Sinclair for suggesting there may be parallels with Digulleville's *Pèlerinage de la vie humaine.* The edition I consulted was J. J. Stürzinger, *Le Pèlerinage de vie* (sic) *humaine de Guillaume de Deguileville* (*sic*), London, 1893, and the line numbers I mention refer to that edition. For an explanation of the differences between the two versions, see E. Faral, "Guillaume de Digulleville, moine de Chaalis," *Histoire littéraire de la*

Digulleville's principal character sees the Heavenly Jerusalem in a dream, and, clothed as a pilgrim, he sets out to reach it (*Pèlerinage*, 201 ff.). After receiving religious instruction and being provided with armour which he carries rather than wears, he arrives at a fork in a road. The right path is guarded by Labeur and the left one by Oisiveté or Idleness. The hedge of Penance runs between the two paths. Labeur toils perpetually, spinning and unravelling a napkin, while Idleness attends to the whims of those who follow her path. In the style of her namesake in the *Déclaration*, she loves dalliance (*Pèlerinage*, 6745-64), and to symbolise her activities, one hand lies across her breast, while the other plays with a glove (cf. 6702, 6847).

The author of the *Déclaration* names the second person encountered by Christ the Pilgrim on His way to the City of Religion in Ruin; she is Wrath, described merely as an old grimacing hag, who holds two pebbles with which she sparks off disputes. This simple profile echoes details of the complete portrait of Wrath who assails the pilgrim in the *Pèlerinage*, 8809-10 and 8883-8. No attention is paid by the expositor to the principal traits that make her such a formidable adversary of Digulleville's pilgrim.

Having entered the path of dalliance, the pilgrim is soon assailed physically and morally by each of the seven Vices in turn (*Pèlerinage*, 6483-7032). Their weapons and instruments of character assassination are at times precisely recalled by the expositor in the *Déclaration*, but in his eyes, he needs to speak only of five Vices, since that is all the artist depicted in the right panel of the illustration. The textual descriptions that are sources for the *Déclaration*'s remarks about Envy, Lechery, Sloth, Pride, and Avarice may be summarised conveniently as *Pèlerinage* 8195-796, 10529-684, 7087-292, 7345-8190 and 9059-10230 respectively.

A close study of the *Pèlerinage* and *Déclaration* statements about Envy, for example, allows us to comprehend the expositor's method of handling his source material. We must also bear in mind that his range of comment is determined to a certain extent by the icon in front of him, it can be appreciated that it behoves him to confine himself to these details alone.

Envy is viewed in the portraiture of the Brussels *Horloge* as proceeding on all fours (cf. *Pèlerinage*, 8199). From her eyes dart two arrows (cf. 8197-8 where the Cistercian poet calls them lances and glaives), which represent False Glance and Contempt for Honour. The explanation of the two shafts proffered by Digulleville is quite different: one is named Anger at Others' Joy and the second is Joy at Another's Adversity. Envy carries two daughters on her back (cf. 8329-30). The first has a false face that seems to bear a smile and she holds a box of pomade (cf. 8340-1). The daughters are Flattery and Fair Countenance which mask her true mien of ugliness and spite (cf. 8386-97). The pointed dagger held behind the back in readiness to strike recalls the *Pèlerinage*, 8398 and 8416. Denigration, called Detraction by Digulleville and the expositor, gnaws at a bone (cf. 8510, 8532) and grasps a brooch encrusted with ears (cf. 8611) for persons who may be ready or disposed to hear anything or everything (cf. 8613-18).

There is no Battle of the Vices and Virtues in the *Pèlerinage*, each trait is separated so that the Vices attack the pilgrim on his journey to the Heavenly Jerusalem, while the Virtues are introduced at different points during the pilgrimage.

When recalling the Virtues of the *Pèlerinage*, the author of the *Déclaration* follows the same procedure as he did for the Vices. Charity in the poem explains that she caused Christ to become incarnate, to suffer the Passion for the sake of mankind and to bequeath to mortals the gift of Peace

France, XXXIX (1962), pp. 1-47. Iconographic programmes for the text were studied by Marion Lofthouse, "Le Pèlerinage de vie humaine by Guillaume de Deguileville with special Reference to the French MS. 2 of the John Rylands Library," *Bulletin of the John Rylands Library*, XIX (1935), pp. 170-8, and by L. M. J. Delaissé, "Les Miniatures du Pèlerinage de la vie humaine de Bruxelles et l'archéologie du livre," *Scriptorium*, X (1956), pp. 233-50, pls. 17-26. The illustration subjects of the Vices and Virtues in the Brussels *Horloge de Sapience* show no affinity with pictorial models in the Manchester or Brussels copies of Digulleville's text.

(cf. 2427-52). This is explained in Christ's Last Testament which Charity reads out to the pilgrim (cf. 2453-598). The wording explicitly mentions the letter P.A.X., meaning Peace, arranged on the L-shape of the carpenter's square (cf. 2513-76). Charity carries the Testament on her person (cf. 2616). However, there is no suggestion in Digulleville that a corner-stone keeps the walls of the Heavenly Jerusalem from collapsing nor that Charity carries a hod for cement on her shoulders.

Towards the conclusion of the dream Digulleville's pilgrim sojourns in one of the castles aboard the Ship of Religion. There he meets several handmaids employed in all manner of tasks. They bear the names of Virtues in some instances. In the *Déclaration* we learn of Chastity, a dormitory maid (cf. *Pèlerinage*, 10579-80) with a rod, symbol of bed-makers (cf. 12662, 12715-16). The sketch of Obedience in the *Déclaration* is inspired by the portrait of a similar figure in the *Pèlerinage* (cf. 12736-8). Each holds fine thongs and threads, but the expositor's symbolic interpretation of their use is a mere recollection of the statement in the poem: Obedience the prioress leads her cloistered flock by these threads and bonds (cf. 12739-44).

The attribution of the rank of prioress to Correction in the Brussels *Horloge* illustration is not substantiated by the *Pélerinage*. Moreover, the attributes of Correction can be aligned with those of Discipline in the poem (cf. 12751-4). The latter uses a file to scourge and purify (cf. 12755-8), whereas Correction wields rods.

The expositor's sketch of Voluntary Poverty as a crowned lady, empty-handed, an anvil on her back, is certainly not based on Digulleville's vision of Voluntary Poverty. To be sure, she had nothing in her hands in the dormitory where she attended and sang happily about her poverty (cf. 12717-22). The reference to the anvil is possibly a reminiscence of its depiction on the back of the allegorical figure of Patience whom the pilgrim met with Grace Dieu long before he had encountered any of the Vices (cf. 3898-952).

The author of the *Déclaration*, clearly, was well acquainted with the Cistercian monk's work. At times the detail he assigns to the Vices and Virtues of the illustration are identical to those of the poem. At other moments, he is less attentive and is confused, especially in his portraiture of the Virtues. Some of his interpretative statements may be construed as personal to him, as he endeavours to adjust the *données* of the painting to Digulleville's textual commentaries on the Vices and Virtues. According to the expositor, Voluntary Poverty is provided with a crown in order to symbolise her inheritance of the Kingdom of Paradise; Charity for her part carries a mortar tray on which lies the cement that binds all Virtues.

The tableau of the money-changer in the lower centre panel of the illustration relating to the City of Religion in Ruins is viewed last of all by the writer of the *Déclaration*. His comment is a short paraphrase of Suso's views about the corrupt alms-gatherers, but modified so as to inculpate prelates and church-rulers specifically.

Let us now turn to the pictorial statement by the artist. In places it is a literal reading of Suso's narrative. The walled city commanding the observer's attention is dominated by a church whose fabric displays neglect; the roof above the nave is open to the skies; large cracks run up the walls. On the periphery no section of the circumvallation is shown in a cast-down state, or leaning, or rent asunder. In the right foreground two malefactors hold a corner-stone labelled PAX which they have removed from the city wall. Humanoid beasts watch from the battlements as a pilgrim approaches the main gate. He is taunted by two women named Ire (Wrath) and Oiseuze (Sloth), the first holds two stones, the second parts her gown at the front while her other hand toys with a glove.

The holy men and women mentioned collectively by Sapientia are presented in four segments of a left vertical panel in the order of apostles, martyrs, confessors and female saints; each is given the traditional attribute. One can distinguish clearly St. Peter holding a key and book, St. Andrew a saltire cross, and St. John the Divine a poison cup with snake; St. Leonard carries a book and fetters; St. Lawrence a grid-iron; among the confessors are St. Gregory with mitre and pastoral staff, and St. Benedict with book and abbot's crozier; the holy women include St. Catherine of Alexandria with her wheel, St. Mary Magdalene with an ointment jar and St. Barbara near a tower.

The allegorical statements about five Virtues are pictorially phrased in the right vertical panel. They appear in the order Charity, Chastity, Obedience, Chastisement and Poverty, and as women dressed in noble attire, except for one person. They are in the company of five Vices, respectively, Envy, Lechery, Sloth, Pride and Avarice. Confrontation between the Vices and Virtues was already by 1450 a well known iconographic theme.[142] Among the range of didactic texts circulating in the Rolin Master's day was a richly illuminated one that will assume great significance in the present discussion. When first composed, it gave rise to an iconographic vogue involving not merely pictorial formats but also instructions for executing them. Its name was the *Somme le Roi*, an epitome of knowledge for the monarch. It was in fact completed in 1279 by Friar Laurent du Bois, the Dominican confessor to Philippe III, King of France (1270-1285). Helped no doubt by royal patronage, the work enjoyed wide circulation and considerable popularity right to the end of the Middle Ages. Countless manuscripts survive containing copies that show contamination and reworkings and confusion of title with another equally popular compendium, the *Miroir du Monde*.[143]

Broadly speaking, the *Somme* is a treatise on the Christian Faith, divided into sections about the Ten Commandments, the Creed, the Seven Sins, the Art of Dying Well, the Garden of Virtues, the Pater Noster and the Seven Gifts of the Holy Spirit.[144] In the last part Friar Laurent selects the Virtue each Gift nourishes, he sets a Vice over against each Virtue, and then suggests a Beatitude he feels is appropriate. The seven Virtues singled out here by Friar Laurent are not the series of three Theological and four Cardinal ones so frequently encountered, but a series compiled from the extensive range of moral attitudes and principles that the Middle Ages inherited from Antiquity. Thus, the Gift is Scientia or Knowledge, the Virtue is Equity, the Vice is Ire and the Beatitude is the Weepers.

Although early copies of the work are illustrated, and a regular series of portraits and vignettes becomes established for decorating the manuscripts,[145] a few decades pass before sets of instructions about them appear in ateliers. Art historians owe a debt of gratitude to the renowned paleographer and historian, Léopold Delisle, for editing an almost complete group of these directions from a manuscript dated 1373.[146] In 1953 E. G. Millar added to the conspectus of the codification of descriptions when he published a detailed account of a *Somme le Roi* in his possession.[147] In a more recent paper Rosemond Tuve unveils the panoramic sweep of the typology and symbolism attached to this contrasting Virtues-Vices programme in the Middle Ages, particularly in manuscripts produced in English and French ateliers.[148] Moreover, she prints the 'lost' directions

[142] See Male, *Art religieux*, pp. 309-28; Tuve, *Virtues and Vices*, two parts (details in my Bibliography) and her study *Allegorical Imagery*, pp. 57-79.

[143] See Edith Brayer, "Contenu, structure et combinaisons du *Miroir du Monde* et de la *Somme le Roi*," *Romania*, LXXIX (1958), pp. 1-38, 452-70. There is no critical edition of the French *Somme*.

[144] The best overview of the work that is readily accessible is by C. V. Langlois, *La Vie en France au moyen âge*, Paris, 1928, IV, pp. 123-92.

[145] Cf. Delisle, *Librairie de Charles V*, I, pp. 238-42 publishes details of the series of fifteen tableaux that occur in Paris, Bibliothèque de l'Arsenal, MS. 6329, dated 1311; Paris, Bibliothèque Mazarine, MS. 870, *c.* 1295 and Paris, BIbliothèque Nationale, MS. fr. 938, *c.* 1294. The list can also be read in Tuve, *Virtues and Vices*, p. 53.

[146] Delisle, *Librairie de Charles V*, I, pp. 244-6, extracted from Paris, Bibliothèque Nationale, MS. fr. 14939.

[147] See E. G. Millar, *An Illuminated Manuscript of La Somme le Roy, attributed to the Parisian Miniaturist Honoré*, London (Roxburghe Club), 1953, pp. 2-6. Colour reproductions of several illustrations enrich his later study for the Faber Library of Illuminated Manuscripts and entitled *The Parisian Miniaturist Honoré*, London, 1959, pls. 3-8. The manuscript owned by Millar was eventually acquired by the British Library, London and given the shelfmark Additional 54180.

[148] Rosemond Tuve, *Virtues and Vices*, in two parts; I give details in the bibliography. Some of the information about the *Somme le Roi* formats was repeated in her monograph, *Allegorical Imagery*, pp. 79-102.

for two of the paired Virtues and Vices. If the Brussels *Horloge* series had been more widely known at the time, she would have most certainly discussed them.

Friar Laurent linked the Virtues and Vices as follows: Humility-Pride, Friendship (Amitié)—Envy, Equity—Ire, Prowess—Sloth, Misericordia—Avarice, Chastity—Lechery, Sobriety—Gluttony. Artists decorating the early copies of the *Somme* aligned these confrontations in pairs or groups of four, but some went a step further to reinforce the morality of the Virtue and the Vice by painting an exemplification of the abstraction. Other miniaturists presented the two systems at one and the same time. Thus, in a 1295 copy of the *Somme*, the celebrated Parisian illuminator Honoré presents Humility as a woman in one frame and underneath it an exemplum of a sinner praying before an altar. On the same level as the Humility portrait is an exemplum of Pride, namely, Ahaziah falling headlong from battlements, and below his frame is yet another exemplum of a hypocrite.[149] By contrast, the iconographic programme of another manuscript from the closing years of the thirteenth century presents the female Virtue Friendship above a tableau of David and Jonathan, confronting the woman Hatred above a portrait of Saul menacing David.[150]

During the fourteenth century the two major programmes continued to be executed in *Somme* manuscripts. The instructions to artists contained in the 1373 volume already mentioned make this point.[151] A tangential phase was reached when the series was occasionally adapted in whole or in part by illustrators of liturgical books, such as Psalters, Breviaries and Books of Hours. Take the case of the well known Belleville Breviary, whose Use or Rite is Dominican.[152] In the seven days' series of psalms for Matins, one meets a David-picture or equivalent at the beginning of each psalm, an exemplum illustrating a sin, a Sacrament,[153] a Gift of the Holy Spirit and a Virtue. Interestingly enough, the programme of opposing Virtues and Vices is absent.

What surely must be a remnant of an earlier design for the moral parallels is preserved in the Bedford Hours. The Bedford Master, perhaps because of space limitations, presents the hybrid series in marginal medallions for which he is famous. They occur on the first folio of the Seven Penitential Psalms.[154] In the top left corner is an exemplum of Humility and in the lower left another, of Pride. Down the right-hand side, however, one sees paired female figures without exempla: Patience—Ire, Charity—Envy, Chastity—Lechery, Sobriety—Gluttony, Soufisance—Avarice, Diligence—Sloth. It is quite likely that the arrangment of the series of figures in a column could have suggested a similar lay-out for the Rolin Master. Another possibility of course is that both artists derived their lay-outs independently from an unidentified fourteenth-century model.

The inconsistency of the nomenclature employed by painters deserves a comment in its own right. Friar Laurent may have been satisfied in 1279 with his pairing of Misericordia with Avarice, but, clearly, later artists or their clients preferred a different shade of meaning to be expressed by a female figure portraying a Virtue or a Vice. Names were altered even if the portaiture was not. In the Bedord Hours, Soufisance, an obscure face of Misericordia, is opposed by Avarice, and Prowess takes on the new name of Diligence when confronting Sloth.

[149] London, British Library, MS. Additional 54180, fol. 97v.; reproduced in Millar, *Parisian Miniaturist Honoré*, pl. 6.

[150] London, British Library, MS. Additional 28162, fig. 6v.; reproduced in Tuve, *Virtues and Vices*, pl. 7 (d).

[151] See Delisle, *Librairie de Charles V*, I, pp. 244-6, and Tuve, *Virtues and Vices*, pp. 52-5.

[152] In two volumes as Paris, Bibliothèque Nationale, MSS. lat. 10483-10484; see Leroquais, *Bréviaires*, III, pp. 198-210; Tuve, *Virtues and Vices*, pp. 65-72, in particular p. 69, n. 108; and Lucy F. Sandler, "Jean Pucelle and the Lost Miniatures of the Belleville Breviary," *Art Bulletin*, LXVI (1984), pp. 73-96.

[153] On the pairing by St. Thomas Aquinas of the Seven Sacraments and the Seven Virtues (three Theological and four Cardinal) and the way these find iconographical expression in the Belleville Breviary, see Frances G. Godwin, "An Illustration to the *De Sacramentis* of St. Thomas Aquinas," *Speculum*, XXVI (1951), pp. 609-14.

[154] London, British Library, MS. Additional 18850, fol. 96; reproduced in Tuve, *Virtues and Vices*, pl. 12.

If one examines more closely at this juncture the Rolin Master's gallery of portraits, he is seen to follow Suso's nomenclature for the Virtues, placing the labels Charity, Chastity, Obedience, Chastisement and Poverty near the appropriate figures. The case of the Vices is different. This particular Dominican's work could offer little guidance, so the artist turns to a traditional pairing of names, one that derived from the late thirteenth-century copies of Friar Laurent's *Somme*. The following table may help the reader to appreciate the point more readily. The Bedford Master's terms are included for comparison.

Somme[155]	Bedford Hours	Brussels *Horloge*
Amitié — Envy	Charity — Envy	Charity — Envy
Chastity — Lechery	Chastity — Lechery	Chastity — Lechery
Prowess — Sloth	Diligence — Sloth	Obedience — Sloth
Humility — Pride	Humility — Pride	Chastisement — Pride
Misericordia — Avarice	Soufisance — Avarice	Poverty — Avarice

Can the same degree of identity be found among the raiments or physiognomies of the female figures themselves and the attributes each displays? The evidence suggest a negative response, because the Rolin Master is not portraying the actual allegories of the *Somme*, he is interpreting those of the Brussels *Horloge*, especially for the Virtues. The attire of the women is modified in places to reflect more pointedly Suso's condemnation of certain religious. Chastisement is dressed as a meek prioress who is harrassed by Pride and her hand-maid Flattery. The Virtues and Vices are seen in both outdoor and indoor venues, in general accord with Suso's contention that vice and corruption are both within and without the City of Religion. A frame traditionally separated Virtue from Vice, but it has disappeared in this miniature. Each woman now interacts more sharply with the other against a common backdrop. A further visual modification is the disappearance of the exemplum that once accompanied the portraits of each Virtue and Vice.

In depicting the Money-Changer theme, the artist is very direct, one may even suggest he shows a touch of realism. Seven Dominican prelates and friars file into a money-changer's shop and hold out to his waiting scales a chalice, a processional cross and church plate.[156]

XIII *The Ram with the Iron Crown* — Fol. 37v.

Suso relates that in the City of the Church and Religion was a man of God who prayed continuously to Our Lord to hold fast to the City and save it. "And he saw one night while asleep a horned ram appearing in the West and which had left the flock. It had two horns on its head and an iron crown, and it had the strength to rule over the City. Behind it came foxes all belonging to its pack and each had a crown. With them came also a vast multitude of all manner of male and female animals. These kept with the throng, some out of fear, some out of covetousness, others out of ignorance, and others out of naiveté. The ram, which was bent on usurpation and rule of the City for itself and its followers, roamed the region like a tyrannical leader."[157]

[155] From Paris, Bibliothèque Nationale, MS. fr. 938; see above, n. 145.

[156] Spencer, *Horloge*, p. 287 speaks here of the painter choosing to "make simony a Dominican sin," but the comment is inappropriate for the theme depicted.

[157] (Fol. 37v.) Et vit une nuyt en dormant un mouton cornu qui s'apparut par devers occident et departit du tropel. Il avoit deux cornes en son chief et une couronne de fer sur sa teste, et avoit puissance de regner sur ycelle cité. Aprés lui aloient regnars et gouppilz qui tous estoient de sa route et avoit chascun une couronne. Avec eulx avoit grant multitude de toutes manieres de bestes, masles et femelles, les autres par paour, les autres par couvoitise, les autres par ygnorance et par simplesse se tenoient en celle multitude. Le mouton qui vouloit regner et usurper pour lui et pour les siens la dominacion de la cité s'en aloit cerchant toute la region comme ung tirant a grant force.

After describing devastations, Suso announces growing opposition to the creature. "Although some of God's people did not always tread the path of truth and plead their cause as they should have, nevertheless, when they saw the adversity and sorrow that the City suffered, they did all they could to bring it comfort and to resist its adversaries. Then the "duke" or leader of the Children of God who was "duke" and governor of the City came to their aid; he was a courageous and praiseworthy man and a resolute arbiter... When the ram saw it was wasting its time, it and its supporters turned to the Prince of the Multitude, begging him to assist them against the duke and his party. The ram supplied so much false information and advice to the innocent and guileless ears of the Prince that it obtained the domination and lordship of the people."[158] The ultimate fate of the ram was that it fell down and broke a horn, for which the Children of God praised the Lord.

Suso's narrative of events in the vision stands on its own as a masterpiece of veiled allegory about ecclesiastical affairs of the diocese of Constance in his life-time. Much ink has flowed from the pens of modern critics in an attempt to solve the mystery and to identify the leader of the "Children of God" and the "Prince of the Multitude". In fairness to many whose views are now being discredited, it has to be pointed out that they were working with a faulty Latin text of the *Horologium*. Among Suso scholars, the view that has until recently enjoyed credence has centred on Louis of Bavaria's struggle with the Papacy.[159] Spencer followed the discussion of the influential Suso critic, J. A. Bizet,[160] and summarised, conveniently for our purpose, his views on the allegory in these terms: "Louis, excommunicated by the Pope but generally supported by the German princes, had retaliated by accusing the Pope of heresy and of usurping the secular powers of the lesser princes. Therefore the Ram with the Iron Crown of the text would be a symbol for Louis wearing the crown of Lombardy (1327) and the seventy crowned foxes would be the German princes and the cities who supported him. John XXII was finally deposed in favor of Louis's candidate, Nicholas V (1328). Suso refers to the party of the opposition as the "enfants de Dieu". The Ram's broken horn of the text is the sign of defeat, either the abdication of Nicholas V (1329) or Louis's own downfall (1334)."[161]

Künzle, the editor of the first complete critical edition of Suso's *opusculum*, believes that the allegorical veil may be drawn back to reveal the following local dramatic events at Constance in the year 1334.[162] Suso is the man of God who prays that his morally bankrupt City, undermined by vice, should be spared further humiliation and oppression, especially of the kind about to be visited upon them by the arrival of the Ram with the Iron Crown. The beast is from the flock, that is, someone from the religious community at large, and it has the authority to rule in the City. The person symbolised in this manner is Nicholas von Kenzingen whose benefices lay to the West of Constance. He had been a sub-deacon only before being elected Bishop of Augsburg, and then was elected by the chapter of Constance to their See. Nicholas bore ram's horns on his escutcheon. The Constance patricians, about seventy in number, supported him also, but not the populace.

[158] (Fol. 38) "Et combien que les aucuns du peuple de Dieu ne alassent pas tousjours par le chemin de verité et qu'ilz ne feissent pas tousjours raison comme ilz deussent, toutesfois pourtant qu'ilz veoient lé griefz et les adversitez que la cité souffroit, ils se penoient et traveilloient de tout leur povoir de la conforter et resister aux adversaires. Lors leur vint en aide le duc des Enfans de Dieu qui estoit duc et gouverneur de la cité. C'estoit ung homme digne de louenge, courageux et grant justicier... Quant le mouton vit qu'il perdoit sa peine, lui et plusieurs de ses gens se tournerent par devers le Prince de toute la compaignie et lui prierent qu'il voulsist estre en leur aide contre le duc et sa route. Et lui mist au devant toutes fausses informacions et suggestions, et tant lui bouta en ses oreilles qui estoient innocentes et sans barat que il empetra la principalité et la seigneurie du peuple."

[159] G. Barraclough, *The Medieval Papacy*, London, 1968, pp. 146-51.

[160] Cf. J. A. Bizet, *Suso et le déclin scolastique*, Paris, 1946, pp. 132-9. Other critics with similar views about the central theme of the Vision are mentioned by Künzle, *Horologium*, p. 23, n. 1.

[161] Spencer, *Horloge*, p. 287.

[162] I here follow the discussion as set out by Künzle, *Horologium*, pp. 23-6. It is a pleasure to acknowledge the assistance I have received from Madame Lisl Mathew in clarifying my understanding of Künzle's argument.

Nicholas paid a visit to the Pope in Avignon (the text's *princeps totius multitudinis*) in April 1334 to ensure that his nomination was confirmed. So tyrannical was Nicholas's rule in the early days that opposition grew from a rival contender for the episcopal throne, Albrecht von Hohenburg. The honest, God-fearing people of Constance and many members of the clergy sought him out and he soon became the leader of the opponents of Nicholas. He is in Suso's eyes the *dux filiorum Dei*, in the French the words are "le duc des enfans de Dieu" (fol. 38). Suso's vision ends with the ram's broken horn, that is, with Nicholas's definitive loss of one of his two bishoprics.

The author of the *Déclaration* seems just as lost in the true meaning of the allegory as were critics generations after him. He follows his own line of explanation, commencing it once he has related the *Ram with the Iron Crown* tableau with the previous illustration. As already noted, Suso established the relationship of the two themes by telling the story of a man of God of the ruined City who saw a ram in a vision. Thus, the vision was not claimed to occur because of the moral decay of the City's inhabitants. Our expositor, however, declares that an outcome of the decay is a series of dissensions and schisms. He proceeds to situate the religious decline in the period before the election of Martin V, who reigned as Supreme Pontiff from 1417 to 1431. Seemingly committed to this viewpoint, the expositor feels it is necessary to assign the characters and the drama of the painting to the Schism. The ram is said to represent the Pope but the *Déclaration* fails to explain why an anthropomorphic image of a pontiff follows in the ram's train. The content is further clouded by the commentator's description of the ducal figure as the "...King of France...", and the attendants as "...clerks and doctors from all university faculties..." The tribute paid to the King of France, Charles VI, in being a decisive leader in the power play to terminate the Schism is historically correct. the explicitness that the expositor proclaims as a virtue of the illustration must surely have been in the eye of the beholder.

Eleanor Spencer is the first scholar to link the criticism of the *Déclaration* and the elucidation by Suso critics to the subject matter of the painting. It seems that she was swayed by the expositor's statement, and that she did not realise that the miniature is closer to Suso's narrative, however interpreted, than it is to the explanation in the *Déclaration*.

Without actually quoting Suso's text or its French translation, Spencer makes the statement, "Undoubtedly the illuminator's object was to illustrate Suso's words, but other evidence of his independence requires us to explore the possibility of a new fifteenth-century meaning, hopefully more explicit than the *declaration's* vague reference to the time of the Schism and Charles VI. Assuming then the palace crowned with fleurdelis is French and that the king therein is Charles VI "simple innocent",[163] his scholarly advisers would be the committee of the University of Paris (1394)... The Ram would be Sigismund, crowned at Aachen in 1415 but still ambitious for the formal confirmation of his power in Rome... The Pope would presumably be Martin V, supported by the cardinals and by the secular forces in the form of an army. The fine, high-stepping mule suggests a symbol of proud obstinacy rather than humility... The king's party, unarmed, is mounted as a sign of readiness to act."[164]

Spencer's suggestions are plausible to a certain extent, but they contain a disquieting weakness. Perhaps the figure in the loggia is a French king, perhaps the attendant clergy are scholarly advisers, but who then are the kneeling plebeian figures at the monarch's feet? Why should the king's party "be mounted ready to act" when the horsemen are manifestly an embassy on the move? Why is Martin V blessing the ram Sigismund, and why is Sigismund headed for a ruined city.

It is fair to state Spencer's line of reasoning was clearly divorced from a close reading of the French text of the *Horloge*. She relied on the perspective of the Bizet elucidation of Suso's veiled

[163] A characterisation obtained by Spencer from the wording of the *Déclaration*.

[164] Spencer, *Horloge*, p. 288.

thoughts. Bizet's arguments received a setback in 1977 when Künzle published the first scholarly edition of the *Horologium* based on numerous manuscripts and with an impressive *apparatus criticus*. With a reliable critical edition in front of him and access to fourteenth-century chronicles about Constance,[165] he was able to demonstrate the plausibility of a totally different approach to the allegory. It shows that Suso was alluding to regional affairs and concerns.

How faithful is the iconography of the fifteenth-century miniaturist to the vision of the fourteenth-century Dominican, Suso? The four principal elements of the dream, namely, the City, the ram, the Pope and the "duke" are visually depicted in the miniature. The ruined city has the characteristics of the one in the previous illumination, with similar perimeter wall, entry gate, Gothic church with soaring lantern tower, dwellings of all shapes. As the ram advances to the main gateway, with foxes close on its heels, a mounted Pope gives his blessing to the beast's progress. In a loggia sits a duke at whose side are clergy directing his attention to a petition from humble citizens kneeling in front of him. The mounted party, who ride out in the direction of the Pope, are members of a ducal embassy.

It will be appreciated that the vision has been rendered literally with one or two minor liberties. The French text does not specify the number of attendant foxes, although Suso wrote *Quem sequebantur quasi septuaginta vulpeculae*. The French text declares that the foxes are crowned. In the painting we see five animals and they are without crowns. The horde of male and female animals of the pack which follow the ram have been suppressed altogether. Focus on the king-like duke and governor of the City is conveyed by the very popular motif known as the audience. Attention is drawn to the Pope by a totally different pictorial theme, the mounted procession. The Pope's mount, an ass or mule, catches the eye, because its presence is unexpected. As already mentioned, Spencer saw the creature as a "symbol of proud obstinacy rather than humility," and perhaps there is a grain of truth in her idea.

XIV (1) *The Prior's Investigation* — Fol. 39
(2) *Susanna*
(3) *Ahasuerus, Esther and Haman*
(4) *Haman*
(5) *Mordecai Receiving the Signet-Ring*
(6) *Saul Threatening David*
(7) *Saul's Death*

Sapientia gives the Disciple general advice about coping with adversity in whatever station he occupies in life: "Above all, what you have to do, if you are a prelate or a judge, is not to hold shallow opinions nor readily believe those who are accustomed to take sides. If anyone speaks or holds forth against anyone, do not believe him immediately and especially when persons speak with bias by supporting one and condemning the other against all reason. Likewise, do not receive information against a person who is acknowledged to lead a good life, for if you do the opposite, you will destroy discipline and give the wicked boldness to do evil and to resist their sovereign lords. And in this way the strength of religion will be weakened and suffer adversity. And it behoves you to investigate any cause you know nothing about. When you learn of it, examine it promptly and then give to the guilty party what he was going to give to the innocent, and make him turn his sin on himself, so that his sword pierces his own heart and his own bow is shattered. In this way the evil-doers will be caught in their own wickedness. In this way were stoned for their iniquities the

[165] Cf. Künzle, *Horologium*, p. 24.

two elders who condemned Susanna to be lapidated; she was innocent but her trust was in God who saved her. Thus perished Haman who was an adversary of the Jews and who through falsehoods and misinformation had obtained from Ahasuerus that God's people should be destroyed and put to death. Thus acted Saul, the proud king put to death by his own sword; he persecuted David against reason, David the good and noble knight."[166]

It is important from the outset to alert the reader to the translator's inaccurate rendering of Suso's phrase *praelatus enim vel rector fratrum tu cum sis, licet in minimo gradu, discas...*, where it is a question of a prior or leader of Dominican brethren. By rendering the Latin statement as "Et sur toutes choses que tu auras a faire, se tu es prelat ou juge, ne soies pas de legiere creance," the translator has obscured the specific reference to a Dominican Prior.[167]

Carefully reminding the reader that the illustration is still in the same chapter as the two preceding prospects over the City of Religion in Ruins and the Ram with the Iron Crown, the author of the *Déclaration* suggests that honest citizens cannot strive for perfection without undergoing trials and tribulations. These include unjust accusation, which Suso talked about and which the artist illustrates firstly by a representation of a 'Chapter of Faults'. Although he does not use this term, the expositor recognises the setting and the actors in the drama, and employs the term "prior". The *praelatus* of Suso and the prelate or judge of the French translator are ignored.

The commentator states that the textual narrative contains "... several *exempla* in which the just are condemned..." It would have been more correct to report that Suso mentions cryptically heroes or heroines of Antiquity who were wrongfully accused. In the first instance, the artist's visual statement conflates Suso's wording, as it were, while the expositor expands the pictorial subject matter by recourse to Biblical narrative. In the case of Susanna, the details he supplies are sparse. If one reads his statement, one understands why the heroine moves towards a stake and why old men are being stoned.

The circumstances of Esther's Banquet and the outcome of Haman's treachery are told with much detail, based less on the painting than on the Old Testament narrative. One notes that the two eunuchs, the agents of death in Esther, VI, 2, are referred to as *deux portiers* (*ms. portieres*) by the commentator. The same reliance on the Old Testament is evident in the exemplum about Saul. The expositor summarises at the end, as if concluding a sermon, but he expresses the moral couched entirely in his own words: vengeance will fall on the heads of those who make unjust accusations.

The artist's iconographic interpretation agrees with Suso's Latin phrase, it would seem, since the first segment of the full-page miniature shows a prior[168] presiding over a 'Chapter of Faults'. He

[166] (Fol. 39v.) "Et sur toutes choses que tu auras a faire, se tu es prelat ou juge, ne soies pas de legiere creance, ne ne adjouste pas legierement foy a ceulx qui ont a coustume de faire partie. Et se aucun parle ou propose contre aucun, ne le croy pas tantost et especialment quant ilz parleront par faveur en soustenant l'un et condempnant l'autre contre raison. Item ne reçoy point informacion contre personne qui est approuvee estre de bonne vie, car se tu fais le contraire, tu destruiras bonne discipline et donneras aux mauvais hardiesse de mal faire et de resister a leurs souverains. Et par ainsi seroit afoiblie et traveillee la vigueur de religion... si te appartient de enquerir la cause que tu ne scez, et quant tu la sauras, si l'examine diligemment et lors donne au coulpable ce qu'il vouloit faire au juste, et luy fay retourner son pechié sur lui afin que son glaive lui perce le cuer et son arc soit froissié. Ainsi seront prins les mauvais en leur malice. Ainsi furent lapidez les deux viellars pour leurs iniquitez qui condempnerent Susanne a estre lapidee, laquelle estoit innocente mais sa fience fut en Dieu qui la sauva. Ainsi perit Aman qui estoit adversaire des Juis, qui par ses faulses mensongieres informacions avoit empetré du roy Assuaire que le peuple de Dieu fust destruit et mis a mort. Ainsi fit Saül, l'orguilleux roy mis a mort de son propre glaive qui contre raison persecutoit David, le noble chevalier et bon."

[167] The gloss on the word *praelatus* is *prior* in Künzle, *Horologium*, p. 415.

[168] I feel that the title *Disciple's Trial*, chosen by Spencer, *Horloge*, p. 281, n. 16, does not accurately capture the spirit of the subject matter, which emphasises the role of the judge, not that of the accused.

listens carefully to complaints by friars who stand and point down at a brother kneeling humbly at their feet. The prior also hears attenuating comments from two other brothers seated near him, and he is attempting to decide guilt or innocence in the accused, and fairness or malevolence in the accusers.

Sapientia's reference to Susanna concerns merely the stoning of the elders. The theme has been given an expanded visual treatment which records three episodes of the narrative in Daniel XIII. She is dramatically portrayed, advancing reluctantly towards a stake that is about to be lit; she is saved from that terror by the elocutionary prowess of young Daniel whose interrogation of the elders (verse 24) reveals their complicity and malice: *Dic ubi vidisti eos colloquentes* (the word in the scroll is *connuscentes*) *sibi* "Say where you saw them making each other's acquaintance." The replies from verse 58 are also supplied in banderoles: *Sub (s)cino* "Under a clove-tree," says one; *Sub prino* "Under a yew-tree" says the other. On the extreme right the viewer witnesses a lively lapidation of the two old men.

Susanna has been a prominent figure in Christian Art since the days of the Catacombs. The popularity of her icon may be related to a passage in the *Ordo commendationis animae* which runs *Libera, Domine, animam eius sicut liberasti Susannam de falso crimine.*[169] Her life as depicted in literature and art reflected not only unjust accusation and calumny but also alleged carnality and lasciviousness. The emphasis placed on the moral or sensual elements always depended on the context. Sapientia's remark implies moral outrage; the episodes in the visual rendering likewise arouse moral indignation rather than any thoughts of her physical charm and voluptuousness.

The deceits of Haman are exposed in the centre register of the *Horloge* painting. Suso's comment is quite cryptic, yet the visual statement is replete with detail derived from the Book of Esther. Esther was once a ward of the Jew Mordecai, and became in due course Queen to King Ahasuerus in Susa. She holds a banquet for her husband and his chief minister, Haman, a persecutor of the Jews. She announces details of the persecution to Ahasuerus and demands the boon from him that all Jews be left in peace (Esther, VII, 1-7). The first panel of the triptych presents the banquet in progress. In an adjoining chamber one sees Haman lying on the queen's bed and about to be seized by servants. The king orders Haman to be hung on the very gallows that had been prepared for Mordecai. The centre compartment depicts Haman's demise by hanging (Esther, VIII, 8-10). The sequel is shown in the next illustration, with a throne-room as a setting, where Mordecai kneels to receive from Ahasuerus the signet-ring of high office (Esther, VIII, 1-2).

Whereas Esther may have been considered a national heroine for the Jewish people, she was often viewed by Christians in the Middle Ages as a prefiguration of the *Stella Maris* of the Liturgy and of the *Virgo Coronata et Mediatrix*. By her intercession with Ahasuerus, Esther could be said to symbolise the intercessionary role played by the Virgin on Judgment Day. She was popular with artists in France and Germany in the fifteenth and sixteenth centuries.[170]

In the lower register of the miniature, Saul, persecutor of David "against reason" is seen in a throne-room aiming a spear in a demented fashion at David, his court entertainer, an episode derived from I Samuel, XIX, 9-10. Saul prefigures Judas in this dramatic act and symbolises misplaced Hatred, an emnity that caused the king to be morbidly suspicious of an innocent man.[171]

For the circumstances of Saul "the proud king, put to death by his own sword," the iconography draws on the narrative of I Samuel, XXXI, 1-9. The venue depicted is a slope, recalling the foothills of Mount Gilboa. The Philistines have already slain the three sons, seen in the centre of the middle

[169] Réau, *Iconographie, Ancien Testament*, p. 394. On pp. 393-8 he reviews the episodes in Susanna's life that became pictorial representations.

[170] Réau, *Iconographie, Ancien Testament*, p. 337. On pp. 335-41 he discusses not only her allegorical role but also the cycles of paintings she inspired.

[171] *Ibid.*, p. 264.

ground. Another corpse may be that of Saul's armour-bearer. The king lies dead on the left, having fallen purposely on his own sword. Enemy soldiers in shiny plate-armour are on the battlefield in the act of stripping the vanquished, and one is about to cut off Saul's head.

XV (1) *Carole* Fol. 40
(2) *Sick and Dead Youths*
(3) *Sapientia with Lilies and the Disciple*

At the beginning of the sixth chapter about the beauty of Divine Sapientia the Disciple reflects on the transitoriness of life: "There was a time when my youth flowered and my heart was lively and fickle, and I took great pleasure in going in spring to pick flowers in the green meadows with my friends. To turn my desires into reality some of them invited me at times to play, amuse myself and have fun in several games. With great joy in their hearts they moved their heads and beat their palms together, and jumped and cavorted in the manner of sheep in the flock, saying out loud: 'Our life is short and we have toil and trouble by day and by night; we have no hope of obtaining riches
or repose in the next life. We know no one who was come back from the dead. We were born and made of nothing, and then later on we shall be just as if we had never existed. Come now, eat and drink, use up and spend extravagantly the riches of the earth, for as long as our youth lasts. Let's drink fine vintage wine; let's put on sweet-smelling perfumes; let's cover our heads with rose garlands before they fade; let's be sure that there is no field or place where we have not taken our pleasure; let's show happiness everywhere, for we can take nothing with us."[172]

The third component of the triptych, *Sapientia with Lilies and the Disciple*, is more directly a pictorial projection of the Disciple's narrative in the first person. In a restless moment during his meditation on the meaning of life, he states: "Then, high in the sky, right in front of me, I saw a flower appear that was beautiful, radiant and bright; it was the renowned lily of the valley, and it was supremely pleasing to behold; it was the most beautiful flower I had ever seen. I ran forward to look at it close up, and when I was near, it suddenly vanished and I saw it no more, but instead, I seemed to be gazing at a goddess, one of great beauty, who was then in front of me. She resembled a crimson rose and was as white as snow, brighter and more dazzling than the sun."[173]

The discussion in the *Déclaration* is not ordered in an expected manner, that is, moving from the *Carole* to the *Sick and Dead Youths* and then to the third segment. A sentence at the end announces the three topics. The early part of the commentary offers a restatement of Suso's theme

[172] (Fol. 40v.) "Un temps fu que ma jeunesse fleurissoit et mon corage estoit fres et volage, si prenoie grant plaisir d'aler ou printemps cueillir les fleurettes en (*MS.* et) la verdure de la prairie avecques mes compaignons. Et les aucuns d'iceulx, pour acomplir mes desirs, me prioient a la fois de moy esbatre, jouer et soulacier a plusieurs jeux. Et par grant liesse de cuer mouvoient leurs chiefz et batoient leurs paulmes, et en la maniere des moutons du tropel saultoient et tripoient, en disant: 'Nostre vie est briefve et si sommes en grant douleur et labeur jour et nuyt, et n'avons point d'esperance d'avoir bien (*MS.* bien bien) ne repoz en l'autre siecle, ne on ne congnoist homme qui soit retourné de mort a vie. Nous somme nez et fais de neant et puis aprés nous serons aussi comme se nous n'eussions oncques esté. Or venons, buvons, mengons, usons et despendons largement des biens qui sont sur terre tant comme nostre jeunesse dure. Buvons bon vin et precieux, et usons des ongnemens delicieux. Couvrons nos chiefz de chappeaulx de roses avant qu'elles soient flaitries et gardons que il n'y ait pré ne place ou nos plaisirs ne soient acomplis. Monstrons partout signe de lyesse, car nous n'en emporterons plus.'

[173] (Fol. 40v.) "Et lors je vis en mon encontre, hault vers le ciel, une fleur apparoir moult belle, clere et reluisant, qui estoit renommee fleur du champ, et estoit souverainement plaisant a regarder, et de toutes les fleurs que j'avoye veu par devant, c'estoit la plus belle. Je couru et m'avençai pour la veoir de pres, et quant je fu pres, elle s'esvanuit soudainement, et ne la vy plus, mais me sembla que je veoye une deesse plaine de toute beaulté qui estoit devant moy, laquelle ressembloit a une rose vermeille et blan[che] comme nege, plus clere et plus reluisant que le souleil."

concerning the shallowness of earthly pleasures. There is also a personal viewpoint given about the Disciple's vision of the lily of the valley which appears to transform itself into the deific Sapientia. The *Déclaration* then identifies the person as the Disciple's object of affection. Thus, the *Carole* as a representation of dancing, and the death-bed sequence do not elicit any comment from the expositor.

In the textual extract cited at the opening of this study of the triptych, Suso reproduces with a maximum of poignancy phrases from Wisdom, II, 1-2 and 6-9. The visual interpretation captures the joyful exhilaration of the young as they participate in a lively carole,[174] set in a pastoral *locus amoenus.*[175] Three youths and three maidens dance around a tree in which a musician friend is perched, marking the tempo for the steps by playing a pipe and beating a *tambourin.* A young couple at a near table is totally absorbed in quaffing refreshments.

While the tableau appears as a literal rendering of the levity and abandonment of youth, the gestures of the dancers and drinkers are symbolically empty and meaningless and spiritually barren. The young have rejected immortality and given themselves over to hedonistic pleasures. There is a certain biographical nostalgia in Suso's remarks, put in the mouth of the Disciple, about his mispent youth before his entry into the Dominican Order. The pictorial image underscores the purely secular nature of the event.

The Sapiential quotation employed by Suso "after this life we shall be as if we had never been here on earth", is suggestive of the intervention of Death in the lives of the hedonists. The pictorial rendering of the idea presents a bed-room with a dead youth on the viewer's left and a moribund one on the right. The scene may be a narrative event with the inert form symbolising the state that the expiring person will ultimately reach. However, because the usual way of reading a narrative sequence is from left to right, the dead youth on the left may be a stark reminder to his dying friend that he is going to suffer to same fate. The companion in death lies on a sombre bed of buff yellow, flecked with gold and canopied with a blue drape in which gold stars twinkle. The moribund juvenile, already bereft of friends, lies on an elegant bed covered with a rich pink counterpane. Strewn over it are herbal quoits, their sweet-smelling perfumes warding off some of the odours of death. He is assisted by a woman of charity in his efforts to hold up the candle whose steady or flickering flame will eventually tell if he loses or wins the battle against mortality.

The visual symbolism of the *Sick and Dead Youths* is direct and poignant, both in its contrast with the preceding testimony to folly in the *Carole* and the ethereal setting of the tableau that follows. It must also be read in relation to the more macabre symbolism of the *Death-Bed Scene with Death Personified* which illustrates further on in the *Horloge* series the broader subject of the Art of Dying.

The elements of the textual vision have been clearly reproduced in the painting, *Sapientia with Lilies and the Disciple*; the visual perception is made all the more credible by the positioning of the 'goddess' on a slope, holding at arm's length the symbol of purity and virginity, the lily of the valley. The wonderment of the Disciple is captured by his half-raised arm and palms turned outwards, while his sudden genuflexion conveys his act of humility and adoration.

Suso's choice of the flower could not have been made at random. Not only was its symbolic meaning known to all Christians, it was more particularly recognised by Preaching Friars. In his celebrated treatise, *Somme le Roi*, drawn up *c.* 1279 for Philippe le Hardi, the Dominican author,

[174] I use the form *carole* and not *carol*, since the former is the exact Medieval French term for a round dance accompanied by singing; see Margit Sahlin, *Etude sur la carole médiévale. L'origine du mot et ses rapports avec l'Eglise*, Uppsala, 1940. The use of the English term *carol* in this sense is archaic, and if employed here, it may suggest the act of standing in a circle while singing.

[175] See Dagmar Thoss, *Studien zum locus amoenus im Mittelalter*, Stuttgart, 1972 for the symbolism attached to idyllic landscapes.

a certain friar Laurent, included a section on the subject of Virginity, symbolised by a lily with six leaves, and described as purity of body, purity of heart, meekness, fear of God, harshness of life, and steadfast will.[176]

In the first half of the fourteenth century a Dominican from Soissons (L'Aisne) composed a *Plantaire* in honour of the Blessed Virgin Mary. The plants are described in terms of their vegetal characteristics, medicinal properties and metaphysical symbolism. The nature of the lily occupies 170 lines of verse and includes reference to its symbolism of the purity of the Virgin as well as to her virtues and prerogatives.[177]

Given that the love and devotion to the Virgin was widely practised by Dominicans in the Middle Ages, it is not surprising that in the Disciple's vision Suso transformed the lily into a goddess-like figure as white as snow. It is also not untoward that this image should be represented as Sapientia, resplendent in white, presenting the symbolic plant to a person in the habit of a Dominican.

But there is much more that can be said about the miniature. A Dominican viewing the act of the Disciple reaching for the lily would relate the gesture to the person of the founder of the Order of Preaching Friars, St. Dominic himself. For many decades preceding the fifteenth century, the saint's iconography showed him often holding a lily. One can point to a late fourteenth-century historiated initial by the Brussels Initials Master, in the first instance.[178] Another portraiture is a woodcut of Flemish-Netherlandish origin, *c.* 1480-1490.[179] One may also point to the representations of the saint in frescoes by the great Dominican artist, Fra Angelico.[180]

XVIa *Last Judgment* — Fol. 46v.

Among the subjects treated in the seventh chapter is the Last Judgment. The textual passage that inspired the miniature occurs on the folio facing the painting, shortly after the storm at sea which is depicted in the illustration that follows this one. The Disciple speaks: "And immediately I was shown a prophetic vision in a wonderful way, and it told this story: the thrones and seats are in place and the oldest man has taken his seat. His raiment is as white as snow, the hair on his head is as pure and unsullied as cleanest wool. His throne was burning fire, and from his face streamed a rapid and turbulent river of fire, and from his mouth emerged a sword with two sharp cutting edges. When I saw him I was greatly terrified and fear struck my heart, so much so that my hair stood on end from the great revulsion I experienced; and all the more so because, in the presence of everyone, I beheld along with everyone else all the sins and peccadillos committed in my lifetime written down at length. And I was constrained by the angelic attendants to appear sad, grieving and fearful before the judge, who seemed to me to be beyond himself with rage and anger. And when the judge was seated on his throne of majesty and saw me in front of him, he raised his head, and fire seemed to issue from his eyes. Then, in great rage and indignation, he spoke and displayed the anger and vexation of his heart. When I heard him speak, I collapsed on the ground from fear, and I put my weak and awe-struck head on my arm as best I could. I was silent for an hour out of the great dread I experienced."[181]

[176] W. N. Francis, *The Book of Vices and Virtues*, London, 1942, pp. 255-8.

[177] See *A Plantaire in Honor of the Blessed Virgin Mary*, ed. by Sister Mary Alberta Savoie, Washington D.C., 1933, pp. 52-7.

[178] The portrait's present whereabouts is not known; it is reproduced in Meiss, *Late XIV Century*, fig. 797.

[179] In the National Gallery of Art, Washington D.C.; reproduced in Stokstad and Stannard, *Gardens*, p. 127.

[180] Cf. Pope-Hennessy, *Fra Angelico*, pls. 14, 92 and fig. 22.

[181] (Fol. 47) "Et tantost me fust monstre[e] une vision de prophecie par une tres subtille maniere, et disoit: Les sieges et les trosnes sont mis et le plus ancien se siet sus. Son vestement est blanc comme nege et les cheveulx de son chief sont netz et purs comme blanche laine. Son trosne estoit feu ardant et de sa face yssoit ung fleuve de feu roide et violent, et de sa bouche ysoit une espee trenchant de deux parties.

The *Déclaration* appears to treat the *Last Judgment* and the *Shipwreck* as one integrated painting. For the first of these subjects the commentator confines his description to essentials, such as the relative position of the Disciple, the Devil, the interceding angel, the Judge, and the fire that is reported to "burn the world and descend over the Damned." He omits pictorial detail, such as the emergence of the resurrected souls or the Leviathan's Mouth, or the Sword of Justice. As on many earlier occasions in the *Déclaration*, the expositor points the moral with a preacher's zeal: "Thus, every creature that is full of sin... should be well and truly terrified, and afraid to die..."

It is of little import if Suso, like numerous medieval writers, quoted *verbatim* from the Judgment visions beheld by Daniel and St. John the Divine.[182] The fundamental consideration to record is that he stopped short of rehearsing in the Disciple's vision what may be termed the plenary Judgment. His concentration's objective is the Disciple's relationship with the Judge of the World and not with other resurrected souls. The visual representation of Suso's text on the other hand, retains the Disciple, the guiding angel and the Devil, but includes them among many figures who are summoned before the Judge, as outlined in Revelation XX, 11-15. The resultant iconography is just as awesome as Suso's narrative.

The absence of trumpet-blowing angels in the *Horloge* painting shows that Doom is no longer being announced. That had occurred in the subsidiary miniature of the *Shipwreck*. Now the Judgment has commenced. In the upper left, gold stars shine out from a familiar blue firmament, while on the right, a blue-grey, cloud-filled sky glowers over the shrieking damned whom the Archangel Michael drives into the Jaws of Hell. His expression is cold and detached, yet his beauty is akin somewhat to that of the striking young dandy in the *Carole*.

Against an orange and gold radiance, Christ sits on a rainbow, in His role of Judge of the World. His face bears a vivid sentiment of poignant anguish. Being the victor over Death on the Cross, He is, here, the Christ of the Passion.[183] A sword, the harsh symbol of Justice, is placed horizontally on the picture plane, its hilt to His lips. From His hand a stream of fire flows towards Hell and carries the message of damnation, inspired by Matthew, XXV, 41: *Ite maledicti in ignem eternum* "Be gone, cursed ones, to the eternal fire". A banderole floating above His right hand, raised in blessing, proclaims the good news of Salvation to the Righteous, quoting Matthew, XXV, 34: *Venite, benedicti patris mei* "Come, you who are blessed of my Father."

In the painting's top-left quarter, the souls of three resurrected persons, represented as small naked figures, with arms lifted in supplication, traverse the heavens and approach the Mercy Seat in the embrace of their guardian angels. Seated below, on a long, high-backed bench, three intercessionary Apostles contemplate their Saviour. The kneeling Virgin, clad in a dark-blue, hooded cloak, intercedes on behalf of the Disciple. Behind her, also on his knees, and hands clasped in prayer, is St. John the Baptist.

Quant je le vy, je fus monlt (*sic*) esbahy et fus feru de paour jusques au cuer en tant que mes cheveulx se leverent de la grant horreur que j'euz; et plus fort car en la presence de tout le monde, je veoie, et chascun le veoit, devant moy en escript tous les pechiez et tous les deffaulz que j'avoye fais tout le tempz de ma vie et fu constrainct par force de messages de moy representer triste et dolent et paoureux devant la face du juge qui me sembloit tout forsené d'ire et de courroux. Et quant le juge fut assiz et son siege de maiesté et il me vit devant lui, il leva la teste, et sembloit que feu lui yssist des yeulx. Lors par grant courroux et indignacion parla et demonstra l'ire et la desplaisance de son cuer. Quant je l'ouÿ parler, je cheÿ de paour, tout pasmé a terre; je mis mon chief feble et esp[a]oury enclin sur mon bras si comme je peus mieulx, et me taisi bien par l'espace de une heure pour la tres grant paour que j'avoye."

[182] See Daniel, VII, 9-10 and Revelation I, 16. These verses are rendered in the translations from the words, "The thrones and seats are in place" down to "with two cutting edges."

[183] See Schiller, *Iconography*, II, pp. 206-8, who discusses the iconographic similarity between Christ in the Last Judgment typology and Christ portrayed as the Man of Sorrows. Often the Instruments of the Passion find their way into representations of the Judgment (cf. Réau, *Iconographie, Nouveau Testament*, pp. 739-40). This is not the case in the *Horloge* painting.

At the base of the composition is the Disciple, head turned to the left, prostrating himself in humble supplication to the Redeemer. A winged devil holds close to the friar's face an open book containing his sins.

In spite of the emotional element conveyed by the wrath registered on Christ's face, He still retains the dignity of His prototype, the Christ in Majesty. Christ the Judge, as is well known, is a modification of the *Maiestas Domini* figure, a change that began to become noticeable in monumental art in the thirteenth century, and which was soon absorbed in the typology of the Last Judgment.[184] The three principal elements of the theme when represented in French art are in evidence in the *Horloge* miniature. They are the Resurrection of the Dead; the separation of the Elect and the Damned; the 'tribunal' composed of Christ, Intercessors and Elders. The figures are usually viewed in a vertical progression. Due to the small amount of space available on the *Horloge* folio, the elements are seen horizontally.[185] The gaping Leviathan belches smoke and flame, recalling the passage in Job XLI, 21.[186]

XVIb *Shipwreck* Fol. 47

The storm scene and dismasted ship are clearly inspired by the opening words of the seventh chapter: "It happened once upon a time that the sun was covered by cloud and it vanished from the southern sky. And because of frost and the great cold, the flowers which were fresh and new became in one night crushed, faded and then died. And the ship which was on the sea, sailed hither and yon because the storm tossed it about perilously. Then I had great fear and trembled a lot, and I was as if blinded and filled with darkness. I sat down all alone, and with a very embittered heart I began to bring back to mind all my past life."[187]

If one reads carefully Suso's words, one appreciates that he is not stating that the Disciple is on the storm-tossed ship. Yet, that is where the artist has placed the Disciple. The *Déclaration* too seems to depart from Suso's intention. The expositor commences the account of the painting by emphasising that anyone who beholds suddenly stormy weather with rain, hail and snow, and a ship in the act of foundering, becomes fearful about his own life and the prospect of facing the Judge of Doom. The *Déclaration* does not indicate at this juncture that the "anyone" is really the Disciple of Suso's narrative.

The expositor then turns to the illustration and declares, as one would expect, that the Disciple is on the distressed ship, the elements of Nature being unchained over land and ocean. The Trumpets of Doom already sound in the Heavens and summon him to Judgment.

Reflecting on Suso's similitude of the power of the elements as expressed in the extract, one realises he does not locate the ship near any particular land formations nor does he comment upon the state of the rigging and masts. Even the placement of the Disciple is not indicated. A reasonable assumption has been made by the artist that the Dominican was on board the vessel.

The iconographic theme of the *Shipwreck* owes much of its visual impact to a skilful interplay between the hostility of Nature as a forerunner of Doom and the mortal friar's trepidation at the prospect. On the horizon, to the left, spires rising on keeps and turrets stab the stormy sky, and

[184] Cf. Van der Meer, *Maiestas Domini*, pp. 383-97.

[185] Réau, *Iconographie, Nouveau Testament*, p. 737.

[186] *Ibid.*, p. 751.

[187] (Fol. 47) "Un tempz advint que le soleil estoit couvert d'une nuee et s'estoit departy de la region de midi; et par force de gelee et grant froidure les fleurs qui estoient fresches et nouvelles furent en une nuyt toutes amaties, flaitries et perdues. Et la nef qui estoit ou milieu de la mer aloit ça et la, car la tempeste la degettoit perilleusement. Lors eus je grant paour et trembloye fort et fu si comme aveugle et plain de tenebres. Je me seoie tout seul et commencé a ramener a memoire toute ma vie passee en grant amertume de cuer."

on the right, a double-towered bastion presses down on the melancholy landscape. A dusting of snow on the barren hills and the bare branches of a solitary tree does nothing to lighten the gloom. Two angels arrive and blow lustily on long, thin trumpets; their right hands point to the top left of the picture—in fact, across the folio to the facing tableau of the *Last Judgment*—and reinforce the effect of the summons that can be read on a scroll: *Venite ad iudicium* "Come to Judgment". At this moment the Disciple kneels in the well of the ship, his habit billowing in the wintry air, his hands held up in consternation as he beholds the unwelcome messengers. The mast and spar of the vessel have snapped, the sail and broken halyards tumble around him. All is chaos.

XVII (1) *The Disciple Offering his Heart to Sapientia* Fol. 49v.
(2) *Sapientia with Faith, Hope and Charity*
(3) *The Disciple Shutting out Sapientia*

In the course of a dialogue with his heart, the Disciple says: "You are the very gift and beloved gem which I presented to Sapientia at the onset of my youth, when I left behind worldly love. Then, I made Sapientia a gift of you, my heart, as the pledge of earnestness, as a love token and as the sign of perfect charity. Around my heart I wrote her name in blood and of you, O heart, I made the seat and throne for her on which to sit and rest. Of you, I made the chamber to lodge and shelter her." The text actually continues without a break: "Now tell me how and in what manner I should receive this spiritual and eternal bride when she comes and knocks at the door of my heart. How should I open it for her?"—The heart answers: "I shall not wait long, I could not bear to keep my beloved waiting at my door; rather, I shall run in haste to her, the arms of my desire stretched out, and I shall embrace tenderly and joyously the one who pleases me so much and does me so much good. And I shall hold her in passionate embrace and welcome her humbly with an all-glorious and loving greeting."[188]

Later in the chapter, Sapientia reproaches the Disciple for his lukewarm devotion: "You say that my comings and goings cause much displeasure to those who love me and whom I love, and you wonder at this. So that you may better understand my answer, you should know that the reason why the soul is sometimes upset with me and about my departure, is because of it and its own shortcoming. You see, when the sun's rays enter the room through the window, he who closes the window causes the ray to leave and brightness to disappear. And when one has received a guest in one's house and one looks crestfallen, he has every reason to leave... Sapientia then observes: And when the soul becomes slow, lazy and tardy in doing good, or when it withdraws and retreats from its good words and beginnings, then it shuts the window to my entry and casts me out..."[189]

[188] (Fol. 50) "Tu es le propre don et le joyel amoureux que je lui presenté au commencement de ma jeunesse quant je me departi de toute amour mondaine. Adonc lui feiz je present de toy mon cuer pour ses erres, pour le joiel d'amours et pour le signe de parfaicte charité. Entour mon cuer escrisi son nom de lettre de sang et de toy lui feis je siege et trosne pour lui seoir et reposer; de toy fu faicte la chambre pour lui hebergier et hosteler. Ores me dys doncques comment et par quelle maniere deveray je recevoir celle espouse spirituelle et pardurable quant elle vendra et hurtera a l'uis de mon cuer. Comment lui deveray je ouvrir?" Cy respond le cuer: "Je n'attendray pas longuement ne je ne pourroye souffrir que m'amie attendist a mon huys, mais lui courray hastivement au devant, les bras de mes desirs estendus, et embraceray tendrement et de cuer joyeux celle qui tant me plaist et qui tant de biens me fait, et si l'estraindray ardamment et la salueray humblement de un glorieux et amoureux salut..."

[189] (Fol. 51) "Tu dis que mes alees et mes venues font grant desplaisir a ceulx qui me aiment et que j'aime si t'en merveilles. Pourtant que tu entendes mieulx ma response, tu dois savoir que la cause pourquoy l'ame est aucunes fois troublee de ma departie et envers moy, c'est par lui et par son deffault. Tu vois que quant le ray du soleil est entré en la chambre par la fenestre, celui qui clost la fenestre en fait departir le ray et la clarté. Et quant on a receu un hoste en son hostel et on lui fait mate chiere, il a cause de s'en aler... Et quant l'ame se rent lente et paresceuse et tardive a bien faire ou que elle recule et se retrait de son bon propoz et son bon commencement (*MS.* commencencement), adoncques elle clost la fenestre de mon entree et me boute hors de la compaignie..."

The purport of the *Déclaration*'s commentary on this triptych is to emphasise the spiritual acts and gestures which Suso's characters make, and to relate the action of each panel to the one next to it. The Disciple's act of surrendering his heart is seen by the donor as a measure whereby a love token changes hands; it is a pledge of earnestness and sign of perfect charity, according to Suso's text. But the author of the *Déclaration* speaks from Sapientia's viewpoint, so that the gift to her stems from her gift of the grace of recourse to God, of the grace of commitment to devotion and of the grace of love for Divine Sapientia. The expositor claims it is grace that assists her to present herself, in the company of Faith, Hope and Charity, at the door of the Disciple's soul. Suso's account of this arrival is presented through the eyes of the Disciple and the emotive register is totally sensual. The Theological Virtues are not in attendance.

The atmosphere of rejection and gloom at Sapientia's departure, so poetically expressed by Suso, is faithfully captured by the expositor in his account of the last panel. But the reason he gives for the pathos is the rise of earthly and worldly thoughts which darken the Disciple's understanding. One observes further that the commentator's form of spiritual narrative has effaced completely the dialogues and monologues that are a feature of Suso's original work.

To a twentieth-century eye the paintings of the triptych assume additional dimensions. The first subject is a pictorial transformation of a monologue which the Disciple addresses to his heart after giving it to Sapientia. The religious format selected by the artist shows the Disciple genuflecting and offering a heart inscribed with IHS to Sapientia who is wearing her usual cruciform halo and standing on an offertory table. Her body effects an S-shaped stance, the head is inclined to the left. In her left hand she holds her doctrine, and with a deliberate gesture she beckons to her pupil. The Dominican theme is forcefully presented. In Spencer's words, "the Dominican devotion to the Sacred Heart of Jesus seems to be forecast in Suso's works, here (f. 49v.) in his descriptions of the Disciple's vision of Sapientia above the altar (*sic*) to whom he offers his own heart inscribed with the Sacred Monogram."[190]

In the painting that illustrates this passage, the humble Disciple peers out from the open door of his cell at four figures who are the allegories of Sapientia and the Theological Virtues, Faith, Hope and Charity, each with her attribute. The presence of the Virtues is not required by Suso's narrative. It is, however, possible that the mention *signum caritatis* "sign of charity", suggested the person of Charity. The carnality and voluptuousness conveyed by Suso's written word have been replaced by simplicity, beauty and a moralising ethereal quality that betoken a transformation of interpretation that is of theological moment.

The visual presentation of Sapientia with the three Virtues is unusual, iconographically speaking, in Western Christendom at such a late date. Theologians and painters had long ceased to juxtapose these icons, but in the Eastern tradition, Sapientia as Sophia is a recognised saint whose progeny were three daughters, Faith, Hope and Charity.[191]

Suso's moral lesson is projected realistically in time and space in the last panel of the triptych. The Disciple withdraws into himself in a chapel; he shuts the window while Sapientia is outside. She shelters the light of a candle and is about to depart the scene. Her eyes are downcast, her countenance shows the pain of her rejection. The unwanted guest leaves its temporary abode, and the Disciple's heart is about to be plunged into darkness again.

[190] Spencer, *Horloge*, p. 280. The Sacred Heart of Jesus as an iconographical theme is discussed in Réau, *Iconographie, Nouveau Testament*, pp. 47-9, 51. He does not speak of the Dominicans at all in relation to the Sacred Heart. The earliest icon mentioned is a wood carving of 1505.

[191] Réau, *Iconographie, Saints*, I, p. 510, and III, pp. 1232-3.

XVIII (1) *The Disciple Offering his Soul to the Trinity* Fol. 56v.
(2) *Life's Pilgrim*
(3) *The Disciple's Contemplation*

In the ninth chapter the Disciple is urged by Sapientia to abandon earthly things: ''Lift yourself then out of these vile and dishonest temporal pleasures. Open the eyes of your understanding and see who you are, where you are and whither you are going; then, for certain, you will have knowledge of what I have said. First of all, if you wish to see who you are, you are the mirror and representation of the deity because of the semblance of the Trinity that you carry in your soul; for your soul has will, understanding and memory. You are also made in the example of the Trinity in that your soul cannot die; and the more so, just as I am in myself infinite, without end and without beginning, so are the desires of your soul and your heart. A drop of water could no more fill the great ditch which contains the sea than the pit and abyss of the desires of the soul and of the heart could be sated and satisfied by all the worldly joys they could find on earth. You must next consider where you are. You are in a valley of.''[192] There is a gap in the French transcription at this point, owing to the removal of a vellum folio, as was noted above in chapter I. The admonition is therefore resumed by recourse to the Latin original: ''valley of distress, in exile, on a pilgrimage where good and evil are continually interwoven. Laughter here is mixed with grief, and the end of bliss is filled with sorrow. Here truly, the world has deceived its lovers from the beginning, promising them prosperity and giving them adversity; and it still does these sorts of things, even today. To be completely rid of all such preordained disturbances, we must withdraw somewhat and return with all our heart to the foundations of the Faith.''[193]

The expositor takes pains to set the background for this triptych and summarises the narrative in the chapter to this point. His statements show a total familiarity with Suso's ideas and with Sapientia's exhortation to the Disciple to reflect and contemplate who he is, where he is, and where he is going. When he views the first picture the commentator delivers a brief homily on man's affinity with the Trinity, on the three powers of the Soul and on the Disciple as an exemplification of God's eternity. The theme of the tribulations that mark corporeal existence is said to be represented in the panel of *Life's Pilgrim*. The commentator recalls the tribulations with terms such as pleasures, sorrows, snares, temptations.

Sapientia's concluding exhortation to the Disciple is then taken up in the *Déclaration*. In answer to the question, whither goest thou? she had suggested that he withdraw from the pre-ordained upheavals of the world and return to the foundation of the Faith. The expositor seizes the opportunity to modify the direction by indicating a bifurcation of the Disciple's future with one path leading to Paradise and the other to Hell. He terminates the theological thought by announcing

[192] (Fol. 56v.) ''Lieve toy doncques de ceste vile et deshonneste plaisance temporelle. Euvre les yeulx de ton entendement et regarde qui tu es, ou tu es et ou tu tens. Et adonc pour certain tu auras congnoissance des choses que j'ay dittes. Premierement se tu veulz regarder qui tu es, tu es le mirouer et la figure de la deité pour la semblance de la Trinité que tu portes en ton ame; car ton ame a voulenté, entendement et memoire. Tu es faitte aussi a l'exemple de la Trinité car ton ame ne peut morir. Et plus fort car, si comme je suis en moy infinie, sans fin et sans commencement, aussi sont les desirs de ton ame et de ton cuer. Car neant plus que une seule goute d'eaue ne pourroit emplir la grant fosse ou toute la mer est enclose, tout ainsi et encore mains pourroient l'abisme et la fosse des desirs de l'ame et du cuer estre assouvis ne saoulez de toutes les joyes mondaines qu'elle pourroit avoir ne trouver en terre. Tu dois aprés considerer ou tu es. Tu es en une valee de...''

[193] Künzle, *Horologium*, p. 453, lines 18-25: *Deinde advertere debes ubi sis: quia in valle miseriae, in exsilio, in peregrinatione, ubi bona permixta malis continua vertigine revolvuntur. Risus hic dolore miscetur, et extrema gaudii luctus occupat. Hic nempe mundus suos ab initio fefellit amatores, dum eis prospera promisit et adversa retribuit, quod et similiter nunc quoque facere non desistit. Sed ad pleniorem omnium praemissorum perturbationum amotionem paulululum nobis secedendum est, et ad fidei principia latius recurrendum.*

peremptorily that the Disciple appears in the painting at the moment of contemplating this particular choice.

As we examine the triptych we read a legend of three words *Hec tria considera*, which are not taken from Suso, but which clearly epitomise Sapientia's opening challenge to the Disciple. The phrases *Qui es*, *Ubi es*, and *Quo tendis*, which proclaim the theme of each panel, are inspired by her words *Vide quid sis, ubi sis et quo tendas*. The wording in the left and right scrolls: *Ymago Dei per intelectum* (sic) *memoriam et voluntatem*, and *Exemplar eternitatis quantum ad perpetuitatem et infinitatem*, seem to recall the Latin of Suso: *Imago trinitatis eo quod eius imago in te resplendeat, exemplar aeternitatis, eo quod inviolabili incorruptione gaudeas.*[194] A rendering of these passages, based on the French translation, was supplied at the opening of this section.

The key themes in the passage for the purpose of a visual presentation are the Trinity and the Soul, and the Soul's longing for communion with the Deity. The relationship is suggested by the placement of the Disciple, kneeling on the open ground and holding upwards the figure of a young child, personifying the surrender of his Soul. The iconography of the Trinity is traditional in that the principal personages are seated on a throne behind which seraphim gather in large numbers. Christ wears a halo while His brow is constrained by a Crown of Thorns. A patriarchal cross rests on His right shoulder. He sits on the right of a bearded God the Father whose omnipotence, omniscience and omnipresence are suggested by a conical tiara. Both figures sit side by side under one raiment, and between them is a fluttering Dove and an open book.

The Disciple is portrayed in pilgrim's garb as he stands in a landscape that includes an estuary and a distant city. In his immediate presence are two swathed corpses and nearby, two lovers in worldly attire about to be ensnared by two devils holding upright a giant net, an evocative symbol of life's pitfalls and ruination. The portraiture suggests, however, not merely the mortality and frailty of life for the average human; there is also the knowledge that a pilgrim's staff was an attribute of St. Dominic, symbol of the Preaching Friar's mission to traverse the world in order to disseminate the Word of God. According to his legend, the saint received the staff from the hands of St. Peter.[195] The other attribute, the Book of Gospels, which St. Paul handed to him on the same occasion, to symbolise the spread of God's Word, is not included in the *Horloge* painting.

The last panel in which the Disciple wears his usual Dominican habit once again and stands all alone in his cell, captures the spirit of contemplation recommended by Sapientia. The words, "Where are you going?" of the legend qualify the mood, while the open door through which he gazes unemotionally suggests realistically his act of concentration on the future and eternity.

XIX (1) *Devils of Hell*
(2) *Cold of Hell*
(3) *Fires of Hell*

Fol. 57 missing

The reader may recall from observations in chapter I that fol. 57 is missing, and with it, of course, the appropriate passages of the French translation around the triptych. Suso's text assists us to repair the loss. The Disciple continues to contemplate possible destinies, and these include Hell. In a vision he thought he saw a region filled with darkness that presented a terrifying spectacle for him. An explanation of its purpose caused him greater terror. He beheld some of the frightening activities in the place. "An overpowering stench arose from the various areas that had been mentioned. The sounds of hammers could be heard; intense gloom and darkness was all pervasive; one could make out horrible faces of devils; wailing and moaning filled the air. Sinners were

[194] For these excerpts, see Künzle, *Horologium*, p. 453.

[195] Réau, *Iconographie, Saints*, I, p. 395.

tormented, now one way, now another, ruthlessly and mercilessly. From icy waters they passed to crushing heat.''[196]

Fortunately, a brief mention of these three lost miniatures was made by the author of the *Déclaration.* His comments link the triptych to the preceding painting in which the Disciple is contemplating his future, Hell included. What the commentator saw in the lost triptych is reported in a descriptive mode which seems close to the wording of Suso's *opusculum.* This suggests that the artist also had been inspired by the textual narrative.

XX (1) *Gallows of Hell* — Fol. 57v. missing
(2) *Pains of Hell*
(3) *Pains of Hell*

Suso's comments about Hell continue. The Disciple next perceives other causes for terror. ''Those who had robbed the poor and friends of God or who had crushed them with extortionate taxes, were put on the Gallows of Hell. They suffered pains beyond human comprehension; they were brutally hung, and pulled and wrenched in the same place; however, they did not die, but continued to be tortured in an unspeakable manner.''[197]

The words of the *Déclaration* are not very expansive about detail in the segments of the lost triptych. Nevertheless, the expositor does indicate that the artist's portrayal of the bodies on the Gallows depicts the tongue, feet and neck as areas for the noose to grip.

XXI (1) *The Slothful, Hedonists and Wasters in Hell* — Fol. 58
(2) *The Fornicators in Hell*
(3) *The Avaricious in Hell*

The legend across the bottom of the first picture *Les oyseux, mondains et vains* ''The Slothful, the Hedonists and the Wasters'' characterises the victims of Hell's fury: ''The drunkards and those who had spent all their time in gluttony and in serving their stomachs went mad with hunger, howled like wolves and had raging thirsts to the point that their tongues were hanging out, seeking a drop of water to relieve the heat, but not finding anyone to give them it. Nearby were frightening and terrifying devils with flaming tubs and buckets filled with molten lead and stinking sulphur, all red and glowing like ovens; and the demons were pouring this searing beverage into their bodies.''

The second painting reproduces imaginatively the continuing drama flowing from Suso's pen: ''Next, one could see the fornicators and those who had indulged their bodies with gratifications and who had wallowed all their lives in their obstination and filth. These ones were attacked and devoured by serpents, scorpions, snakes and toads, creatures that were all bloated, venomous and inflamed, and which injected their poison and venom right into the victims' stomachs. The men and women who had turned their hearts to carnal love ate the earth of Hell; the devils persecuted

[196] Cf. Künzle, *Horologium*, p. 455: *...de praedictis locis foetor intolerabilis exoriebatur. Mallei resonantes audiebantur, tenebrae profundissimae densabantur, et videbantur ibi facies daemonum horribiles. Ibi erat planctus et gemitus, et alternantia mala impios sine pietate discerpebant. Ab aquis nivium transibant ad calorem nimium.*

[197] Cf. Künzle, *Horologium*, p. 455: *Raptores namque et praedones ac eorum complices, et hi qui pauperes et amicos Dei, dum adhuc viverent, spoliaverant, vel per exactiones indebite afflixerant, ad patibula quaedam infernalia, supra humanam aestimationem poenalia, rapti violenter suspendebantur et ibidem torquebantur; nec tamen moriebantur, sed modo indicibili cruciabantur.*

them, seized darts and arrows feathered with scorching flames and wounded them grievously. The man and women who had been in life companions in sin and guilt were in Hell partners in torment."

The third episode relates the fate of the avaricious. Suso's narrative reads in the translation: "Above all others, were tormented the covetous, the avaricious and the deceitful alms-collectors; and they were in deep ditches full of boiling metal, swimming in it; and whenever they tried to get out of the bath, the devils were on hand to push them back in again forthwith."[198]

Even though Suso offers his readers a somewhat graphic account of the punishments undergone by selected sinners, classified according to the type of sin, e.g. slothful, wasters, fornicators or avaricious, the author of the *Déclaration* sees in this triptych further representations of the torments and tortures of Hell. Perhaps he considered that the realism of the segments needed no further comment.

The artist's visual perception of Suso's text directs the viewer to the acts of howling and raging by the victims, rather than to the tubs and buckets of molten metal. The flames in the foreground are the only elements that suggest the fiery environment. The representation of the beasts doing the torturing in the foreground is faithful to the textual models. For the companions in carnal sin, *les luxurieux*, two pairs of men and women are depicted as seated facing each other with circumventing shackles. Already the flames reach their loins and genitals, while darts inflict further pain. The realistic pictorial translation of the text is a masterpiece of devilry, action and human pathos.

It is well known that *Hell* scenes constructed with segments of the *Last Judgment* had, by the late Middle Ages, become vehicles for the portrayal of the Seven Deadly Sins and the punishment each merited, although there was not general agreement on the exact detail.[199] Inspired by Suso's text, where the Damned are characterised by their sins, this triptych presents the three groups of Slothful, Fornicators and Avaricious.

XXII (1) *False Judges in Hell* (2) *Wicked Women in Hell* (3) *Pit of Hell*

Fol. 58v.

Suso continues in graphic fashion his account of the torments of Hell: "Those also who had been judges and rulers in the world were very harshly put to torture and to torment. The most almighty in life was most almightily tortured. Who could say and relate just how cruelly tortured were the false judges, the pitiless rulers, and the clerks and men of the cloth who had basely earned their living and who had been more intent on temporal than spiritual gain? Included with them were

[198] (Fol. 58) Les yvrongnes et ceulx qui avoient tout leur temps despendu en gloutonnie et en servir a leur ventre estoient tous enragiez de fain et ulloient comme loupz affamez et languissoient de soif en telle maniere que ilz gettoient la langue traicte en demandant une seule goutte d'eaue pour attremper leur chaleur et ne trouvoient qui leur donnast. Decoste eulx estoient dyables hydeux et espoventables qui avoient cuves et seilles enflammees, plaines de plomb fondu et de souffre puant, rouge et ardant comme fournaise, et leur gettoient dedens le corpz celui angoisseux buvrage. Aprés veoit les luxurieux et ceulx qui delicieusement avoient leur chair nourrie et qui tout le tempz de leur vie avoient demouré en leur obstinacion et en leur ordure. Ceulx estoient assaillis et devourez de serpens, de scorpions, de couleuvres et de crapous enflez, envenimez et tous ardans, qui leur gettoient leurs santonines (*MS.* saintons) et leurs venins jusques au cueur du ventre. Ceulx aussi qui avoient mis leurs cuers en amour charnelle, hommes et femmes, mordoient la terre d'enfer (fol. 58v.) car les ennemis leur faisoient grans persecucions et prenoient dars et saiettes enferrees de flambe ardant et les navroient douloureusement. Ceulz et celles qui avoient esté au monde compaignons en pechié et en coulpe estoient en enfer compaignons en tourmens. Sur tous les autres estoient tourmentez les couvoiteux, les avers et les faulx questeurs, car ilz estoient en fosses parfondes, plaines de metail boullant et la dedens se baignoient et souvent s'efforçoient d'issir hors du baing, mais les ennemis estoient la qui tantost les y reboutoient.

[199] Cf. Réau, *Iconographie, Nouveau Testament*, p. 752.

covetous and avaricious monks, slothful and troublesome lay people, knights, inn-keepers, overly loquacious women, gossips, dancers, proud and shameless pleasure-seekers, along with several other false Christians—all were harshly punished, worked over, painfully tortured. All wailed and howled like dumb beasts in full cry in such a manner and in such horror that it was great sorrow not only to suffer what they endured but also to hear their clamouring and pitiful lamentations. And it did not seem that in the world there could have been pain similar to this."

The second compartment is inspired by Suso's next passage, which is taken in part from Wisdom XI, 18-19: "The Disciple saw there bears, lions and newly created species of horrible monsters whose mouths spat flames and abominable and stinking clouds of smoke. From their eyes came flashing sparks. The animals devoured and tore at the sinners, filling them with venom and breaking off their limbs."

As a terminal punishment, Suso mentions the *puteus abyssi*: "Finally, all the sinners were gathered up and hurled into a deep, hideous pit filled with putrefaction, from which arose such a stench that no human heart or body could bear or tolerate it."[200]

The *Déclaration* groups this triptych with the preceding one and mentions that the viewer sees in both the tortures and torments of Hell; the final ignominy is the Pit. There is a rejoinder that the series of punishments may be envisaged by anyone who meditates on them.

The artist's source of inspiration was Suso's narrative. The title, *Les Faulx Justiciers*, attached to the first segment of the triptych, is a guide to the type of hapless sinners who undergo all manner of terror and torment. Lucifer, crowned and sitting on a stool placed at an oblique angle, directs the activities of his horde of devils. To the left, a man lies across an execution block, his lacerated body is subjected to repeated blows from a sharp-pointed hammer. In the middle ground, another lost soul is impaled on a stake, and in the foreground, two of the false judges are ripped to shreds by fiends manipulating a spiked wheel. The anguished faces with gaping mouths convey forcefully the idea of "clamouring and loud lamentations". The realistic punishment depicted in the painting may have been inspired by contemporary fifteenth-century methods of torture employed against criminals in castles and prisons all over France.

The viewer's interest is certainly aroused by the presence of a fantasmagoric lion, a bear and deformed monsters which devour or torment naked judges. A hideous giant clutches two screaming humans in its paws, while riding on its shoulders is another tormented being, his hands held to his mouth in horror.

The Pit of Hell is represented by a large round cistern packed with terrified victims. Over their heads floats a small scroll outlined in red suggesting blood, and with a poignant inscription: *Ubi sunt vaniglori(fic)atores mondi* (sic), which loosely translated would read "Where are the vainglorious of the world?" The source for this legend is unidentified. Between the cistern or

[200] (Fol. 58v.) Ceulx aussi qui avoient esté justiciers et souverains au monde estoient moult rigueureusement mis en gehaine et a tourment. Et celui qui avoit esté plus puissant au monde estoit plus puissamment tourmenté. Qui pourroit dire et raconter coment estoient la cruelment tormentez les faulx juges, les gouverneurs sans pitié, les clers et les gens d'eglise qui (*MS.* qui qui) vilement avoient gaigné leur vie et qui avoient plus entendu a gaing temporel que a l'espirituel. Aussi les moynes couvoiteux et avaricieux, la gent laye oiseuse et ennuyee, chevaliers, taverniers, et femmes trop parleresses, jangleresses, danceresses, tres bauderesses, orgueilleuses et despiteuses et plusieurs (*MS.* et plusieurs *twice*) autres faulx crestiens estoient la durement tourmentez et asprement punis et traveillez qui tous ensemble ulloient et crioient comme bestes mues a haulte voix par telle maniere et par telle horreur que c'estoit grant affliction non pas seulement de souffrir ce qu'ilz souffroient mais aussi de oÿr leurs clameurs et piteuses complaintes qu'ilz faisoient. Et ne sembloit mie que au monde eust ne peust avoir peines semblables a ycelles. La veoit le disciple ours, lyons et bestes toutes nouvelles de horrible façon qui gettoient feu par leurs gueules et une fumiere puante et abhominable, et de leurs [yeulx] yssoient scintelles estincelans; et celle[s] bestes les devouroient et desciroient et en mordant les emplissoient de venin et les (fol. 59) despeçoient par membres. Finablement, on les prenoit tous ensemble et les gettoit on en un puis hydeux et parfont, plain de toute punaisie duquel il yssoit une pueur telle que corps ne cueur humain ne pourroit souffrir ne porter.

cauldron and the front of the picture plane two creatures dance a ghoulish jig, their clawed feet stamping on three unbelievers. Iconographically speaking, the physiognomies, corporeal traits and activities of the devils in the cycle of *Hell* scenes of the Brussels *Horloge* conform for the most part to the traditional imagery of the agents of Satan and Lucifer in medieval French paintings.[201]

XXIII *Court of Heaven* Fol. 62

Sapientia describes to the Disciple the various heavens; on reaching the highest one, she says: "And the heaven above it that I tell you about is quite smooth, flat and polished like a mirror ...and this heaven is the chamber and royal abode where live the blessed spirits and God's Children who praise and glorify me without end. There are the noble and lighted celestial thrones from which the cursed company of bad angels were cast out; they are to be filled from the number of good Christians, the Elect of God."[202]

She shows the Disciple more: "Turn your eyes to all the parts and look at the great multitude of people and note carefully how they all drink at the fountain of goodness, and how with great appetite they become intoxicated through the sweetness of the beverage. Observe how they are attentive and gaze into that most excellent mirror in which all things are delineated and clearly represented. In it they discern great bliss and wonder at it. There is no travail there; they behold the sovereign good eternally and draw on it without ever becoming replete or bored with its intoxication. Rather, they desire eternally more and more what they see and taste and savour. Come even closer and hear how the Queen of Angels and of Virgins whom you love and honour so devoutly, is favoured with a singular privilege of love and glory, and how she transcends the highest angels, and how through trust and tender love she is close to and leaning on the shoulder of her dear Son, and how she is surrounded by the quite overpowering scents of violets, roses and fleurs de lis..."[203]

Sapientia continues: "And see how the first and sovereign angels who belong to the Order of the Seraphim and the blessed souls who have been raised to their status are burning with the love of God... how the Order of the Cherubim receives and gathers the abundance and fullness of divine light and reflects it back to the others over a wide expanse; how the Order of the Thrones and all the other blessed ones in their company find repose in God and God in them blissfully..."[204]

[201] Réau, *Iconographie, Ancien Testament*, pp. 60-3.

[202] (Fol. 61v.) "Et le ciel que je te dy, qui est dessus, est tout uny, tout aplanié et aussy polly comme ung miroir... et cestui ciel est la sale et palais royal on habitent les benois esperis et tous les enfans Dieu qui m'en loent et glorifient sans cesser. La sont les sieges celestes nobles et enluminez desquelz tresbuscha celle maldite compaignie d'ennemis et doivent estre rempliz du nombre des bons crestiens esleuz de Dieu."

[203] (Fol. 62v.) "Tourne tes yeulx de toutes pars et regarde la grant multitude de gens et te prens garde comme ilz boivent a la fontaine de tous biens et comment par grant appetit ilz se enyvrent pour la doulceur du buvrage. Regarde comment ilz sont entendans et regardans en celui tres excellent mirouer ou quel toutes choses sont figurees et clerement representees. La voient ilz toute joye et en ont grant merveille. La n'a nul travail car tousjours ilz voyent le souverain bien et en usent sans en estre enyvrez ne ennuyeusement soulez, mais tousjours desirant et plus et plus celui qu'ilz voyent et qu'ilz goustent savoureusement. Vieng encores plus pres et escoute comment la royne des angres et des vierges que tu aymes et que tu honneures tant et si devotement est douee de singulier privilege de amour et de gloire; et comment elle seurmonte et trespasse la haultesse des angelz; et comment par fiance et tendre amour elle est acoustee et apuiee sur l'espaule de son chier filz; et comment elle est environnee de toutes precieuses oudeurs de violettes, de roses, de fleurs de lis..." Suso is not as specific as the French translator about the Virgin's position in relation to Christ: *...qualiter regina virginum, quam summo venerare affectu, privilegio singulari honoris et gloriae transcendit culmen ordinum, innixa prae teneritudine super dilectum, inter areolas aromatum, circumdata floribus rosarum et liliis convallium*; cf. Künzle, *Horologium*, p. 463.

[204] (Fol. 63) "Et voy comment les premiers et les souverains angelz qui sont de l'ordre de Seraphin et les benoites ames qui sont eslevees en leur compaignie sont ardans en l'amour de Dieu... comment aussi l'ordre de Cherubin reçoit et recuelt l'abondance et la plenitude de la lumiere divine et respendent et envoyent aux autres largement; comment aprés l'ordre des Throsnes et tous les autres bieneurez qui sont en leur compaignie se reposent en Dieu et Dieu en eulx joyeusement."

"Further, you must see how the Apostles and principal followers of God are honourably seated on judgment seats and how they have supreme power to judge and pass final sentence; and how the martyrs are bright, lit up by rubescent light; the confessors cast rays as flamboyant as flames; and virgins are pure and white as snow. In this way you can see and know how the whole Court of Heaven is resplendent, overflowing and suffused with bliss and every happiness."[205]

The Trinity is named in a still later passage: "...Yet, it is true that when the Soul comes into Glory and beholds the Blessed Trinity which is unity itself, the Soul is completely and ecstatically transported and transformed in God and beautified by the realisation of such ineffable pleasure."[206]

The positioning of the painting by the author of the *Déclaration* is made correctly and its content is contrasted with the infernal machinations that were the subject of the Disciple's contemplation. The Disciple now comes to wonder at "...the beauty and refulgence of the city and glory of Paradise." The expositor associates this beauty with the vision of God. The simile of the stars of varying brightness and the differences in the merit of individuals belongs to the commentator. He is also thinking for himself when he writes that the presence of so many kinds of holy men and women in Paradise enables the viewer to comprehend better its glory. His remaining comment is for the figures contemplating the celestial throne of the Trinity. They correspond to those of Suso. Yet neither the painter nor the expositor mention the various kinds of Heaven, the mirror image or the Virgin leaning on the shoulder of her Son, which are all elements of Suso's narrative. On the last detail, the expositor follows the portraiture in the miniature, he states that the Virgin is close to the Trinity. However, when the commentator observes that St. John the Baptist is above all the Orders of Angels, he is expressing a traditionally held view, one that is not illustrated by the *Court of Heaven* painting.

Examining more closely the artist's achievements, we appreciate how difficult was his task of representing visually all the details of Suso's Heavenly Court. The design chosen for their execution was nine compartments with frames partly effaced to give the appearance of a single painting. The focal point is, not unexpectedly, the centre of the top row, and it is occupied by the Trinity surrounded by seraphim.

The Virgin, wearing a crown of foliate conformation, sits on the steps of the dais, in close proximity to the three persons, but at a lower level. Her crowned figural representation recalls a tradition in both monumental art and manuscript illumination for the Court of Heaven as a setting of the theme, the *Coronation of the Virgin*.[207] The French translator's words describe her as "leaning on the shoulder of her dear Son." A literal interpretation of the statement would have meant placing the Virgin behind the mantled Christ and into the confines of the throne itself; but such a visual presentation would surely be suggesting that the Trinity was composed of more than three beings.

The "overpowering scents of violets, roses and fleurs-de-lis" are symbolically conveyed by the presence of the lily which the Virgin holds. The reader will recall the flower's symbolism of purity and virginity displayed in the earlier miniature of *Sapientia with Lilies and the Disciple*.

[205] (Fol. 63) "Encores te couvient regarder commet les apostres et les principaulz amis de Dieu sont honnorablement assiz sur les sieges du jugement et comment ilz ont souveraine puissance pour jugier et pour rendre sentence diffinitive; et commet les martirs sont clers, enluminez de couleur vermeille, les confesseurs gettent rays flamboians come feu; et les vierges sont nettes et blanches comme la neige. Et par ceste maniere peus tu veoir et sçavoir comment toute (*MS.* toute toute) la court du ciel est habondant, reluisant et plaine de la doulceur divine et de toute lyesse."

[206] (Fol. 63v.) "...Pourtant est il vray que quant l'ame vient en gloire et elle voyt et regarde la benoite Trinité qui est simple unité, elle est tant parfaittement esjoÿe et transportee et transformee en Dieu et beatifiee par fruicion et plaisance qui ne se peut dire ne raconter."

[207] The subject is exhaustively treated in P. Verdier, *Le Couronnement de la Vierge*, Montreal, 1980. For some pertinent remarks about miniatures, see p. 12, n. 15 and p. 138, n. 96.

One observes that the portraiture of the Virgin contains yet another symbolic feature, the exposed breasts. In this case the Virgin presses her right nipple between the thumb and forefinger of her left hand. The gesture is seen in the *Virgo lactifera* tradition when its object is to direct the nipple at the Holy Child. But the Child is not present in this *Court of Heaven* in the role of a suckled infant. It seems that the symbolism of the naked breasts in this particular painting is not to be interpreted in a naturalistic function but rather as a gesture of intercession before God. In the context of fourteenth-century manuscript illustrations, Meiss observed that the "Virgin merely displays her uncovered breast, often when accompanied by Christ displaying His wounds, both of them thus demonstrating their power as intercessors before God."[208]

No attempt has been made to represent the hierarchy of angels that Suso mentions. Seraphim cluster behind the throne, and other angels in small groups display musical prowess. At the top left, a celestial trio plays animatedly, one angel on a *viele*[209] and his two companions on pipes. At the top right three others intone the *Sanctus*, while a pair at the lower right join in the same melody. In the foreground, lower left, another angelic musician in a rose-coloured cloak plucks a lute with a plectrum. Opposite him, another plays a wind instrument, but it is almost obscured.

As the viewer beholds the multitude of seated figures in the eight compartments, he is reminded of Suso's *immensa multitudo* and his manner of categorising some of the throng as apostles, martyrs, confessors and virgins. The serried ranks have been arranged and positioned in a traditional fashion. In front of the nimbus and in the company of their attributes, a winged lion and a winged bull, sit St. Mark and St. Luke. Behind them, very much reduced in scale, are their companions St. Matthew and St. John the Divine, also attended by their symbols, an angel and an eagle. On the two front benches placed to the right and left of an oval, one can distinguish St. John again among his brother disciples, Saints Peter, Paul, Andrew, Bartholomew, James and Thomas. The forerunner of Christ, St. John the Baptist, sits among them, his diminutive lamb tucked into the crook of his left arm.

The middle stratum, divided by burnished gold bars into three segments, presents firstly the Doctors of the Church among whom it is not difficult to identify St. Gregory the Great wearing a papal tiara, St. Augustine and St. Ambrose, both mitred. Next to them is a grouping of female saints of whom St. Catherine of Alexandria and St. Agnes with her lamb are readily discernible. Catherine's left hand grasps a broken wheel; she is clad in a green jacket trimmed with ermine, and she alone of all the holy women is crowned with an elegant coronet. The other virgin martyrs wear white undergarments, pink blouses and white and grey cloaks lines with gold. The third compartment is a gathering of men who had been martyred and one can easily recognise St. Lawrence and St. Stephen in the front row.

The far left segment of the lower register gives prominence to St. Francis of Assisi and St. Anthony of Padua who congregate with other beatified monks and friars. The adjoining panel is occupied by the Magdalene, St. Elizabeth and possibly St. Anne. Spencer was inclined to believe that the figure of Elizabeth may well be that of Elizabeth of Hungary, a tribute perhaps to Elsbeth Stagel of Töss, Suso's friend and biographer.[210] Finally, we see the crowned figure of a King of France, doubtless St. Louis IX, in his gold and blue robe of State bespangled with fleurs-de-lis. At his side are an ermine-clad noble along with magnates or lords who achieved sainthood.

The iconographical representation is not just a matter of an ordered multitude attendant on the Trinity, it also displays that abstract, spiritual and sublime quality that is the very prism of Suso's

[208] Meiss, *Late XIV Century*, p. 128. He also refers to a seminal paper on the topic by E. Panofsky, "Imago Pietatis," in *Festschrift für Max J. Friedländer zum 60. Geburtstag*, Leipzig, 1927, pp. 261-308, in particular, p. 292 ff.

[209] See Hutois, *Iconographie musicale*, p. 122.

[210] Cf. Spencer, *Horloge*, p. 291.

vision: Light, that mighty omnipresent force which enables the multitude's bliss to be renewed eternally at the fountain of goodness. In this particular version of the theme of the Court of Heaven, a transfer of symbols is achieved by a highly original panoramic flood of luminescence. Rays emanate from the Trinity's mandorla and strike deep into each of the groups in the eight compartments. The spiritual plane transcend the visual, when almost all the Elect of God turn their faces to the fount of light. The countless images are as if transfixed by the darting beams so that the viewer beholds one ineffable moment of radiant bliss and everlasting peace. He would also recognise the symbolism attaching to the blaze of light as the effusion of the Holy Spirit from the Trinity over the universe in general and the Elect of God in particular.

XXIVA (1) *Absalom Becoming King* Fol. 72
(2) *David Fleeing from Absalom*
(3) *Absalom Slain by Joab*
(4) *Sons of Jacob*
(5) *Isaiah, Jeremiah, Habakkuk and Daniel*

Sapientia counsels the Disciple in chapter XIII of the *Horologium* to look back over the eons of time and reflect on the countless blessed holy men who are tried and tested in adversity. She begins by quoting King David's lament in Latin from Ps. LXXI, 20: *Quantas ostendisti michi tribulationes multas*, which the translator presents along with his rendering. This in turn may be understood in English as "Thou hast made me pass through bitter and deep distress."

Sapientia's references to persons tested by adversity include Joseph: "Joseph who was to be lord of the land of Egypt was sold by his brothers, and was so defamed by a woman that he was put in prison and held there wrongly for a long period."[211]

Sapientia stresses how perfunctory the Disciple's sufferings have been, compared to those of the saints: "If you wish to weigh your travail and your adequacy against the horrible and frightening terrors which your predecessors endured, you will find that all your tribulations are merely a trifle and a mockery. Behold Isaiah, who was nobly adorned with every grace as everyone knows, yet he was sawn and split lengthwise by a log-saw. Jeremiah was sanctified in the womb of his mother, yet he was stoned to death and then thrown into into a fire and burned... Daniel who was so amiable and loved by everyone was put into a pit with lions in the hope that they would strangle him."[212]

After stating the theme of the chapter, namely, the Disciple's need to endure trials and tribulations with the same fortitude as did the saints of old, the writer of the *Déclaration* seems to apologize for the artist's inability to portray all of the worthies. The point that is left unstated is that Suso merely named the holy men and announced their fate. The Dominican did not speak of Absalom at all, yet the expositor supplies from the Bible a detailed narrative of three tribulations David suffered because of Absalom. The commentator next reports facts from the tale of Joseph that are not recorded in Suso or in the picture: "his brothers... pretending to their father that a wild beast had devoured him...", and "...his brothers worshipped him..."

[211] (Fol. 72v.) "Joseph, qui devoit estre sire de toute la terre d'Egipte, fut vendu par ses propres freres et fut tellement diffamé d'une femme qu'il en fut enchartré et detenu longuement en prison et a tort."

[212] (Fol. 72v.) Et pourtant, se tu veulz peser ton labour et ta souffisance contre les hydeuses et horribles angoisses que tes predecesseurs ont enduré, tu trouveras que toutes tes tribulations ne sont que une truffe et derrision. Vecy Ysaie qui fut tant noblement adornez de toutes graces comme chascun scet et toutesfoiz il fu sié et fendu au long d'une sie de fust. Jeremie fut sainctifié ou ventre de sa mere et toutesfoiz il fu lapidé et occis de pierres et puis fut getté en un feu et ars... Daniel qui tant estoit aimable et amé de toutes gens fu mis en une fosse avecques les lyons en esperance que les lyons l'estrangleroient."

The account of the third panel could be seen as based entirely on the miniature, or as a paraphrase of Suso's information about three Old Testament worthies. Whatever the source, the expositor does not report the presence, in the illustration, of Habakkuk bringing food to Daniel.

Viewed through twentieth-century eyes, the artist's first panel is manifestly more than a representation of the abstract idea enunciated by Suso concerning unspecified tribulations that David endured. Absalom is represented as his father's antagonist on three occasions. In the left section of the upper register, Absalom and mounted escort ride up to the gates of Jerusalem where numerous citizens and elders come forward to offer him a crown.[213] On the right of the miniature, David and followers flee through a rear gate, their procession being led by two Levites, the traditional bearers of the Ark of the Covenant. The shrine is aptly named *archa federis Domini* on a scroll hanging beneath it. From an elevation, Semei, so named in the painting, showers stones on David's retinue and shouts abuse: *Egredere, vir sanguinum* "Begone! you man of blood."[214] The third tragic event in David's life is Absalom's death. The pictorial representation contains enough detail to suggest that the Old Testament story was being followed. On the extreme left, Absalom is portrayed with his head caught in the boughs of an oak tree, while his mule stands patiently behind him; his own soldiers ride out of sight, oblivious of his fate. When Joab sees this, he rides up and drives his lance through the body. In the right-hand corner, David sits in his high-backed throne in front of the city walls. A banderole carries the famous cry of anguish: *Absalon, fili mi* "Absalom, my son!"[215]

Although the visual rendering of the theme of distress and adversity follows closely the Biblical passages, the realistic nature of the depiction tends to detract from the precision of Sapientia's references and conflate the idea where no elaboration was intended by Suso. The thirteenth-century *Biblia Pauperum* interprets the lapidation of David by Semei as a prefiguration of Christ, crowned with thorns, being mocked by the Jews.[216]

The visual record concentrates the viewer's attention on the acts of persecution of young Joseph by his brothers rather than on his imbroglio with Potiphar's wife.[217] The persecution is episodic, represented in three phases, all within the same pictorial unit. The youth is cast into a cistern, withdrawn, and offered to passing Ishmaelites. His coat is stained with the blood of a newly slaughtered goat.[218] A fifteenth-century observer, regular, secular or lay, might well see the symbolism of Christ's treatment at the hands of the Jews in the persecution of Joseph by his brothers. For, when he recalled the violence done to Joseph, Stephen reminded the Sanhedrin that "God was with him and rescued him from all his troubles."[219]

The martyrdom of the three holy men in question has been represented once again by an episodic segment without specific narrative from Biblical sources. Suso's words were adequate inspiration. He may well have had in mind the words St. Paul penned about prodigies of faith in his letter to the Hebrews.[220]

The theme of adversity and the manner in which exemplary humans have coped with it and even been punished for their resistance to it receives a sympathetic and evocative treatment. In places the iconography is visually shocking because of the stark and gruesome imagery, and in places it is moving because of the steadfastness of the victims in their resolve and their indifference to their fate.

[213] II Samuel, XV, 10.
[214] II Samuel, XV, 24, 30 and XVI, 7.
[215] II Samuel, XVIII, 9, 14, 33.
[216] See Réau, *Iconographie, Ancien Testament*, p. 279. The best iconographical study I have consulted is by Henrik Cornell, *Biblia Pauperum*, Stockholm, 1925.
[217] Genesis, XXXVII, 12-28.
[218] Réau, *Iconographie, Ancien Testament*, pp. 160-1.
[219] Cf. Acts, VII, 9-10.
[220] Hebrews, XI, 32-8. The presence of Habakkuk providing food and refreshments to the prophet in the lion-pit is not an element of the narrative in Daniel, VI, 16.

XXIVB (1) *Three Youths in the Furnace* Fol. 72v.
(2) *Job on the Dunghill*
(3) *Antiochus Torturing the Maccabees*

Sapientia continues to quote the exemplary holy men who suffered all manner of adversities: "The three Youths beloved of God were cast into a burning furnace full of flames."

As she cites further models of conduct, Sapientia reminds the Disciple of Job, describing him as "exemplary in patience, delivered through God's will into the hands of the Devil."

Sapientia considers the seven Maccabees as "good and worthy men, harshly tortured and killed by their adversaries."[221]

The *Déclaration*'s observations are confined to the focal point in each segment of the triptych, but there is little interpretative colour.

The artist's visual statement, on the other hand, contains much that impacts on the viewer's gaze. The first panel offers the spectacle of two soldiers stirring the embers around Ananias, Misrael and Azarias.[222] Nebuchadnezzar, seated in a high-backed throne, looks on incredulously as they move about, unharmed by the lambent tongues of fire. In the art of the Catacombs, the three men are seen to represent God's Elect, protected from the flames of Hell.[223] The present illustration may be said to continue the iconographic tradition, even though the text of Suso appears to recall their ordeal as a symbol of adversity.

In the second compartment Job exposes his almost naked and ulciferous body to the view of his wife as she approaches the dunghill. He calmly announces: *Dominus dedit, Dominus abstulit. Sit nomen Domini benedictum!* (Job, I, 21) "The Lord gave and the Lord took away. May the name of the Lord be blessed!" She replies in a mocking tone, employing words from Job, II, 9: *Benedic Deo et morere* "Bless God and die." God in a mandorla hears the Devil's condemnatory phrase (Job, I, 9): *Frustra Job timet Deum* "Job fears God to no purpose".

Job may be said to prefigure here Christ's suffering, in the sense that the Devil tried to tempt and demoralise Christ, so the Jews thought to mock and deride Him in adversity. There is no trace in this painting of the symbolic three-pronged fork carried by the wife in the fourteenth-century *Speculum Humanae Salvationis*, whose imagery of iconographic parallels influenced several illuminators in later periods.[224]

When we behold the mayhem caused to the Maccabees in the last panel of the triptych, the principal character is clearly the tormentor, King Antiochus, who watches impassively from his throne. Severed limbs are roasted in a cauldron, the bodies of the living are subjected to quartering, eye-gouging, decapitation and flaying. The torture thus represented could be likened to the graphic depiction of human suffering in the earlier *Hell* scenes of the *Horloge* series, but on this folio, the slaughter is all the more poignant and pitiful because humans are the tormentors of their own kind, just as Herod, Pilate and Caiaphas sat in judgment on Christ and witnessed His moral derision, His physical torture and spiritual buffeting. Antiochus was often viewed as the Antichrist by Church Symbolists.[225]

[221] (Fol. 72v.) "Les trois enfans amis de Dieu furent mis en une fournaise ardant et enflammee. Job qui estoit exemplaire de vraye pacience par la voulenté de Dieu fu mis es mains de l'ennemy... Les Machabeans qui tant furent bons et preudommes furent durement tourmentez et mis a fin par leurs adversaires."

[222] These are the names the painter has placed near each youth. In modern English BIbles they often bear the names Shadrach, Meschach and Abednego.

[223] Cf. Réau, *Iconographie, Ancien Testament*, p. 399.

[224] *Ibid.*, p. 316. The bibliography for the *Speculum* is extensive. An overview of the work's influence in medieval art, including manuscript painting, is in Mâle, *Art Religieux*, pp. 233-46. See also the scholarly study by J. Lutz and P. Perdrizet, *Speculum Humanae Salvationis. Kritische Uebersetzung von Jean Mielot 1448*, 2 vols, Mulhouse, 1907-1909; and the study by E. Breitenbach, *Speculum Humanae Salvationis. Eine typengeschichtliche Untersuchung*, Strasburg, 1930.

[225] Réau, *Iconographie, Ancien Testament*, p. 306.

XXV *Christ Offered the Sop* Fol. 80

The painting is placed at the end of the fourteenth chapter in preparation for the fifteenth one which commences on the next side of the sheet of vellum fol. 80v. The new chapter emphasises the point that the Disciple's conduct must always be inspired by the events of Christ's Passion. Sapientia speaks in the first person in the guise of Christ: "When the Jews, who were and still are Children of Darkness, had nailed me to the Cross, the terrible torture they made me suffer was not enough. Like wild creatures they stayed in front of me at the time I was in death's travail and they mocked me by shouting blasphemies, they shook their heads out of derision, and they tormented me harshly and pitilessly. And without perturbation, while bearing my grief patiently, I said to my Father 'Forgive them for they know not what they do.' Notwithstanding my patience, I, who was and am the Son of God, was hanging between two thieves as the most guilty and most worthy of punishment to their way of thinking. One of the robbers mocked me while the other asked for help. I immediately took pity on him and pardoned his misdeeds and gave him my blessing and promised him the Kingdom of Heaven."[226]

Here the *Déclaration* reasserts the text's demand for the Disciple to be inspired by the pains of Christ's Passion. Even so, there is some ambiguity in the expositor's employment of 'he'. The first 'he' refers to Suso, the second, however, could allude to Suso, the Disciple or the illuminator. To define for the viewer the exact moment of the Passion depicted in the illustration is no of concern to the expositor; it is more important to set the scene, as it were, in the context of the whole narrative about the Disciple.

The miniature depicting the Crucifixion is a sympathetic narrative rendering of the text. It is not, however, a typical fifteenth-century representation of the subject, for, if it were, Golgotha would be densely crowded with figures and almost every event that occurred there would have found visual expression.[227] The *Horloge* painting concentrates on a particular happening, the Offer of the Sop to a parched and enfeebled Christ.

The delicately penned words on the swirling banderoles indicate a sequential account. From left to right across the composition, the viewer reads firstly a short passage from Luke, XXIII, 43: *Hodie mecum eris in paradiso* "Today you shall be with me in Paradise"; then: *Memento mei Domine dum veneris in regnum tuum*, cf. Luke, XXIII, 42 "Remember me when you come into your kingdom, O Lord". The celebrated *Scitio* "I thirst" from John, XIX, 28 is then recorded. There follow words spoken by one criminal to the other: *Neque tu times Deum qui in eadem dampnatione es* cf. Luke, XXIII, 40 "Do you not fear God? You are under the same sentence as He". Christ had previously uttered the compassionate and forgiving words, here placed on His left: *Pater, ignosce illis quia nesciunt quid faciunt* cf. Luke, XXIII, 34 "Father, forgive them, for they know not what they do". In Suso's text the word *dimitte* replaces *ignosce*. The next scroll contains the taunt from the second robber, as reported by Luke, XXIII, 39: *Si filius Dei es, salvum fac temetipsum et nos* "If you be the Son of God, save yourself and us." The last two banderoles are appropriately positioned over the heads of the soldiers and witnesses, since incredulity and ridicule were on their lips: *Si filius Dei est, descendat nunc de cruce et credimus ei* "If He is the Son of

[226] (Fol. 80v.) "Quant les Juifz, qui estoient et sont encores enfans de tenebres, m'eurent atachié a l'arbre de la Croix, l'orrible tourment qu'ilz me firent souffrir ne leur souffisoit pas, mais comme enragiez ilz se tenoient devant moy a l'eure que je traveilloie a la mort, et faisoient leurs moqueries en disant blasphemes de moy, et mouvoient leurs testes par derrision, et sans pitié me tormentoient laide ment. Et sans moy (*MS.* moy moy) troubler, en portant paciemment ma douleur, je disoie a mon Pere: 'Pardonne leur, car ilz ne scevent que ilz font.' Nonobstant ma pacience, je, qui estoie et suis filz de Dieu, fu pendent entre deux larrons comme le plus coulpable et plus digne de punicion a leur semblant. L'un des larrons se moquoit de moy et l'autre demandoit mon ayde, et je le prins tantost a merci et lui pardonné ses meffaiz et lui donné ma grace et si lui promis le royaume du ciel."

[227] Schiller, *Iconography*, II, p. 158.

God, let Him come down from the Cross and we shall believe Him"; *Vath qui destruis templum Dei et in triduo illud reedificas*? "You would pull down God's temple, would you, and rebuild it in three days?" These quotations are from Matthew, XXVII, 42 and 40.

To the onlooker's left, the three Maries weep and mourn. The Virgin has turned her face away from the obscenities around the Cross and makes a gesture of lamentation by crossing her arms over her breast. She is held upright by St. John the Divine. The landscape in the background is uncluttered, Golgotha being at a slightly lower elevation than the towers and ramparts on the left of the picture plane. They may be intended to represent the City of Jerusalem.

Part of the sky has been darkened and the light of the sun and the moon partially obscured, to suggest the impending eclipse. Although here depicted naturalistically, the cosmic symbols have a long association in Art with the Crucifixion. As Schiller records, "The main function of cosmological symbolism is to show that Christ's victory on the Cross embraces the whole world and He is Lord of the Cosmos."[228]

XXVI *Virgin Suckling the Holy Child* — Fol. 83v.

The principal section of chapter XVI is taken up with an exaltation of the Virgin by the Disciple. He commences with her face and works downward over her form, pausing to reflect on the qualities of the major parts. Having reached her chest, he is inspired to observe: "...the breasts with which you suckled your sweet Child, who is King and Lord of every creature, give us greater savour. For through them life is returned to us and from them pours forth the sweetness by which all who languish are cured and restored to good health. And truly, the woman is greatly blessed who said to Our Lord *Beatus venter qui te portavit et ubera que suxisti* "Blessed is the womb which bore you and blessed are the breasts which nourished you." She spoke to Our Lord Jesus Christ in public in front of everyone."[229]

The author of the *Déclaration* has read Suso's anatomical description of the Virgin and recalls the praises heaped, in particular, on the breasts and womb. Suso's quotation from Luke, XI, 27 is repeated *verbatim*. Yet, when he comes to state precisely the content of the painting, the expositor declares it portrays the blessed matron who uttered the words. If one did not have access to the miniature by which to verify his perspective, one might be led to think that no one else is present. It is, indeed, a distinctive oversight by the commentator when he fails to mention the salient theme and dominant portraiture of the Virgin Suckling the Holy Child.

The Virgin, surrounded by a rayed nimbus, and wearing a small coronet, is seated on a wide Gothic throne, offering her naked left breast to the Holy Child and pressing milk from the right one. He is clothed and sitting upright on her lap. The kneeling Disciple in the right foreground intones the *Ave Maria* whose opening words are in a scroll. On the left, a worshipping matron repeats the celebrated phrase already mentioned and it is contained in another scroll. Both banderoles twist, fern-like, to form a heart that frames the Virgin and Child. The matron wears a habit with cloak; around her neck is a wimple and on her head is the hood of widows, as Joan

[228] *Ibid.*, II, p. 109.

[229] (Fol. 84v.) "...toutesvoies les mamelles dont tu alaittas ton doulx Enfant qui est roy et seigneur de toute creature nous donnent plus grant saveur, car par elles nous est la vie rendue et d'elles vient la doulceur par qui tous langoureux sont gueris et mis en bonne santé. Et en bonne foy la femme est bieneureuse de laquelle il fut dit a Nostre Seigneur *Beatus venter qui te portavit et ubera que suxisti*, 'Bieneureux est le ventre qui t'a porté et bieneureuses sont les mamelles qui t'ont alaittié,' dist elle, a Nostre Seigneur Jhesucrist en commun devant chascun." The Latin quotation is from Luke, XI, 27.

Evans terms a similar head-covering.[230] Spencer calls the Rolin Master's figure "an abbess who kneels beside the Disciple to adore the Virgin and Child and who utters the words of Elizabeth (Luke 11:27), "Blessed is the womb..."[231] There are two points of contention: the first is that Luke does not name the woman about whom he speaks. Secondly, an important attribute of an abbess in miniatures is a crozier,[232] but it is not represented in the *Horloge* illustration.

The textual theme of the exaltation and adoration of the Virgin through all her physical attributes has been pictorially narrowed to focus on the tender humanity of her nurture of the Holy Child. A male and a female representative of mankind extol the virtues of the Virgin, he by the recitation of the *Ave Maria*, and she with the glorious words reported by the Evangelist. That the male figure is the Disciple is merely of moment for the textual narrative. What is more striking in the presentation of the theme is the garb of the Disciple. The beholder is led to register a higher level of symbolism. In the presence of the Holy Child, the humble Preaching Friar warmly and adoringly intones the *Ave Maria* before the sublime imagery of the Virgin's own great act of self effacement and humanity.[233]

XXVII (1) Stabat Mater (2) *Deposition* (3) *Christ Appearing to His Mother* — Fol. 87

At this place in the text much of the narrative is supplied allegedly by the Virgin. It includes the record of her sorrows: "...as I was lamenting and grievously sorrowing from the heart, my Son forgot His own torment, and, as if He had my sorrow in His heart more than His own, He began to console and comfort me as His mother, saying: Mother, I am dying and I depart. I take leave of you and I commend you to John, who is my brother, cousin and friend."[234]

As one reads further in the chapter, the Virgin's account of the Passion remains the focus of attention. "Finally, when my Son had rendered and commended His soul to God the Father, and when His body was taken down from the Cross, I held His arms lovingly and embraced Him with a mother's love. I washed His face with my tears and wiped His wounds by dint of kissing Him."[235]

Further on, the Virgin is still speaking: "...when He was buried, I was led back home in great sorrow and suffering. But when it came to the day of His blessed and glorious Resurrection, I saw Him resplendent and so truly risen that all my grief vanished and was completely forgotten, and my joy was increased a hundred thousand fold. My joy was founded in His blessed vision. I then had total solace, being entirely transported into spiritual contemplation."[236]

[230] Cf. Evans, *Dress*, p. 68 and pl. 74. The illustration of the habit is from London, British Library, MS. Additional 39843, a copy of the *Sainte Abbaye*. A different reproduction of similar clothing from the same MS. (fol. lv.) can be studied in Sandler, *Pucelle's Belleville Breviary*, p. 89.

[231] Spencer, *Horloge*, p. 291.

[232] A Benedictine Abbess and an Augustinian one, each with crozier, are portrayed *c.* 1440 by the Cleves Master; see Plummer, *Catherine of Cleves*, pls. 153 and 154.

[233] A detailed study of the theme's appearance in medieval European Art was published by M. Meiss, "The Madonna of Humility," *Art Bulletin*, XVIII (1936), pp. 434-64. An epitome is presented in his *Late XIV Century*, pp. 126-8. See also Réau, *Iconographie, Nouveau Testament*, pp. 96-8, and Schiller, *Iconography*, I, pp. 47-8.

[234] (Fol. 88) "Si comme je me complaignoie [et] en griefve douleur de cuer me garmentoie, lors mon chier filz oublia sa propre douleur et come s'il eust eu plus au cuer mon tourment que le sien, il me commença a consoler et conforter comme sa mere en disant: 'Mere, je meurs et m'en vois. Je pren congié de toy et te recommande a Jehan mon frere, mon cousin et mon amy..."

[235] (Fol. 88v.) "Finablement, quant mon filz ot l'ame a Dieu son pere rendue et commandee, et que son corpz fut despendu de la Croix, je soustenoie ses bras par grant amour et l'ambraçoie par amour de mere; je lui lavoye la face de mes lermes et par force de lui baisier, ses plaies lui torchoie..."

[236] (Fol. 88v.) "...quant il fu enseveli et je fu remenee en mon hostel a grant paine et a grant travail. Maiz quant vint au jour de sa benoite et glorieuse resurrection, que je le vy si glorieux et si veritablement ressuscité, toute ma douleur fu passee et tout entie[re]ment oubliee et ma joie fu cent mille fois doublee. Ma liesse fut fondee en sa benoite vision. Lors (*MS.* lors lors) eus je parfaitte consolacion et fus toute transportee en esp[i]rituelle contemplacion."

In one sense the *Déclaration*'s statements about the triptych are brief and to the point, and therefore iconographically weak. No interest is shown in the fluttering banderoles or the distinctive arrow that pierces the Virgin's heart. The icon that is displayed in the last panel is not thought to be in conflict with the Gospel Canon. The expositor's concerns lie, true to much that he has expressed already, with the general spiritual and theological movement of the narrative and with the extent of pictorial conformity.

The modern art historian is free to raise broader issues as he surveys the iconographic panorama of the triptych. The first compartment is devoted to the text's *Stabat Mater* theme. The name is derived by Suso from the opening words of a Lament attributed to a thirteenth-century Franciscan poet, Jacopone da Todi, in which he alludes to the Virgin's station at the foot of the Cross.[237] In Western Art before the end of the thirteenth century, the event had been mainly represented by the portraiture of the Virgin standing on one side of the Cross's vertical shaft, facing St. John the Divine, positioned on the other side. Early fourteenth-century icons of the event show a displacement of the balanced grouping and the positioning of St. John next to the Virgin on her side of the upright.[238] The iconographical tradition has been observed in the *Horloge* painting, to the point of the exclusion of other motifs such as the two crucified thieves, the mocking Jews and Roman soldiers. The effect is to concentrate all attention on Christ and the Virgin. While the centurion Longinus holds the lance firmly implanted in Christ's side, she stands beside the upright, sorrowing deeply. John also is smitten by grief and pain as he attentively puts his arms around the Virgin in a realistic interpretation of the emotional mood. The wording on the two scrolls on the viewer's left explains the reason for this show of tenderness, mingled with sorrow. The dying Christ, whose wounds ooze blood, has just spoken the words recorded in John, XIX, 25-7: *Mulier, ecce filius tuus* "Mother, behold your Son" and *Ecce mater tua* "Behold your Mother."

The Virgin's icon is also noteworthy for the long arrow reaching from Christ's gaping wound to her breast. It symbolically recalls Simeon's prophecy (Luke, II, 35) on the day she presented the Holy Child in the Temple, that a sword of grief would one day transfix her soul: *Tuam animam pertransibit doloris gladius.*[239]

To the viewer's right, that is, Christ's left, flutter three more banderoles recording utterances in the Final Agony. These are from Luke, XXIII, 46: *Pater, in manus tuas commendo spiritum (meum)* "Father, I commend my spirit to your hands"; from Matthew, XXVII, 46: *Hely, hely, lama zabathani* "My God, my God, why hast Thou forsaken me?"; and from John, XIX, 30: *Consummatum est* "It is consummated". God's hand could be said to be raised above the Crucified Christ to denote His recognition of His Son's debasement, His reception of the Sacrifice, and Christ's elevation to the position of God.[240]

The scene in the centre panel of the triptych really illustrates the words "when His body was taken down from the Cross," the Lamentation being unrepresented. The realistic *Deposition* is replete with iconographical detail that matches the sorrowful occasion. The incidents that unfold explain why two ladders lean against the arms of the Cross.[241] On the viewer's left, Joseph of Arimathea stands half way up one of them and holds a linen sheet under Christ who is bent double as He is lowered by Nicodemus. Nicodemus has left the other ladder and stands behind a servant who levers out gently from the upright the large nail piercing Christ's feet. Already the Magdalene acts according to tradition and clasps His gory feet in readiness to kiss them. Meanwhile, the Virgin

[237] See Réau, *Iconographie, Nouveau Testament*, p. 499.
[238] *Ibid.*
[239] See Réau, *Iconographie, Nouveau Testament*, p. 108.
[240] This interpretation is given for the hand of God, not the Almighty in person, by Schiller, *Iconography*, II, p. 107. The context for her remarks is Carolingian and Ottonian icons.
[241] Coincidentally, the events are very similar to the standard typology of the theme, as explained in Réau, *Iconographie, Nouveau Testament*, p. 515.

lovingly presses His hand to her cheek and raises a cloth, preparing to wipe His lifeless face. St. John the Divine, glimpsed on the left, holds solemnly the Crown of Thorns. The visual representation of the *Deposition* in the *Horloge* belongs to the archaic group in which the crowded Crucifixion imagery is absent.[242]

The third and last panel of the miniature contains all three textual elements, the Resurrection, the Virgin's vision of her Son and her own resultant joy. The second of these themes is the dominant one, the concept of joy is conveyed by the Virgin's physiognomy and by the wording on the scrolls. Christ's *Gaude, mater virgo* "Rejoice Virgin Mother" causes her to exclaim: *Gloria tibi, Domine, qui natus es de virgine* "Glory be to Thee, O Lord, who art born of a virgin."[243] The chapel at whose entrance the Virgin experiences the vision is a dove-white, and one may hypothesise that the whitish architectonic motifs symbolise the Virgin's purity and constitute an allusion to her as Ecclesia. The Resurrection is recalled by an empty sarcophagus, its lid half drawn back, with an angel kneeling with his back to the viewer at one end, and a second angel standing with arms crossed on his breast. Christ is portrayed with the winding sheet, gathered loosely about His loins, exposing in places the wounds, as He reaches out, as if to embrace the Virgin. His left hand steadies the Resurrectional banner.

The subject matter of *Christ Appearing to His Mother* is apocryphal, in the sense that it is not substantiated in the Gospel Canon of the Western Church. The apparition was made popular in Europe by an anonymous thirteenth-century author known as the Pseudo-Bonaventura. His *Meditationes Vitae Christi* which included the story was soon translated into the major languages of the Continent, and the theme enjoyed a vogue in Western Christian Art in the Late Middle Ages, as James D. Breckenridge has shown in his study of the subject.[244]

XXVIII (1) *School of Theology* Fol. 90v.
(2) *Seven Liberal Arts*

The reader has now reached the first chapter of the second Book in which both spiritual and worldly learning comes under close scrutiny. Questing after knowledge about true wisdom, the Disciple "...went from *studium* to *studium* seeking purposefully what he wanted. Once it happened that he thought he was shown in a vision a sphere, that is a wheel, quite round, of fine gold, and very large and very beautiful. It was studded and garnished with fine precious stones. Within lived all the masters and doctors of the seven sciences called the seven arts, and with them was a large number of students. In the habitat there were two abodes and in each lived certain doctors of certain sciences with their students. In the lower one were old masters of all kinds of trades and a large number of scientists. There were astronomers who busied themselves with astrolabes; other scientists studied the nature of the things of creation; geometricians measured and calculated things needing computing and quantifying; musicians sang in fine tune and in great harmony. And each of the others studies his art and his science; for instance, the physicians studied medicine, the blacksmiths studied the art of the forge. And it seemed that each had a veil or towel over his eyes. Because they worked at a time of great heat, they had a sweet and pleasant tasting beverage which kept them cool and which refreshed them. They were revived and relieved by it, but not completely:

[242] The history of the typology is set out in Schiller, *Iconography*, II, pp. 164-8.

[243] I have not been able to determine a source for these utterances.

[244] See James D. Breckenridge, "*Et prima vidit*: The Iconography of the Appearance of Christ to His Mother," *Art Bulletin*, XXXIX (1957), pp. 9-32. The theme is also included for mention in the repertoire of Réau, *Iconographie, Nouveau Testament*, pp. 554-6.

their thirst was not entirely slaked, rather the said drink engendered a greater thirst afterwards than before."[245]

Having tasted the potion in the *studium* of the Seven Liberal Arts, the Disciple was revolted and nauseated by it, so he left and "went off to the place above. It was very beautiful and grandiose, decorated in many colours and with numerous icons and paintings. When he reached the door, he saw written above it "Here is the School of Truth about Theology; its mistress is called Eternal Sapientia; its teaching has the name of Truth; the goal of its study and toil is known as Everlasting Glory." As soon as he had read and understood the wording, the Disciple immediately rushed inside and thought he would become a pupil so that he could reach the goal he had so long sought. In the School there were three kinds of students and three sorts of doctors, and others sat on the floor close to the door and had no taste or liking for what they heard, but they looked always at the door and wasted time. The mistress, Sapientia, poured sweet and honeyed words into their mouths but they were quickly changed into bile and bitterness.[246]... In their presence and right in front of them were devils blacker than Nubians who jumped all over the place, making ready and offering the students *cathedras* and thrones of honour, or displaying gold and silver in great quantities, along with great landed wealth.[247]... Then the disciple looked around and saw a wondrous thing worthy of great mockery: a silver ball was in the air on high and when it landed, it began to roll between the students, and because of its great beauty each was pleased to behold it. Each one liked it and wanted it because it was of such a nature that it bestowed honour and glory on its holder..."[248]

The *Déclaration*'s exposition of the diptych is structurally in two parts, a rehearsal of Suso's narrative in so far as it is echoed in the two registers, then, a summary of them both, stressing the morality and religious aims they contain. The reader or viewer receives no insight into the Disciple's participation in the teachings. The name over the School of Theology, reproduced from Suso by the artist, is not mentioned, although the expositor alludes to the subject matter of the inscription.

[245] (Fol. 91) Et si comme il s'en aloit de estude en estude enquerant partout diligemment ce que il desiroit. Une foiz il advint que il lui sembloit que une espere, c'est a dire une roe toute ronde, de fin or, lui fut monstree en vision mont (*sic*) belle et monlt (*sic*) grande, estoffee et garnie de belles pierres precieuses, en laquelle estoient et habitoient tous les maistres et les docteurs des sept sciences que on appelle les sept ars; et avoient avec eulx grant foison d'escoliers. En celle habitacion avoit deux demourances et en chascune demouroient certains docteurs de certaines sciences avec leurs escolliers. En l'abitacion de dessoubz estoient les anciens maistres de toutes manieres de mestiers et grant foison de philozophes la estoient: les astrologiens qui se occupoient en l'astralabre; les autres estudioient en la nature des choses cree[e]s, les geometriens mesuroient et compassoient les choses qui sont a mesurer et a compasser; les musiciens chantoient par grant melodie et par grant proporcion. Et chascun des autres se occupoit en son art, en sa science, c'est assavoir les phisiciens en medecine, les fevres en favrerie. Et sembloit que chascun eust un voile ou une touaille sur les yeulz. Pourtant qu'ilz labouroient ou tempz de la grant chaleur ilz avoient un beuvrage doulz et plaisant qui les refroidoit et rafreschissoit dont ilz estoient recreez et soulaciez mais non pas plainement, car leur soif ne estoit mie du tout estainte, ainçois ledit beuvrage leur engendroit plus grant soif aprés que devant.

[246] (Fol. 91) ...et s'en ala en celle de dessus moult belle et monlt noble, painte de diverses couleurs et de diverses ymages et hystoires. Quant il vint a l'uys de l'ostel il vit en escript sur la porte "Vecy l'escolle de la verité de theologie dont la maistresse est nommee Sapience pardurable et la doctrine est nommee Verité; la fin de ceste estude (fol. 91v.) et labeur est nommee Gloire pardurable." Si tost comme il ot leu et entendu ceste escripture il courut tantost et entra en l'escolle et pensa que il en seroit escollier pourtant que il peust venir a la fin que il avoit si longuement desiree. En celle escolle ilz estoient trois manieres d'estudians et trois manieres de docteurs et les autres seoient a terre tous pres de la porte et n'avoient goust ne saveur en chose qu'ilz ouÿssent maiz regardoient tousjours par devers la porte en perdant leur tempz. La maistresse, dame Sapience, leur couloit en la bouche parolles doulces et pleines de miel, mais elles estoient tantost converties en fiel et en amertume.

[247] (Fol. 91v.) En leur presence et tout en leur face estoient dyables plus noirs que nulz Ethyopes qui saultoient et menoient grant feste et leur presentoient et appareilloient chaieres et sieges d'onneur, les autres leur monstroient or et argent a foison et grans possessions par leurs seigneuries terriennes.

[248] (Fol. 92) Lors le disciple getta ses yeulz et vit en celle escolle une chose merveilleuse et digne de grant risee, car il vit venir une pelote de hault qui estoit d'argent, et quant elle vint a terre, elle commença a courir et a rouler entre les escoliers, et pour sa grant beaulté chascun regardoit a elle par grant plaisir. Chascun l'amoit et la desiroit a avoir, car elle estoit de telle nature que elle donnoit gloire et honneur a celui qui l'avoit.

He spells out the scholastic concerns of the three kinds of students and how their zeal for study waxes and wanes according to their perceptions of the spiritual or worldly rewards. Even the role of the devils is characterised in harmony with the illustrator's details. However, the employment of the terms 'right' and 'left' does not conform to the modern viewpoint when subjects are being located in the painting. This usage was mentioned above in chapter II, § 3.

The *studium* of the Seven Liberal Arts is said by Suso to hold a large number of students who are taught by scientists and skilled tradesmen. The expositor has little time for the activities of this School, where study is undertaken to gain "riches or renown or honour." These Arts only research earthly things and seven ladies serve as teachers. The commentator speaks of the painting here and not of Suso's description, wherein the implied gender of the mentors is masculine.

The visual statement by the artist abandons Suso's account of the wheel of knowledge,[249] and presents a two-tiered hierarchical tableau with the activities in the *studium* of the Liberal Arts subordinated, spatially speaking, to those of the School of Theology. The upper register's arched frame contains the words *L'Escole de theologie dont le maistre est Sapience, la doctrine verité et la fin gloire pardurable* "The School of Theology whose teacher is Sapientia, the doctrine is Truth and the goal is Everlasting Glory". The lower compartment's inscription reads *L'Estude des sept ars liberales* "The Studium of the Seven Liberal Arts".

The artisans, blacksmiths and physicians are not included in the visual programme for the Liberal Arts, prominence being given instead to seven women clad in fashionable gowns. In their presence are four students whose number is probably a token one. The imposing centre-piece is a hexagonally-shaped book-stand.

Spencer's account of the venue and the figures' activities is apposite and worth quoting *in extenso*. "*Logique* controls her snakes in the manner of a Minoan goddess; *Musique* listens to her notes; *Rhétorique* writes like a young evangelist; *Arithmétique* works at a slate on which numbers are half seen; *Astrologie* examines an astrolabe; *Grammaire* teaches the children to read; *Géométrie* is ready to measure with three tools. The emphasis given to the mathematicians in the foreground invites the thought that Nicolas of Cusa's figure for Wisdom, "the intellectual image... of the divine and infinite form which is God," was the infinite straight line, the *actus infinitus*, the form of all possible figures. The room, lighted by a scholar's lamp, is open to a sky blazing with stars, sun and moon, inscribed with the names of six figures of the Zodiac, Aries to Virgo. Their position points to early June, a period midway between Gemini and Cancer associated in mediaeval calendars with the planet Mercury. It is indeed a pity that the stars form no constellation by which the occasion may be dated more closely. But Mercury is the planet associated with the second age of Man, the time of learning, the schoolboy, and the protégés of Mercury are those who paint, carve or write, men of the arts among which has been counted clock making."[250] In a gloss to Spencer's description, P. Verdier links the scholar's lamp of the painting to Mercury. "Il faut voir dans ce lampadaire Mercure, la planète du dieu protecteur des arts qui prétend à la main de Philologie."[251]

A point seemingly overlooked by Spencer and Verdier, when viewing the *Horloge* allegories, is the 'modern' dress imagery. Given the long iconographic tradition of the theme, there is every reason to expect that the seven women would be portrayed as they were in earlier forms. Instead, their outward appearances have been modernised, suggesting thereby that their activities could be topical. The beholder is, moreover, invited to contemplate the zodiacal signs in the heavens and to relate their presence to the context of the planetary symbolism of Mercury, itself representing the time of leaning.

[249] By a curious coincidence, one of the most important configurations of the Seven Liberal Arts in medieval iconography, preserved in the twelfth-century *Hortus Deliciarum*, of Alsatian origin, presents the seven women in medallions within a wheel.

[250] Spencer, *Horloge*, pp. 289-90, and pl. 30.

[251] Cf. Verdier, *Arts Libéraux*, p. 334.

The format for the School of Theology is equally realistic and singularly dramatic. The eye is drawn instinctively to the top of the picture frame to read the information about the studies being conducted inside the auditorium. The teacher, Sapientia, is clad in doctoral robes, as befits the head of a school of Theology. She retains three elements of her standard portraiture in the Brussels *Horloge*, the cruciform halo, the open book and the rays of enlightenment, here placed as if to emanate from her *cathedra*. The field of *rayonnement* of her intellect is pointedly delineated. The rays only strike the benches where the dedicated students are, and fall short of the right-hand group, who are distracted and tempted by the worldly objects shown to them by devils, and by the bouncing silver ball on the tiled floor. There are the students whom Spencer categorises as "the ambitious worldly ones, greedy for honours and appointments, wanting only to be rich, well dressed and comfortable."[252] One of the devils is seen in the act of nailing a piece of richly designed tapestry to the back of a bench, presumably to make it more comfortable.

The illustration can be viewed in another light by regular or lay. The School of Theology evokes the aura of the Paris *Studium Generale* of the Order of Preaching Friars,[253] whose standard of scholarship and commitment to learning was renowned throughout Europe in the thirteenth, fourteenth and fifteenth centuries. One of its greatest teachers and scholars was St. Thomas Aquinas (1225-1274). The bouncing ball, an image traditionally linked to his name,[254] recalls his presence there. The tableau could be said to be infused with the spirit of theological learning, so cherished by the Dominicans.

XXIX (1) *Death-Bed Scene with Death Personified* Fol. 95v.
(2) *Battle for the Soul*

The second chapter in Book II treats the subject of the Art of Dying Well, and part of the Disciple's instruction in this art includes Sapientia's purposeful remark: "Now think about and recall to mind a man who is in bed at the point of death, and act as if he were talking to you at that instant of demise.' When the Disciple heard this example quoted, he began to withdraw his heart and mind from worldly things, and his thoughts turned inwards, and he began to consider most earnestly the tableau of a man who would be dying in the manner suggested and devised by Dame Sapientia. Then he had a vision that in front of him was a very handsome youth who had been overtaken by a fatal sickness and who was about to die unexpectedly and immediately. There was no order made for his salvation. He lamented most piteously, saying: 'The weeping and sorrowing of death have assailed and surrounded me and the terrors of Hell have attacked me..."[255] The young man again utters a soulful cry, as the text relates: "I hear the voice of Death saying: You are the son of Death. Neither wealth nor logic nor earthly friends can release you from my hands. Your end has come, here is your end."[256]

[252] Spencer, *Horloge*, p. 290.
[253] Künzle, *Horologium*, p. 49.
[254] See Mandonnet, *Order of Preachers*, p. 361.
[255] (Fol. 95v.) "Ores donques pense et te souviengne d'un homme qui est au lit et a l'eure de la mort, et fay tout ainsi comme s'il parlast a toy tout sur le point de mourir." Quant le disciple ouÿ cest exemple, il print a soubztraire son cuer et son entendement de toutes choses mondaines et reprinst sa memoire en soy mesmes, et prist a considerer moult diligemment la semblance de l'omme qui vouloit mourir en la maniere que dame Sapience l'avoit dit et devisé. Lors lui vint une vision que il avoit devant lui ung tres bel jouvencel qui estoit seurpris du mal de la mort et le couvenoit mourir soudainement et tantost, et si n'avoit quelconque (*MS.* quelconques) ordonnance faicte pour son sauvement. Et se complaignoit monlt (*sic*) piteusement en disant: "Les pleurs et les douleurs de la mort m'ont assailli et environné et la paine d'enfer m'a fait assault."
[256] (Fol. 96) Je oy la voix de la mort qui me dit: 'Tu es filz de mort! Richesses ne raison ne amis charnelz ne te pevent delivrer de ma main. Ta fin est venue. Venue est ta fin.'

The *Battle for the Soul* appears to be a pictorial elaboration of a short utterance by the Disciple to the moribund youth: "it is true that daily all of us shall go from life to death, just as water flows always downward and never returns; yet God does not want a soul to be lost, He draws it up to Him, in as much as He knows that our frailty cannot be redressed towards good deeds without His help..."[257]

While correctly stating the subject matter of Suso's chapter, the author of the *Déclaration* speaks of the death-bed scenes in broad terms only. Sapientia presents to the Disciple the idea of a man who is on the verge of death and whom the Disciple beholds in a vision as a "...very handsome youth..." The expositor appears disinterested in such a specific characterisation, and refers merely to "a person who has led a life filled with joy, pleasure and carnal delights..." He declares that "...the wretch only thinks of what he loves, his body or his wealth..." The mood of Suso's narrative and of the illumination is one of remorse and contriteness, an image reinforced by the inscriptions on the banderoles, but seemingly unnoticed by the expositor.

The subject matter of the lower register receives much more attention, with a sensitive analysis of the symbolism of the bridge and combat for the soul, due emphasis being placed on the Art of Dying Well. The rhetorical question "Who is there who has lived in such a way..." causes the listener to believe he hears the words of a preacher.

There is close affinity between Suso's narrative and the visual rendering. It is achieved through the deployment of scrolls to convey the dramatic utterances, and through a faithful adherence to specific details of the Disciple's vision. Moreover, the theme of the Art of Dying Well is interpreted in harmony with traditional and popular conceptions of the Late Middle Ages.[258]

A youth is in his own bed-chamber and on the point of death. One banderole carries the Latin of Ps. CXIV, 3: *Circumdederunt me dolores mortis et pericula inferni invenerunt me* "The sorrows of death have assailed me and the terrors of Hell have attacked me"; yet another scroll near the young man proclaims the lamentations from Job, XIX, 21: *Miseremini mei, (miseremini mei), saltem vos amici mei* "Pity me, pity me, you who are my friends." The faces of those attending him are impassive, their feelings of concern are registered by the tilt of the head or the movement of the hand. The grieving woman, no doubt his mother, shrouded in a dark grey mantle, wrings her hands in anguish. To her right is a middle-aged man, probably the father, clothed in red; he places his right hand over his heart as if to reduce the pain of his impending bereavement. Between these two figures peers a male face, while to their left, stand two younger women, the shorter of whom raises her hands, slightly open and level with her chest, in a gesture of anguished futility. Her face is turned towards Death, but still her expression is blank. Death is portrayed as a desiccated cadaver of decay and putrefaction, half-draped in a shroud and brandishing a javelin high over the bed. The banderole nearby registers the doleful message: *Filius mortis es. Finis venit, venit finis* "Thou art the son of Death. The end is nigh, the end is nigh", the first three words of the Latin are from I Sam. XX, 31 and the others from Ezechiel, VII, 6. Behind Death, against a side wall, is a small domestic altar with a free-standing crucifix. The Disciple kneels in the left foreground, saying *Omnes morimur et quasi aque in terram dilabuntur* "We shall all die, we shall be like water spilt on the ground". The passage is from II Samuel XIV, 14, where the last word is given as *dilabimur*.

[257] (Fol. 97) "C'est chose vraye que nous yrons tous de vie a mort de jour en jour, si comme l'eaue qui court tousjours aval et ne retourne oncques amont, mais nonpourtant Dieu ne veult point que l'ame soit perie, mais la trait a lui pourtant que il scet que nostre fragilité ne se peut adrecier a bien faire sans son ayde."

[258] The theme is extensively discussed in Sister Mary C. O'Connor, *The Art of Dying Well*, New York, 1942, pp. 48-60. Also appropriate, since they give wider dimension to the subject, are Mâle, *Art Religieux*, pp. 382-88; A. Tenenti, *La Vie et la Mort à travers l'art du XVe siècle*, Paris, 1952, pp. 48-68, 97-120; H. Zerner, "L'Art au Morier," *Revue de l'Art*, XI (1971), pp. 7-30, and T. S. R. Boase, *Death in the Middle Ages*, London, 1972, pp. 119-26.

The focus of the drama in the lower panel of the miniature is of a very different nature. Enticed by volumes of darkness and light, as it were, the eye passes from the doom of the condemned at the bottom of the tableau to the radiance of the righteous in the upper part of the painting.

The small naked figure on the bridge that extends over the boiling Stygian waters represents the soul of the dead youth striving for eternity in Paradise. It looks very much like an enclosed garden on earth, where the Blessed, already beyond the cares of mortals, watch the soul's struggle to join their midst. The demon which assails the stretched figure's ankles and the angel holding its wrists leave one in no doubt about the subject matter of the drama.

The torments suffered by the damned are reflected in their faces: with mouths agape, they scream piteously. A writhing lacertilian form seems to be entering the rump of one of the damned, his fate perhaps being to change into the shape of his malevolent torturers.

The universal theme of the Soul's Immortality and the Bridge of Death are strikingly represented in this almost literal rendering of Suso's narrative. The fate of the soul and the battle for possession of it by the forces of Light and Darkness was the subject of depiction in Books of Hours.[259] Numerous images exist that represent the Battle for the Soul in the most literal terms. More often than not, St. Michael's combat with the Dragon or Devil is presented to the viewer to illustrate the theme.[260] On this occasion, however, it is merely a guardian angel on high who tries to rescue the soul from the clutches of a demon.

The Bridge of Death[261] is rarely presented concurrently in the paintings. The *Horloge* tableau contains one such bridge, not mentioned by Suso, and what is of the greatest interest, is its functionality. The soul has just stepped off it, after leaving the graveyard. The Bridge becomes the venue of the Battle itself, but it remains a symbol of the passageway leading from the Earthly Realm to the Divine Kingdom.

XXX St. Arsenius — Fol. 104v.

The painting is placed in the middle of Book II's third chapter in which Sapientia instructs the Disciple on the art of living a good Christian life. One doctrine she espouses is that of the hermit Arsenius: "Dear son, here is the path of salvation which your Arsenius, of whom you spoke at length earlier, followed all his life and made his adherents do the same. And he used to say to them: Shun people, be silent and keep your peace. These are the three tenets of salvation."[262]

The Disciple later had a revelation in which he saw "a very handsome youth approach him, take him by the hand and lead him to a church where there were many sanctuaries. In one a lone man lived out a holy life; he was already very old and hoary, and he wore a long beard. It seemed that he was endowed and filled with God's grace. Next to his small abode stood a ladder which the young man ascended and descended as if for pleasure. Then the youth called the Disciple over and said: Come hither and listen to the lesson I shall read out..."[263]

[259] See Male, *Art Religieux*, p. 388; and Boase, *Death in the Middle Ages*, pp. 39-56.

[260] Cf. Réau, *Iconographie, Ancien Testament*, pp. 44-51.

[261] See G. A. F. Knight, "Bridge," in J. Hastings (ed.), *Encyclopaedia of Religion and Ethics*, London, 1909, II, pp. 848-57. Texts, but rarely art work, that include the theme, were studied by P. Dinzelbacher, *Die Jenseitsbrücke im Mittelalter*, Vienna, 1973.

[262] (Fol. 104) "Beau filz, veci la voie de salut que ton Arseny, duquel tu as tant parlé cy devant, garda toute sa vie et fist garder a ses disciples. Et leur disoit: 'Fuy toy, tais toy et te repose!' Ce sont les trois principes de sauvement."

[263] (Fol. 104) ...et vit par une maniere de revelacion un tres bel jouvencel qui venoit vers lui, lequel le print par la main et le mena en une eglise en laquelle il avoit grant foison de petites chambres entre lesquelles en y avoit une ou demouroit un homme tout seul qui menoit moult sainte vie et estoit ja ancien et chanu et avoit une longue barbe, et sembloit bien qu'il feust doué et rempli de la grace de Dieu. Decoste sa maisonnette avoit une eschielle dreciee en hault, et lui sembloit que ledit jouvencel montoit et avaloit celle eschielle si comme par esbatement. Adoncques le jouvencel appella le disciple et lui dist (fol. 104v.) "Vien cy et escoute la leçon que je vueil lire..."

The commentator of the *Déclaration* sets the picture in the context of the chapter, describing how in the chapter preceding, Sapientia had instructed the Disciple in the Art of Dying Well, and in this new chapter her advice concerns the Art of Living an Unblemished Life. The model life style, we are told by both Suso and the expositor, is that of Arsenius whose three tenets constitute perfection itself. When the commentator comes to discuss the actual miniature, he thinks its contents are worthy of one sentence only.

To the left of the painting a chapel is placed, and at its portal the seated Arsenius can be seen deep in thought and contemplating the book of instruction resting on his lap. The handsome youth of the text has been interpreted sympathetically as a boyish angel. He is presented standing halfway up a ladder, facing outwards in the direction of the Disciple who, in a sitting posture, with his chin cupped in his right hand, listens attentively to a lesson read to him from the book in the angel's hands. The Latin wording of the three tenets is placed in the order *Fuge homines* at the bottom rung, *Tace* halfway up, and *Quiesce* at the top. God the Father in Heaven is surrounded by a nimbus of clouds and a cherubic host; he turns His face to Arsenius whilst blessing the saint with His right hand and carrying an orb in His left.

Although Suso's narrative suggests that the youth moves up and down the prominently placed ladder, the scene presents a totally static atmosphere and reflective mood. Meditation is the underlying theme. The three guiding principles, noticeably emblazoned on the ladder rungs, invite not only the attention of the viewer within the frame, namely the Disciple, but also the beholder who lights upon the miniature on the manuscript's folio. The choice of the griseous tones augments the reflective mood, as do the two seated figures on either side of the point of focus, and the beatific angel with curly locks of hair in the act of reading from an open book. The elements of concentration thus reinforce each other. A touch of whimsy is contributed on the one hand by the intrusive quality of the Disciple's right shoe, seen peeping out from under his white cassock, and on the other, by the irresolute scale of the chapel and its wandering vanishing points.

XXXI (1) *The Last Supper* Fol. 105v.
(2) *Elevation of the Host*
(3) *Tantalus*

In chapter IV of the Second Book of the *Horloge*, Sapientia prescribes the Seven Sacraments for the Disciple when he feels he is in adversity: "But in truth the main one is the Sacrament of the Altar, for, from it above all, issued forth the ray and foundation of divine love."[264] Its most important feature, she maintains, is the presence of Christ. Information about the Mass is imparted to the Disciple by both Sapientia and Christ. At one juncture He says, "For this reason, I gave myself to my disciples the evening of my Last Supper..."[265]

Sapientia next instructs the Disciple in the celebration of the Eucharist, but the moment of the Elevation is not singled out for special mention.

The third panel presents a version of the legend of Tantalus. Suso's reference to him is fleeting. Towards the end of the chapter, after the Disciple has listened to Sapientia's sustained discourse on the mental and spiritual attitudes and expectancies he should entertain and adopt before attending Mass, he is moved to exclaim "I had a wealth of delights, but I was unhappy and wretched;

[264] (Fol. 105v.) Mais a voir dire le sacrement de l'autel est le principal car de lui principalment est venu et yssu le ray et le fondement de l'amour divine.

[265] (Fol. 105v.) "...Pour ceste cause me donnay je a mes tres amez disciples le soir de ma derreniere cene..."

in a flood of grace, I was desiccated and hardened. I resembled Tantalus, who, among riches, suffered great privation.''[266]

The *Déclaration* offers here yet another example of a carefully worded statement about the illuminations position within a chapter. It is as if the commentator foresees the possibility that the owner of the volume will light upon the painting first and then wish to pass to the textual narrative it illustrates. The expositor echoes Suso by naming the Sacrament of the Altar as the most important of the Seven Sacraments, and by proclaiming Christ as the person who instituted the Sacrament at the Last Supper. But the painter's skill in presenting a high point of the meal elicits no personal interpretation from the expositor. In his eyes, the next panel of the triptych, the Elevation of the Host, is apparently so banal a pictorial statement that it warrants only a single sentence.

As for the Ovidian exemplum of Tantalus, the commentator alludes to its classical tradition, and cannot resist retelling the tale. The *Déclaration* contains at this juncture a moralisation worthy of a preacher: ''In the same mould are those who covet this world but cannot lay their hands on it.'' However, Suso's own simile linking Tantalus' predicament to the spiritual state of the Disciple is not forgotten.

It is time to review the artist's achievements in more detail. The first segment of the triptych presents the twelve apostles seated at a round table in the company of Christ, who is in the most visually prominent position, that is, facing the viewer and high up the picture plane. He holds up a piece of bread over the cup and announces: *Hoc est corpus meum* ''This is my body.'' Another scroll on His left continues Luke, XXII, 19: *Hoc facite in meam commemorationem* ''Do this in remembrance of me.'' St. John the Divine leans forward and rests his head on the folds of His Saviour's garments. Judas bends over the table and thrusts his hand into the dish in the centre. This action has the effect of breaking symbolically the harmony of the circle. It constitutes a literal rendering of the passage from Matthew, XXVI, 23: ''He that dippeth his hand with me in the dish...'' The dish in the miniature contains a lamb which characterises the occasion as the Passover meal. Round loaves of bread and ewers are eucharistic symbols. A room or chamber is suggested by architectonic motifs.

This image of the Last Supper is in the form of an episode of the Passion. It is not in the liturgical form, common after the thirteenth century, where Christ administers bread and wine to the Disciples.[267] The painting is also unusual for the typology of the subject in Western Art in that it does not contain the motif of Christ pointing out the traitor. Schiller observed, ''...among Western representations there is only a small group which depicts the institution of the Last Supper alone.''[268]

The appearance of the sacrificial lamb on the dish was customary in late medieval images of the subject. A model of the motif that is later than the Brussels *Horloge* painting by some fifteen or so years, can be seen in the *Last Supper*, *c.* 1464-1467, by Dieric Bouts, in the centre panel of the altarpiece for St. Peter's, Louvain.[269]

The Host held on high by the celebrant parallels visually Christ's action in the adjacent segment of the miniature. White-robed altar-boys or youths flank the dais and two Dominican acolytes support the heavy chasuble of their centrally placed brother who officiates. A similar scene, where the participants are three-quarters facing away from the viewer and are not regulars, can be seen in a liturgical manuscript by the same miniaturist, Lyons 517, a Missal for the Use of the Autun Diocese.

[266] (Fol. 111) ''Je habondoie en delices et si estoie maleureux et chietif, et en affluence de grace je suis sechié et endurci. Je ressemblé au Tantalus qui entre les richesses souffroit de grandes povretés.''

[267] Schiller, *Iconography*, II, p. 28.

[268] *Ibid.*, p. 32.

[269] Reproduced *Ibid.*, II, fig. 99; in colour by H. Guratzsch, *Dutch and Flemish Painting*, New York, 1981, pp. 46-7.

Tantalus is presented as constrained by fetters in a cage with open views over the rolling countryside and on the two objects that torment him ceaselessly, and which are out of reach: food laid out on a table and water flowing over the bowl of a pedestal fountain. He wears clothes that are of the most recent styling in the painter's day. One would estimate him to be in his late twenties. The ubiquitous Disciple is within the picture space. The portraiture of both figures is eye-catching. The theme is Ovidian in origin,[270] and while textually it functions as an exemplum, its visual representation has the role of symbolically exposing the threads of frustrated desire that envelop the frail character of the young man. The Disciple is present, but detached, confronting spiritually the same problem, but free to resist and move away.

XXXII *The Spiritual Marriage of Sapientia and the Disciple* Fol. 127v.

The seventh chapter of Book II has the title *Qualiter multi fideles possint sapientiam divinam desponsare, et quomodo per cotidiana quaedam levia exercitia debeant se continue in eiusdem amore renovare*[271] "How numbers of the Faithful may marry Divine Sapientia, and how through daily good works they should renew themselves continually in that love". The chapter opens on the preceding recto side and includes an account of the spiritual marriage of Sapientia and the Disciple. After the union he exclaims: "...the Son of God, who is the great and sovereign Emperor and King of Glory in this new Easter feast, has celebrated a royal wedding, for He has given me in marriage His dearly beloved Sapientia and made me His son-in-law."[272]

The Disciple is next shown a vision. "In the vision that the Disciple experienced, he was called a new name by his bride when they were in their secret retreat, in the epithalamium, in silence and in seclusion. He had fallen asleep in her arms out of great trust and bliss, although his heart kept vigil over his salvation and the salvation of other creatures. His bride called him 'Frater Amandus', and spoke intimate words that were wise, subtle, trustworthy and very spiritual, and which passed all human understanding."[273]

The Spiritual Marriage of the Disciple is the climax to Suso's mystical treatise. The expositor's commentary in the *Déclaration* makes the point adequately but conveys little of the emotive stimulus that is evident in both textual narrative and illumination. His remarks, for example, do not register the Disciple's ecstasy. They seem to imply that the Disciple presents himself to Sapientia in a chamber, whereas Suso's account suggests she appears in a vision to the Disciple in his cell. The amazement shown by the Disciple on learning he becomes the son-in-law to the King of Heaven, and the slumber in Sapientia's arms are not considered worthy of elaboration by the commentator.

The artist's visual statement is in fact more interesting than the *Déclaration*'s account and invites close study. Three strata of interpretation are possible: the human, the spiritual and the divine. To the viewer's left is a chapel on whose threshold a High Priest joins a young woman and a man in marriage, pronouncing through the device of a scroll St. Paul's words to the wayward Corin-

[270] See Ovid, *Metamorphoses*, IV, lines 1458-9; one may also consult the English translation by F. J. Miller, London, 1916, I, p. 210.

[271] Künzle, *Horologium*, p. 590.

[272] (Fol. 127) "...le filz de Dieu, qui est le grant et souverain empereur et roy de gloire en ceste nouvelle feste de Pasques, a fait unes nopces royaulx, car il m'a donné en mariage sa tres amee Sapience et m'a fait son gendre."

[273] (Fol. 127v.) En celle vision qui lui fu monstree, fu appellé par ung nouvel nom de sa chiere espouse quant ilz furent en leurs secrez retrais, en la chambre des nopces, en silence et en secret; et il fut endormy entre ses bras et par grant fiance et par doulceur. Mais il veilloit de cuer en traittant de son sauvement et du sauvement des autres creatures. Son espouse l'appella 'Frere Amant' et lui dist secrettes parolles, saiges, subtilles et fiables et moult espirituelles, lesquelles trespassoient engin humain.

thians (II Cor. XI, 2): *Despondi vos uni viro virginem castam etc.* "I betrothed you as a chaste virgin to her one and only husband etc." The officiant blesses the union which has the standard form for medieval betrothals. The contract was solemnised at the door of the church or chapel. One is reminded of many scenes of the marriage of the Virgin and Joseph in medieval French art.[274]

Evidence for a spiritual union is suggested by the bride's elegant white and gold raiment, symbolising her moral and physical purity and noble bearing. Her spiritual quality is enhanced by the stellar circlet of glory that she wears. The Disciple kneels before his bride in a gesture of spiritual sublimation.

The divine allegory is conveyed on the wings of the simple bird which hovers over the joined hands. It announces the presence of the Holy Spirit, just as the triple tiara on the bearded officiant transforms his figure into that of God the Father. Suso's readers have known since the beginning of the *opusculum* that Sapientia may represent, according to the context, the person of Christ, whose cruciform halo she wears continually. Thus the Disciple may be said to be joined divinely with the Trinity, just as earlier in the *Horloge* series he was portrayed in the act of offering up his soul to the Trinity.

The setting for the adjoining vignette, presented visually as if occurring simultaneously, is the Disciple's cell. He kneels in all humility at the feet of Sapientia and, paraphrasing I Samuel, XVIII, 18, says: *Quis ego sum ut sim gener regis?* "Who am I that I should become the son-in-law to the King?" Sapientia addresses him with a term of affection: *Frater Amandus*[275], as recorded in a second scroll that is near her hand.

XXXIII (1) *God in Majesty on Mount Sinai* Fol. 157
(2) *Moses climbs the Mountain*
(3) *Moses descends with the Tables*
(4) *Moses casts down the Tables on seeing the Worship of the Golden Calf*

Gerson's text, known more generally to scholars as the *Miroir de l'Ame*, is composed principally of a discourse on the Ten Commandments. The scribe has reproduced in the Brussels volume the opening of the *Miroir* down to the conclusion of the Commandments' section.[276] Reading Gerson, one finds no mention of the circumstances surrounding the origins of the Tables of the Law nor of the principal actors of the drama on Mount Sinai. The question then must be asked as to the narrative source of the large illustration on this folio.[277] A Bible story has already been proved on more than one occasion in the course of this study to be mirrored in both an illustration and the *Déclaration*, and so it is again in this instance.

The expositor's account is purely descriptive and based on Exodus. God on a Mount Sinai in flames and Moses ascending are the the principal actors in chapter XIX, 17-20; the Golden Calf with Aaron and the dancers are the subject of XXXII, 1-6; the descent of Moses, his wrath and destruction of the Tables derive from the narrative of verses 15-19.

Although the commentator's terms 'right' and 'left' do not conform to modern usage, as was pointed out in chapter IV, § 3, his statement reflects a total comprehension of the contents of the composition. Perhaps a personal touch is applied when he attributes respect and reverence to the people when they behold their leader ascending and descending the mountain.

[274] See Réau, *Iconographie, Nouveau Testament*, pp. 171-2.

[275] Suso himself had this name; see above chapter III, § 1.

[276] Cf. above, chapter I, § 2 for an edition of Gerson's work.

[277] Comment by other scholars on this art work is rehearsed above, chapter II, § 3.

The Bedford Master's Chief Associate has chosen to telescope events so that one reads a narrative of several episodes confined to one spatial area. The focus on Moses as the central figure in the drama is weakened by the presence of Joshua on each occasion. Joshua is only briefly mentioned in the context of Exodus, XXXII, 17 as the person who draws the leader's attention to a disturbance in the Israelite camp. Perhaps the artist felt that two figures were required at three places on the semi-circle around the mountainside to balance the groupings of figures in four places equidistant from each other in the foreground. God in Majesty is centrally placed, high up the picture plane, and surrounded by flames to enhance His radiance. Each of the four principal action scenes has a typological history of its own. Models occur in other mediums, such as sculpture and wall painting, as far back as the ninth century.[278] In manuscript illumination cycles for the life of Moses are known by the early fifteenth century. One recalls the series of vignettes executed by Paul de Limbourg for the Exodus chapters I-XV in a *Bible Moralisée.*[279] The artist in the Brussels volume was illustrating later chapters of Exodus so his paintings do not derive from the imagery of Paul de Limbourg. One venerable tradition, however, that the Chief Associate of the Bedford Master did follow, was the portraiture of Moses with horns on his head.[280]

XXXIV *Court of Heaven* — Fol. 199v.

Gerson's *Mendicité Spirituelle* is an intimate dialogue between man and his soul on the subject of recourse to God and His Saints through prayer so as to be able to receive Grace and Wisdom from the Most High.[281] In the first part the soul declares itself "impoverished, sick, imprisoned, wounded, naked and stripped of everything."[282] It proclaims that it and its mortal host are in the dark abomination of a prison that is their exile. The discussion includes mention of the Saints and of the Church of Paradise "...where there is the Treasurer of Grace, the Queen of Mercy, the Mother of the Poor and the Orphaned. Moreover, there is found the Redeemer of Mankind, Our Lord Jesus Christ, who, before God the Father, is intermediary and advocate for sinners... if you have the desire, you may always speak to God, and from Him and His Saints seek help and succour."[283]

The second part of the *Mendicité Spirituelle* presents the soul in the guise of a beggar seeking solace and sustenance. Gerson instructs it in formal exercises, among which are the seven petitions of the Pater Noster, the Nine Orders of Angels and the Reception of the Holy Spirit.

The *Déclaration*'s author reveals in the second sentence of his account of the painting that he has personal knowledge of the broad thrust of Gerson's treatise. The dialogue mode, however, is not mentioned. The expositor's summation of the illustration is cursory but his identification of the principal actors is flawless. One may have expected the words "of Theology" to accompany the term "Doctor", since both the role and attitude of the 'healer' towards the soul sustain the imagery and the allegory.

[278] See Réau, *Iconographie, Ancien Testament*, pp. 203-6.

[279] Paris, Bibliothèque Nationale, MS. fr. 166, reproduced in Meiss, *Limbourgs*, II, pls. 314-35.

[280] The origin of the horns is explained in Réau, *Iconographie, Ancien Testament*, p. 177, but see also the monograph by Ruth Mellinkoff, *The Horned Moses in Medieval Art and Thought*, Berkeley, 1970.

[281] An edition is indicated in the textual note for this work, above in chapter I, § 2.

[282] (Fol. 199v.) "...je suis povre, malade, emprisonnee, bleciee et navree, nue, sans vesteure et si n'ay riens."

[283] (Fol. 200) "Ilec est la tresoriere de grace, la royne de misericorde, la mere des povrez et orphenins. Et qui plus est, la est trouvé la Redempteur de l'umain lignage, Nostre Sauveur Jhesu Crist, qui envers Dieu le Pere est moyenneur et advocat pour les pecheurs... par desir tu peus tousjours a Dieu parler, a lui et a ses sains, leur ayde et secours demander."

The eye of the modern critic readily perceives St. Peter in an entranceway at the lower right and the Doctor facing, in the opposite corner of the foreground.[284] The choice of an infant to portray a soul was of some antiquity by the fifteenth century. A similar configuration, it will be recalled, was executed in this Brussels volume by the Rolin Master in his painting of *The Disciple Offering his Soul to the Trinity.*[285] Responding no doubt to Gerson's symbolism, the Bedford Master's Chief Associate places the infant on a bier, recalling the posture of a truly sick person. The role of doctor and patient thus represented is as close as the artist comes to the dialogue of the textual narrative. Without the treatise as a guide, one may not easily realise that the doctor instructs the soul how to win favour, assistance and succour from the inhabitants of Paradise. Behind the symbolic walls of the Heavenly City stands a cross section of the Blessed, both lay and secular, holy and saintly.

The arrangement of the figures in concentric semi-circles is reminiscent of the construct of the previous composition whose theme was Moses on Mount Sinai. This time a sense of space is achieved by the positioning of single figures in the picture plane close to the Throne of Judgment.[286] Thus, the *Court of Heaven* illustration lends a totally different emphasis to Gerson's work. One could be excused if one thought that his treatise concerned only the Heavenly Paradise.

XXXV (1) *Agony in the Garden* Fol. 229
(2) *Betrayal*
(3) *Christ Nailed to the Cross*
(4) *Crucifixion*
(5) *Deposition*
(6) *Entombment*

The words of John, XIII, 3 are the *textus* for the great Easter Sermon Gerson preached on Holy Friday, 1403 in the Eglise Saint-Bernard, Paris. As the rubricator's introduction copied on fol. 228v. clearly observes, the sermon is in two parts, each of twelve sections. In the morning Gerson expounded the Passion narrative commencing with the Garden of Gethsemane and concluding at the point where Christ enters upon the Via Dolorosa. The 'collation' was delivered "aprés disner" and the congregation composed of the King, Charles VI, members of the Royal Family and dignitaries of the Realm, were conducted by the preacher through all the Stations of the Cross to the Entombment. The narrative of the Gospels was the inspiration for Gerson's commentary.[287]

This source is reiterated by the author of the *Déclaration*, who then proceeds merely to announce the subject matter of each of the six compartments of the miniature. He follows the chronological order of the Passion events, but offers on this occasion no spiritual or exegetical comment. The placement terms 'right' and 'left' when employed here by the expositor are not in line with modern practice, as was observed above, in chapter IV, § 3.

The artist has selected six scenes from the Passion Cycle and arranged them structurally in two registers of three each. We read them as top left, lower left, lower centre, upper centre, top right, lower right. Each of the scenes recalls an event among Gerson's twenty-four, but we remain in the dark about the choice. The availability of stock models in the atelier of the Bedford Master's Chief Associate could have been an influencing factor.

[284] Earlier comment on the painting is to be found above, chapter II, § 3.

[285] See above, the commentary on picture XVIII (1).

[286] It is an interesting exercise to compare this construct with the Rolin Master's one in picture XXIII of the Brussels volume.

[287] An edition for the Sermon is supplied above, chapter I, § 2.

The first segment is, in fact, a traditional rendering of the *Agony in the Garden*, the area here being defined by a fence of wooden palings. While His three companions slumber, Christ addresses to God in a distant mandorla the celebrated words, modelled on Matthew, XXVI, 39: *Mi si fieri pater transfer calicem istum a me* "If possible, O Father, let this Cup pass me by." The segment placed beneath the *Agony* depicts the next momentous event, the *Betrayal*. Soldiers surge into the narrow confines of the Garden while Judas carries out this treacherous gesture.

In the lower centre the artist presents *Christ Nailed to the Cross*, in which workmen actively pierce His flesh with nails under the watchful eye of two overseers. The *Crucifixion* which occupies the upper compartment is excessively crowded, with the traditional figures and numerous scrolls. The thief on the viewer's left pleads *Memento mei, Domine, dum veneris in regnum tuum* (Luke, XXIII, 42) "Remember me, when you come into your kingdom, O Lord." Above the head of the other felon is a banderole bearing the inscription from Luke, XXIII, 39: *Si filius Dei es, salva temetipsum et nos* "If you be the Son of God, save yourself and us". Two of Christ's cherished thoughts are set out in vertical scrolls, on His right, and both are from John, XIX, 26: *Mulier, ecce filius tuus* "Mother, behold your son" and *Ecce mater tua* "Behold your Mother."

A short scroll placed to the left of the head of the soldier holding the Sop records Christ's anguish, expressed by the one word from John, XIX, 28: *Sitio* "I thirst". Fluttering vertically beside it is another banderole proclaiming the centurion's profound words, recorded in Matthew, XXVII, 54: *Vere filius Dei erat iste* "Truly, this man was the Son of God."

The *Deposition* in the top right of the polyptych is a successful pictorial rendering of a great moment of shock and sorrow. Similar emotions cloud the faces of the onlookers as Christ's body is about to be lowered into a sarcophagus.

Conclusion

Let us recall that MS. IV 111 of the Bibliothèque Royale, Brussels, is a unique treasury of mid-fifteenth-century Parisian book illuminations. Of the 35 illustrations, 3 were painted by the Bedford Master's Chief Associate to embellish works by the Chancellor of the University of Paris, Jean Gerson (1363-1429). The remainder constitute the most sumptuous iconographical statement known for the *Horloge de Sapience*, a translation by a Franciscan of the *Horologium Sapientiae* by Heinrich Seuse O.P., *alias* Suso. The dramatic narrative of the mystic treatise is dominated by the allegorical persons of Sapientia and her pupil, the Disciple. The all-pervading presence of the Logos and the Humanity of Christ, and the aura surrounding the progress of the principal character towards understanding and enlightenment contain Dominican attitudes.

The Brussels manuscript is unique in another way among the extant copies of the *Horloge de Sapience.* Its illustrations became the subject of comment, description and explanation shortly after they had been completed. The *Déclaration des Hystoires*, as the tract is called, may be considered as a precursor of modern iconographical studies. One hesitates to say ''model'' or ''prototype'', since the aims and objectives of its author were not pictorial but didactic and religious.

The art work viewed in its entirety is seen as an outstandingly successful programme whose constituant elements effect an integration of text and illustration. The single extraneous element, that is, one that is unrelated to Suso's or Gerson's works, can be excused because it is traditional. It is the portrayal of an owner as a knight in armour, kneeling in supplication to Sapientia in the frontispiece. Even so, the figure could be an overpainting and may not have belonged to the artist's original scheme of illustrations.

The Rolin Master's illustrative programme offers a whole range of responses to the textual material. As rehearsed in chapter III, he employs pictorial cycles such as those of the Passion, the Stations of the Cross, the Humanity of Christ, the Relationships between Sapientia and the Disciple and so on, all in keeping with the spiritual emphases imposed by Suso in his own treatise. The groupings of the tableaux may be by polyscenes or monoscenes, full-page or reduced single unit format.

Although many of the 108 pictorial units executed by the Rolin Master are based on stock patterns e.g. *Nativity, Adam and Eve, Trinity*, one is nevertheless impressed by the artist's ability to be imaginative. He enhances the immediacy of his distinctive response to the textual narrative.

One of the painter's more noticeable iconographic traits was seen to be the expansion of the thematic statement, coupled with an enlargement of the textual narrative. The example that was highlighted in chapter V was Tantalus, a name mentioned in passing by the Dominican Suso, yet the artist presents a portraiture that depicts the subject's notorious dilemma. Again, an oblique reference by Suso to an incident in the Old Testament—a routed leader was given milk when he asked a hospitable female for water—was transformed by the miniaturist into two splendid illustrations relating the essentials of the tale of Jael and Sisera.

Enlargement of textual detail occurs visually on the occasion when Sapientia is said by Suso to knock at the door of her Disciple's heart; the pictorial response presents Sapientia at the cell door of her pupil, but she is not alone, being in the company of the three Theological Virtues.

The Rolin Master's capacity to conflate at times the textual narrative is well documented by the *Battle for the Soul.* Two phrases by Suso: "God does not want a soul to be lost, He draws it up to Him" become an inspiration for a visual graveyard, bridge, aerial combat to possess a soul, beheld by onlookers standing on the high ground of Paradise.

Two traditional themes may be pictorially combined in a single format to enhance and enlarge a conceptual statement by Suso. A moribund, wailing youth is said to hear the disembodied Voice of Death speaking to him. Instead of presenting a young man in bed with banderoles conveying the message of Death, the Rolin Master portrays Death as a desiccated cadaver brandishing a javelin in the direction of the juvenile, along with a riot of scrolls. The personification of Death from a traditional icon is combined with a standard Death-Bed scene.

Textual detail may be observed to the letter by the artist. It is always a matter of knowing the identity of the source or sources of inspiration. The reader was often instructed in chapter V about the bifurcation and plurality of sources: Suso for the mainspring, the Bible or another text for detail. A significant example of this process was the portrayal of the Vices and Virtues alluded to individually by Suso, then paired by the artist according to an earlier pictorial tradition, then garbed and accoutred in line with *minutiae* modelled on Digulleville's *Pèlerinage de la vie humaine.*

One technical accomplishment in the Rolin Master's painterly arsenal is the method of pictorialising dreams or visions. Only once does he rely on the traditional means of placing the actor of the vision sequence above the head of the dreamer. This is the celebrated *Sapientia with Lilies and the Disciple.* The more usual pictorial practice for the artist when required to represent the thoughts or dreams or visions of the Disciple is to paint the action of the vision in realistic terms and to place the Disciple in a corner near the frame in the act of witnessing the drama. Notable examples are *Christ Nailed to the Cross, Court of Heaven, Epicureans* and *St. Arsenius.*

In the matter of written metaphors, the Rolin Master attempts to reproduce them realistically, but often without understanding allusions and concomitant emphases. For the exemplum of a lover paying no heed to thorns provided he can hold his beloved rose, the artist paints a rose-briar in the hand of the Disciple's beloved, Sapientia. When Suso employs the *Ram with the Iron Crown* to convey allusions of a libellous nature about rulers in the city of Constance, the illustrator presents his viewers with a zoomorphic creature, one of whose horns is broken, while the other supports a crown. But the artist loses the thread of the metaphor when he places foxes and a Pope on a mule in the Ram's retinue. Suso saw the foxes in the Pope's entourage and said nothing of the Supreme Pontiff's procession on a mule.

A measure of implied symbolism that may have been known to artist and viewer alike is contained in the depiction of a bare cloister in shadows around a central fountain to symbolise the place where members of the Order of Preachers re-enacted the Passion procession to Calvary. Another symbolic *fantome* may have been the womanly figure demonstrating the mechanism of time-pieces in the *Clock Chamber.* A fifteenth-century viewer would most likely have recognised the symbolic act of measure and cadence in all things requiring regulation just as Temperantia had done in art for centuries. Even Sapientia seated on a throne of majesty and looking totally deific, could suggest above all a typology that associated her in the viewer's mind with God the Father in the *Maiestas Domini* theme. The artist, by portraying Sapientia in this guise, was also portraying the Logos as Wisdom in female garb.

With only three paintings to serve as a basis for comment about the iconographic respect for texts, we hesitate to state anything categorical about the Bedford Master's Chief Associate. He appears to employ one or two of the procedures that were evidenced in the Rolin Master's iconographic art. He finds inspiration in Gerson's text, but the development mode is determined by another text altogether. For example, he places four scenes concerning the origins of the Tables of the Law at the head of Gerson's tract on the *Dix Commandements*. However, the theologian says little or nothing about the sources of the Decalogue. The artist present on his own initiative a visual statement taken from chapters of Exodus. Likewise, the *Court of Heaven* miniature, although recognisable thematically and iconographically, does not convey the import of Gerson's *Mendicité Spirituelle*; it emphasises an element that was not primordial to the writer's argument.

Further, one must ask the question as to why the same painter gave prominence to the six Passion scenes he selected to illustrate Gerson's Sermon *Ad Deum Vadit* when the textual narrative offered twenty-four incidents to choose from. There are innumerable occasions in the *Horloge* series where the same question could be asked of the Rolin Master. Suso mentioned three similitudes in chapter II of the *Horloge*; why did the artist decide to leave one aside, the one concerning the paramours, when he composed *Sapientia and the Disciple behold Christ Carrying the Cross*?

Another unanswered question for the historians of Iconography and Text concerns the emphasis given to the whole subject matter of the *Horloge de Sapience* in the Brussels volume. Why did the illuminator not conform to the patterns of earlier illustrators and execute a tableau as frontispiece and another as the keynote visual statement at the head of Book II?

Lastly, our study is not able to respond to a request for information on the effect that such a revolutionary iconographic programme had on the reception of the text. It clearly enhanced it as far as the owners and readers and viewers of the Brussels manuscript were concerned, but no reliable statement can be made about the text's renown in succeeding generations until all the extant copies of both the Latin *Horologium* and the French *Horloge*—nearly five hundred were mentioned in chapter III—have been examined from an iconographic viewpoint.

Plates

AND

The Déclaration des Hystoires: *Brussels, Bibliothèque Royale, MS. IV 111, fols. 3-11v*

Notes on the Edition of the Text with Translation

In the edition and translation of the *Déclaration*, the needs of the art historian have been constantly in view. The fifteenth-century expositor refers to the paintings as numbered by a contemporary hand, and in the translation of the text this number is observed.

The picture-titles at the head of each iconographic form are identical to those employed earlier in this study. This uniformity will enable the critic or reader to be confident that the same subject matter is being referenced at all times.

The French text is presented in accordance with the editorial conventions adopted in 1926[1] and universally accepted by Medieval French scholars since that date. This agreement provided for certain letter transcriptions and word divisions to conform to accepted patterns, and for the modern style of punctuation and capitalisation of letters to be introduced. The points before and after Roman numerals are not reproduced. The acute accent is used only for final stressed *e* and a cedilla is placed under 'soft' *c* before *a* and *o*. Abbreviated Latin words except *etc.*, are given in their full form, while rejected readings or textual obscurities are mentioned in the notes. Square brackets are placed around words not in the scribes' copy and around the folio number indicating where a fresh folio begins.

Two idiosyncrasies of our copyists have to be observed. The form *payens* for the 'pains' of Hell is very common, while the more easily recognizable *paines* is rare. The other spelling inconsistency concerns the gerund and present participle. For all verbs the scribes employ *-ens*, *-ent* more frequently than *-ans*, *-ant*. The effect is that the modern translator needs to be on the alert constantly for the correct sense of the form. For example, *voyent* may mean 'seeing' as well as 'they see'.

A modest Glossary is supplied mainly as a guide to unusual graphies and unusual meanings. It does not purport to be exhaustive.

In the translation process, it has been felt necessary to produce an English text that follows closely the sense of the original and at the same time conforms to English stylistic conventions. The result is that, in the translation, punctuation, length of sentences and capitalisation of terms differ from the fifteenth-century French practices, such as they were.

[1] See *Romania*, LVII (1926), pp. 242-56. I should like to thank the staff of the Cabinet des Manuscrits at the Bibliothèque Royale, Brussels, for allowing me to obtain a microfilm of the manuscript.

Whereas personifications may attract a capital letter in both languages: Paresce — Sloth, Charité — Charity, Longue Vie — Longevity, the French term *sapience* invites special consideration. It is employed by the French author with several meanings, and disentangling the sense of each has been fraught with difficulty. The upshot of a careful study of his usage is that *dame Sapience*, the name of the female personage who is mentor to the Disciple, may be consistently rendered *Sapientia*; *divine sapience* will be named in English *Divine Sapientia*, thus avoiding any interpretation of meaning by the translator; lastly, *sapience* without an epithet will be called *wisdom*, or *Wisdom* if personification is present.

One small detail remains to be aired. It is the French author's use of *il* for either the compiler of the Latin treatise, Suso, or the artist responsible for the illustrations. In each case, the English translation records *il* as 'he', and the reader is left to identify for himself the person intended.

[Fol. 3] S'ENSUIT LA DECLARATION DES HYSTOIRES DE L'ORLOGE DE SAPIENCE

Here follows the Exposition of the Illustrations of the Horloge de Sapience

I *Sapientia in Majesty*

Fol. 13

Premierement au commencement du livre est dame Sapience en forme et figure de femme signifient Jhesus Nostre Sauveur qui est dit et appellé vertus et sapience de Dieu le pere[2], laquelle est assise en throne de magesté comme vray Dieu. Et tient ung livre a sa main dextre, et a la senestre, le monde, signifiant que par elle et d'elle est ysseue toute science et sapience par laquelle est le monde gouverné et reparé. A la dextre de ycelle dame Sapience est le disciple aucteur et composeur de ce livre estant en chaire comme preschant a divers estas du monde que ilz sentent et gouttent de Dieu en bonté, et le quierent en simplesse de cuer, et dit: *Sentite de Domino in bonitate, et in simplicitate cordis querite illum*[3], qui est le theume et principe de sa doctrine.

First of all, at the opening of the Book is Sapientia in the image and likeness of a woman, representing Jesus, Our Saviour, who is called the strength and wisdom of God the Father. She is seated on a Throne of Majesty like the true God; she holds a book in her right hand and the orb of the world in her left, signifying that through her and from her has emanated all knowledge and wisdom by which the world is governed and reconstituted. To the right of Sapientia is the Disciple, author and composer of this Book, standing in a pulpit, seemingly preaching to persons of different classes, exhorting them to set their mind sincerely upon God, and seek Him out with simple heart. The words he speaks: Sentite de Domino in bonitate, et in simplicitate cordis querite illum *constitute the subject matter and essence of his teaching.*

[2] The Pauline source for this statement is given in Latin and French in the section on illustration VI (4), below.

[3] Sap. I, 1. The French translation that precedes the Latin quotation is quite free.

La premiere hystoire.

Cy commence le livre appelle horologe
de sapience lequel fist frere Jehan de sou
haube de la nacion dalmaigne de lordre des prescheurs

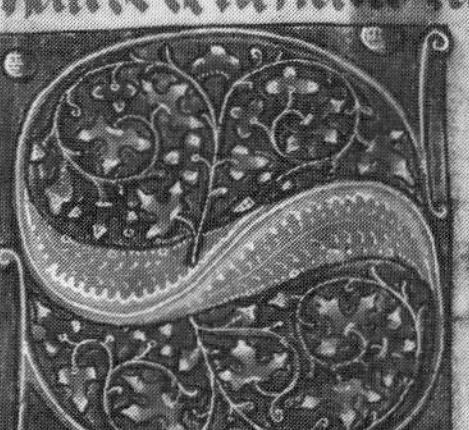

[S]Almon en son livre
de sapience ou premier
chappitre dit. Sentite
de domino in bonitate.
et in simplicitate
cordis querite illu[m]

[S]Entez et entendez de dieu en bonte
confermez vous en son ordonnance
et a sa voulente. querez sa presence en
simplesse de cuer et en purte de pensee.
Car ceulx le truvent qui ne le temptent
pas. qui font leur devoir de bien faire.
selon leur puissance sans atendre
laide particuliere de dieu. Et il se mon
stre a ceulx qui ont fiance en lui.
qui apres ce quilz font leur povoir et
leur devoir de bien faire attendent au
seurplus la grace et laide de dieu [A]u
commencement de la fondacion de saincte
eglise la sapience divine en plusieurs
guises et en plusieurs manieres se mo
stra et sapparut a ceulx qui estoient es
leuz a gloire qui sont des esleus pour
estre sauvez et de sa lumiere enlumina
et endoctrina leurs pensees Lesquelz en
ferveur et en ardant desir desperit servoi
ent a dieu en saintete et vraye iustice de
conscience se representoient a lui. Il
en avoit envoye le feu du ciel en les osse
mens de quoy leurs cuers furent enflammez
du feu de charite. [C]Et ceste flamme
ne se peut celer ains se monstra et espa
di par dehors largement par les exemples

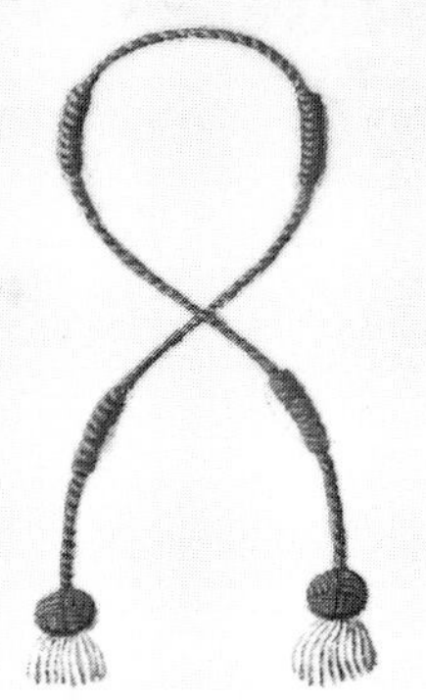

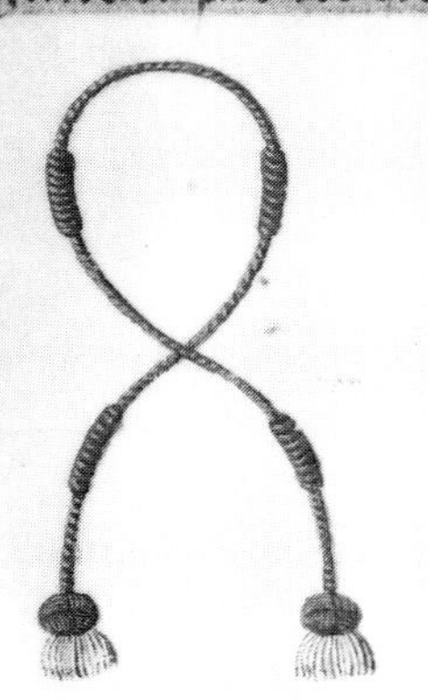

II *Clock Chamber* Fol. 13v.

[Fol. 3] La seconde hystoire est ou mesme prologue dudit livre a l'endroit du parafe qui commence: *Pour ceste cause.* Pour ce que, comme il dit pou devant, la Divine Sapience est tousjours soigneuse du sauvement de toute humaine creature, elle ralume les estains, elle rechauffe les refroidiz et resveille les endormis. En ce signifiant est nommé ce livre *Horologe de Sapience.* Et en ceste figure est dame Sapience comme ordonnant et gouvernant une horologe et ung resveil disent au son[4] de plusieurs clochettes: *Ante secula qui Deus et tempora homo factus est in Maria*[5], lequel verset est approprié couvenablement a Nostre Seigneur Jhesucrist qui est l'eternelle sapience et filz de Dieu le pere et a prins char humaine et nasqui en la Vierge Marie. Et a ses piés est le disciple contemplant le maintien de sa maistresse[6].

The second illustration is located in the same Prologue of the said Book and at the place of the paragraph which begins Pour ceste cause. *Because, as he stated a little earlier, Divine Sapientia is always careful to save every human creature, she therefore revives the dead at heart, she warms those who are benumbed, and she awakens the somnolent.*

To represent this, this Book is called the Horloge de Sapience. *And in the picture Sapientia is portrayed appearing to control and regulate a clock and an alarm, declaring to the tune of many small bells:* Ante secula qui Deus et tempora homo factus est in Maria. *The verse is suitably appropriate for Our Lord Jesus Christ who is the eternal wisdom and Son of God the Father, and who became incarnate and was born of the Virgin Mary. And at Sapientia's feet is the Disciple contemplating the conduct of his mentor.*

[4] The scribe has clearly written *en son* and not *au son.*

[5] For *et* the copyist wrote *est.* The word *Maria* in the *Déclaration* is not abbreviated by the scribe.

[6] Spencer, *Horloge*, pp. 282-3, n. 19 prints this commentary.

de leurs bonnes euures de leur vie par
faitte z de leur saincte conuersacion.
Ceulx ci queroient dieu en grant
bonte z en parfaitte simplesse de cuer.
Car toute leur entente z leur estude
estoit les occupacions mondaines ou
blier. leurs pechiez gemir z pleurer.
les choses diuines assauourer. Eulx
dedens eulx contenir. et garder le corps
et a seruitude ramener. Et seulement
a la diuine sapience acordre. et leur estim
occuper. Mais las dolent au jourdui
le monde est enuielli en malice et lamour
diuine est es cuers de plusieurs tellement
refroidee quelle est z peu sen fault estainte
Car on en treuue pou qui estudient
en deuocion ne qui aient diligence de
renouueller de la grace de dieu pour
eulx eschauffer ne qui par desplaisance
de leurs faultes aient souuent la ler
me a loeil ne qui desirent la presence
la visitacion la doulce alloqucion
et plement de la grace diuine. Ains
se occupent z estudient en vanitez en
narracions en genealogies et hystoires
vaines z temporelles et en delices cor
porelles esquelles ilz sont endormis
par vng grief z pesant sommeil.
Et pour tant la diuine sapience
qui tousiours est songneuse du sau
uement de toute humaine creature
en desirant damender la vie de ses
esleus et en voulant oster z aneantir
tous vices de leurs consciences veult
et entent en ce liure principalment
allumer les estains z enflammer les
les refroidiez. les pecheurs resmouuoir
les mal deuots a deuocion rappeller et
promouoir. Et les endormis par negligence
a lestude des euures vertueuses esueiller

La .ij.e hystoire

III (1) Ecce Homo — Fol. 15v.
(2) *Christ Carrying the Cross*
(3) *Christ Addressing the Daughters of Jerusalem*
(4) *Cloisters with Fountain*
(5) *The Disciple Lamenting before the Crucified Christ*

[Fol. 3] La tierce hystoire est en ce mesme prologue a l'endroit du parafe qui commence: *A savoir est*. Et est la cause de l'ystoire pour ce que le disciple souvent nommé de dame Sapience avoit de coustume toutes les nuys aprés matines de faire en par soy une procession entour le cloistre et revenir devant le crucifilz en remembrance de la procession que fist Nostre Seigneur Jhesucrist depuis qu'il fut condampné a mort jusques a rendre l'esperit a Dieu le pere en l'arbre de la croix. Et ceste procession est figuree en l'ystoire presente qui est faitte en maniere d'ung cloistre ou milieu duquel est le praiau et la fontaine.

Ou premier parquet est le pretoire de Pylate ou Nostre Seigneur fut jugé et condampné et est vuit par Pylate en saillir pour presenter Nostre Seigneur flagellé et couronné d'espines aux Juifs qui l'escrient: "A crucifier!" Dehors ce pretoire est Pylate presentent Nostre Seigneur ainsi flagellé et couronné, et porte on aprés lui une robe de pourpre en derision, Pylate, disent aux Juifz: *Ecce rex vester*![7]

Aprés, en ung autre parquet, est Nostre Seigneur portant la croix, et le mainne on hors de la cité.

Aprés, est Nostre Seigneur lïé et mené ou lieu de Calvaire et on lui porte la croix aprés lui. Et les devotes femmes qui lui avoient servi et ouÿ sa predication siuwent plourans; auxquelles il dist: *Filie Jherusalem etc.*[8]

Aprés est le crucifiement comme en une esglise, comme on a a coustume de mettre a chascune esglise ou milieu, et la, le disciple finoit sa procession en se complaignant [fol. 3v.] que il n'avoit pas la devocion ne la compassion a Nostre Seigneur comme il desiroit et come il appartenoit.

The third illustration is in the same Prologue at the place where the paragraph A savoir est *is located. And the reason for the illustration is that Sapientia's oft-named Disciple was wont, every night after Matins, to perambulate on his own around the cloisters and to come back in front of the crucifix, in remembrance of the procession made by Our Lord Jesus Christ from the time He was condemned to death until He rendered unto God the Father His Spirit on the Holy Rood. And this perambulation is represented in the present illustration which is structured like a cloister with the stretch of grass and fountain in the middle.*

In the first segment is Pilate's praetorium where Our Lord was judged and condemned; it is empty because of Pilate's hurried exit to present Our Lord flagellated and crowned with thorns to the Jews who shout: "Crucify Him!" Outside the praetorium Pilate is presenting Our Lord thus flagellated and crowned. Behind Him they carry a robe of purple in mockery and Pilate is saying to the Jews Ecce rex vester!

Then, in another segment, Our Lord carries the Cross and is led out of the city.

Next, one sees Our Lord bound and led to the site of Calvary and the Cross is carried behind Him. The devout women who had served Him and hearkened to His sermons follow, weeping. To them He says: Filie Jherusalem etc.

After that, the Crucifixion is depicted as if in a church, just as it is customarily set up in the centre of each church; and there, the Disciple concluded his perambulation, lamenting [fol. 3v.] that he did not have the devotion or the compassion for Our Lord that he wanted to have and should have.

[7] John, XIX, 14.

[8] Luke, XXIII, 28.

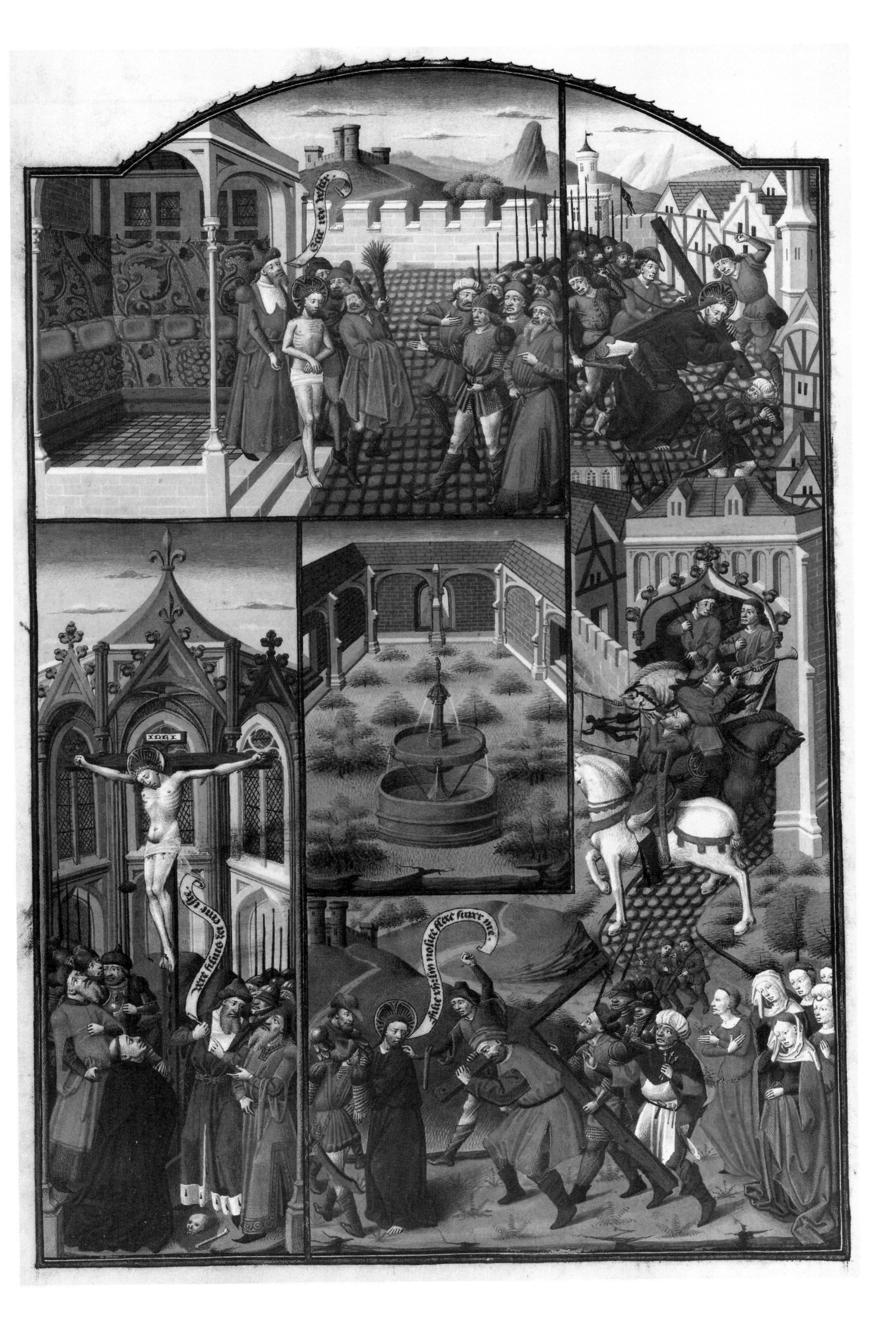

IVA *Youth in Dalliance* Fol. 16v.

IVB (1) *Reception of a Postulant* Fol. 17
(2) *Studying in the Cell*
(3) *Lessons in the Refectory*
(4) *Tree of Life*

[Fol. 3v.] La quarte hystoire est au commencement du[9] premier chapitre qui commence: *Hanc amavi, etc.* et est ou livre aprés le prologue. Et est ceste hystoire de la conversion du disciple de Sapience.

Premierement, en ung petit parquet hors de l'ystoire, est comment le jouvencel en sa fleur d'aage estoit habandonné aux vanitez et voluptez du monde et de la fut visité de l'inspiration de Divine Sapience.

[Fol. 3v.] Aprés, en l'autre page, ou corps de l'ystoire, est comme le jouvencel entre en religion et est en chapitre recevant l'abit de religion. Et est de l'Ordre des prescheurs, car celui qui fit ce livre estoit de celle ditte Ordre, et d'ores en avant sera en cest habit.

Aprés est[10] le souvent nommé disciple en chambre, estudient en la Bible et specialement aux livres de Salomon et de Sapience auxquelz, sur tous autres livres, est enseigné a aymer sapience. Et ce que il en trouvoit, il mettoit en son cuer; et pour la conservation de memoire, il le escrivoit en cedules seloncq que dame Sapience lui inspiroit.

Aprés est le dit disciple en refectouoir, et tout ce qu'il ouoit lire esmouvent a l'amour de sapience, il escripvoit en ses tables et le mettoit en memoire; entre lesquelles retient une auctorité qui disoit que Sapience mieulx vault que toutes choses que on puet desirer. Elle tient a sa dextre Longue Vie et a sa senestre Gloire et Richesses, et ce verrez en ce parquet come le disciple escript et dame Sapience tient a sa main dextre ung souleil qui signifie la vie, et a la senestre une couronne et des joyaulx.

Une autre auctorité le dessus dit disciple ouit lire qui dit: *Lignum vite est hiis qui apprehenderunt eam, et qui tenuerit eam beatus.*[11] Pour laquelle est ou darrenier parquet de ceste hystoire dame Sapience tenent une croix et la baillent au disciple, laquelle est signifiee par l'arbre de vie qui estoit en paradiz terrestre, de laquelle qui mengeroit du fruit, vivroit perpetuellement; aussi, par le fruit que l'arbre de la croix a porté nous est rendue la vie pardurable.

The fourth illustration is at the beginning of the first chapter which commences Hanc amavi etc. *and is located in the Book after the Prologue. And the illustration concerns the conversion of Sapientia's Disciple.*

Firstly, in a small segment, separate from the illustration, is depicted how the youth in the prime of life was given to the vanities and sensuous pleasures of the world until receiving the revelation of Divine Sapientia.

Next, on the following page, in the body of the illustration, can be seen how the youth enters religion; he is in chapter receiving the habit. And he belongs to the Order of Preachers, because the writer of this Book belonged to the said Order. Henceforth, he will be wearing this habit.

Then, the oft-named Disciple is in his cell, studying the Bible and especially the Books of Solomon and Wisdom, where, more than in all the other Books, he is taught to love wisdom. And

[9] MS. *ou premier.*
[10] MS. *apres que le souvent*
[11] MS. *tenuerūt.* The scribe of the *Déclaration* does not appear to recognise the quotation from Proverbs, III, 18 employed by Suso, and where the verb is *tenuerit.*

Quant il ot bien conceu entendu
considere et incorpore toute la revelacion
et il fu revenu a lui il estudia et pensa
moult diligemment a tout ce qui
lui avoit este monstre et trouva et
vit clerement que cestoit pure saincte
doctrine et divine revelacion Et quen
tout le proces navoit chose qui feust
contraire ne discordant aux dis des
docteurs et des sains peres mais tout
ce quil avoit veu et entendu par divine
revelacion estoit tesmoigne autorise
et conferme par la saincte escripture.
Car il avoit si grant et si propre
semblant entre la revelacion et la saincte
escripture que cestoit tout un en sens
et entendement en parolles et en choses
tres semblables par tout Apres
laccomplissement de cest labour celui
qui le livre avoit escript fut en grant
doubtance en pensant que sil publioit
et reveloit ce quil avoit veu et escript.
que les envieux tourneroient tout le
fait a moquerie et a truffe et se peneroi

De la grante hystoire.

ent de lempeschier et de lam
Ilz ne appreuuent riens bon
font et quilz dient Et so
suspecion si affaittiez a leur
trop souvent ilz soustiennen
et leurs contraires mespris
et repudient et condempnen
les secrez misteres et revel
comme silz naient onques
a savour les inspiracio
et quilz nappartenissent o
dons du saint esperit. Et
et desprisent le sens des sa
gnes et les attribuent a m
mensongables et magin
sciences qui sont doubteu
fondees et tres perilleuses
afferment et soustiennen
veritables Si doubta
que telle maniere de gens
despecier et destruire tou
leurs dens agues et venim
ceste cause pensa il quil cel
prendroit son courage
divine sapience par certa
voult que le livre feust m
et en appert Car la vierg
avec son filz sapparut a v
personne et lui comman
livre feust publie et den
toutes personnes aman

Cy commence le premier c
ce livre ouquel il parle co
de sapience divine sont tr
merveilleusement. Et co
de ce livre fut trait a sa di
De le commencement de s

Tant amain et exquis
mea et que suis mis

what he found therein, he took to his heart; and to retain it in his memory, he wrote it down on small scrolls according as Sapientia inspired him.

Next, the said Disciple is in the refectory; all that he heard read out that moved him to the love of Wisdom, he used to transcribe on to his tablets and commit to memory. He retains one saying in particular that declared that Wisdom is worth more than anything one can desire. It holds Longevity in its right hand and Glory and Riches in its left. You will see in this compartment how the Disciple does his writing and Sapientia holds in her right hand a sun which represents life, and in her left a crown and jewels.

Another saying which the Disciple heard said aloud runs Lignum vite est hiis qui apprehenderunt eam et qui tenuerit eam beatus. *This is why in the last segment of the whole composition Sapientia is handing over a cross she holds to the Disciple. The Cross is represented by the Tree of Life which stood in the Earthly Paradise; he who would eat of its fruit would live for ever. Thus, through the fruit borne by the Holy Rood, life everlasting is returned to us.*

lignum vite est hiis qui apprehenderint
eam et qui tenuerit eam beatus.

Ca. iiii^{e}

VA (1) *Temptations of the Flesh* Fol. 18
(2) *Repentant in his Cell*

VB *Epicureans* Fol. 18v.

[Fol. 3v.] La cinquiesme hystoire est devant le parafe[12] qui se commence: *Advint que l'ennemy*, et est ceste hystoire des temptations du disciple: comment le dyable luy monstre les plaisances, richesses et voluptez mondaines.

Et aprés est ledit disciple en sa chambre, se reduisant a soy et repentent des cogitations qu'il avoit eues et des temptations que il avoit receu; et par l'inspiration de dame et saincte Sapience, il en ploure et gemit.

[Fol. 3v.] Aprés en la page ensuivant est une hystoire comment il advint au disciple par l'instigation de l'ennemy une ymagination de l'oppinion des epycuriens[13] qui tiennent que la felicité de l'omme est en suivant ses plaisances et desirs. Et est le deable qui monstre au disciple les gens buvans et mengans et menans[14] joyeuse vie; et aprés, les bestes mues aux champs, mengans et dormans, et ne pensent point qu'il advenra ne que il sera aprés ceste vie.

[12] MS. *la parafe*; elsewhere in the copy, the word is masculine.

[13] MS. *epycurions*.

[14] MS. *menons*.

leuure de sapience louoit z comandoit
monlt sapience en disant Sapiencia
est speciosior sole et super omnem disposici
onem stellarum luci comparata inue
nitur prior. Sapience est plus bel
le que le soleil cest la plus belle et la plus
clere et premiere de toutes les estoilles
plus reluisant que lumiere du jour.

ste lecon. Il se commenca a embraser
en lamour de sapience et si disoit
de dens lui. Certainement se tu
cerchoies tout le monde. tu ne pour
roies pas trouver une dame qui
feust semblable a ceste tresnoble da
me tant en grace et en vertus.

La cinq. hystoi

The fifth illustration precedes the paragraph commencing Advint que l'ennemy; *it concerns the temptations of the Disciple: how the Devil shows him the delights, wealth and sensuous pleasures of the world.*

And then, the said Disciple is in his cell, turning inward, and repenting his thoughts and the temptations he had received. Inspired by holy Sapientia, he weeps and laments out loud about them.

Next, on the page that follows, is a picture illustrating how there came to the Disciple, under the incitement of the Devil, a mental picture of the aims of the Epicureans. They claim that Man's happiness comes from following his own pleasures and desires. The Devil is there showing the Disciple humans drinking and eating and leading the good life; then, he shows dumb creatures in the fields eating and sleeping, not thinking about what will happen nor what there will be after this present life.

VI (1) *Sapientia in a Terebinth Tree* Fol. 20v.
(2) *Sapientia asks the Disciple for his Heart*
(3) *Lessons in the Refectory*
(4) *St. Paul Addressing the Disciple*

[Fol. 3v.] La vj^{e} hystoire est a l'endroit du [fol. 4] parafe qui commence: *Le terebint est.* Avecques les autres auctoritez dessus diz ouyt le disciple une belle auctorité moult recommandent sapience, qui se commence: *Ego quasi therebi[n]tus expandi ramos meos etc.*[15] Et est en l'ystoire dame Sapience en une abre[16] appellee therebintus qui est moult vertueuse et souef fleurent, et tient dame Sapience a la main dextre une bouette plaine de ongnement souef fleurent, et ung brain de baulme, et a la senestre ung livre, signifient que sa doctrine est doulce, souef fleurent, plaine de grace et d'onneur, come dit l'auctorité dessus dite.

Aprés est une autre auctorité que dit meisme Sapience a cellui qui l'aime: *Prebe fili cor tuum michi*[17], et est le disciple a l'inspiration de ceste belle requeste de dame Sapience, la voulent embrasser et baiser; et elle, le recepvoir. Et est a dire: "Mon filz, donne moy ton cuer ou met ton cuer en moy!"

Ainsi que le devot disciple de Sapience ouoit lire ces belles auctoritez, louant la benoite sapience de Dieu, lui, estant ung jour ou refectuoir, ne se peut contenir, mais, de grant amour esmeu, frappa du doit sur la table et s'escria: *Verum est, verum est,* "Il est vray ce que on en dit." Et est ce qui est figuré en ce tiers parquet ou a ung refectuoir.

Une autre foiz estoit ledit disciple amoureux de la Divine Sapience estudient en sa chambre, et estoit fort desirant [savoir] qui estoit ceste dame Sapience si belle, si noble, si haulte, de quoy il avoit tant ouÿ de louanges. Et lors lui vint au devant le dit de saint Pol qui dit: *Christus est Dei virtus et Dei sapiencia etc.,*[18] c'est a dire que "Nostre Seigneur Jhesucrist est la vertu et la sapience de Dieu, qui contient en soy tous les tresors de sapience et de science." Et est ce qui est figuré en ce quart parquet ou le disciple est interrogant, et saint Pol ly respont.

The sixth illustration is where [fol. 4] the paragraph Le terebint est *begins. Along with the other sayings already mentioned, the Disciple heard an expressive one strongly recommending wisdom. It runs* Ego quasi therebi[n]tus expandi reamos meos etc. *And in the illustration, Sapientia sits in a tree called terebinth which has powerful properties and is sweet-smelling. She holds in her right hand a box full of sweet-smelling ointment and a sprig of balm, while in her left, there is a book. All this means that her teaching is gentle, fragrant, full of grace and esteem, in line with the wording of the afore-mentioned saying.*

Then, there is another saying which Sapientia utters to anyone who loves her: Prebe fili cor tuum michi. *And the Disciple is inspired by her loving request, and desires to embrace and kiss her. She wishes to receive him. And the words of the saying run: "My son, give me your heart or place your heart in me."*

So it was that the pious Disciple of Sapientia used to hear these fine sayings read out loud that praised God's wisdom; and one day, when he was in the refectory, he could not restrain himself. Carried away by overwhelming love, he struck the table with his finger and cried out Verum est, verum est: *"What is said is true!" And this is the subject of the third compartment where there is a refectory.*

On another occasion the said Disciple, enamoured of Divine Sapientia, was studying in his cell. He was most desirous of knowing the identity of the woman Sapientia who was so fair, so noble,

[15] Ecclesiasticus, XXIV, 22, where the verb is *extendi*.
[16] The form *arbre* is the scribe's preferred one.
[17] Prov. XXIII, 26.
[18] Cf. 1 Cor. I, 24.

so distinguished and about whom he had heard so many praises. There came to him the saying of St. Paul which runs: Christus est, Dei virtus et Dei sapientia etc., *in other words "Our Lord Jesus Christ is the strength and wisdom of God who contains within Himself the complete treasury of wisdom and knowledge." And this is the subject depicted in the fourth segment where the Disciple questions St. Paul and the latter answers.*

VII (1) *God the Father*
(2) *Man of Sorrows*
(3) *Christ Crowned by God the Father*
(4) *Adam and Eve*
(5) *Miracles of Christ*
(6) *Christ Rejected at Nazareth*
(7) *Sapientia and the Disciple*
(8) *Nativity*
(9) *The Temptation of Christ*

Fol. 23

(*Reproduced also as Frontispiece*)

[Fol. 4] La vij^{e} hystoire est au commencement du secont chapitre qui se commence: *Venerunt*[19] *michi omnia bona.* En celui chapitre le devot disciple desire moult de venir a la congnoissance de la Divine Sapience. Et est ce qui est figuré en ung petit parquet en bas a la dextre, ou ledit disciple prie sa bien amee, dame Sapience, que elle lui vueille donner congnoissance de soy. Et elle lui respont au cours du chapitre que on vient a congnoissance de choses haultes et souveraines par les creatures basses et inferiores (*sic*). Et est ce que il met en ceste hystoire, car il met le paradiz terrestre ou furent mises toutes les creatures, c'est assavoir l'omme et la femme, et la serpente, et les bestes et les arbres qui furent creez pour la consolation de l'umain lignaige. Et dessus la terre: le ciel et[20] les estoilles, et le souleil et la lune, et les angelz et archangelz et par la congnoissance de ycelles creatures on vient a la congnoissance de Dieu le createur. Et est ce qui est figuré en la renge du costé dextre.

Et pour ce que en la Divine Sapience Nostre Seigneur Jhesucrist est humanité et divinité, pour venir a la congnoissance de la divinité on doit premierement congnoistre l'umanité. Et est ce qui est figuré ou moyen parquet d'embas ou est l'enfance de Jhesucrist. Emprés est la temptation de Jhesucrist; dessus, sont les miracles a la predication Nostre Seigneur et comment les Juifz le vouldrent trebucher par la mon—[fol. 4v.]—taigne et il eschappa des mains des Juifz comme l'evangile dit: *Jhesus autem transiens per medium illorum ibat.*[21]

Au dessus du parquet du milieu est la passion, sepulture et resurrection en laquelle Nostre Seigneur se monstra vray homme et vray Dieu le createur. Ainsi par la congnoissance des creatures on vient a la congnoissance de Dieu le createur[22]. Et par la congnoissance de l'umanité de Nostre Seigneur Jhesucrist on congnoit ses miracles; et par ses miracles, sa divinité.

Aprés en hault, au darnier parquet a la senestre, est figuree la gloire de Nostre Seigneur Jhesucrist que Dieu le pere lui donna aprés l'Ascention, et le couronna de gloire et honneur[23], et le constitua seigneur et juge sur toute creature, et fist seoir a sa dextre perpetuellement.

The seventh illustration is at the beginning of the second chapter whose opening words are Venerunt michi omnia bona. *In this chapter the devout Disciple deeply desires to attain knowledge of Divine Sapientia. This is what is depicted in a small compartment on the lower right: the said Disciple asks his beloved Sapientia to give him knowledge of herself. She answers him in the course of the chapter that one comes to a knowledge of high and mighty things through lowly and base creatures. And this is what he puts in this illustration, for he includes the Earthly Paradise where all creatures were placed: man, woman and the serpent, and the beasts and the trees created for the delight of mankind. Above the earth are the heavens, the stars, the sun and the moon, the angels and archangels. Through knowledge of these created things, one comes to know God the Creator. And this is depicted in the row on the right side.*

[19] MS. *voverunt.*
[20] MS. *Et dessus le ciel et la terre.*
[21] Luke, IV, 30.
[22] MS. *dieu et le createur.*
[23] MS. *gloire de honneur.*

La. vij^e. hystoi

Le second chapitre · Comment on peut venir a la congnoissance de la diuinite par la passion de Jhūcrist / et quelle forme dieu prist par la ditte passion de Jhūcrist ·

And because in Divine Sapientia, Our Lord Jesus Christ is humanity and divinity, one must first know humanity in order to come to knowledge of divinity. This is represented in the middle segment at the bottom where Christ's Infancy is depicted. Then comes the Temptation of Jesus Christ; above, are miracles that occurred when He was a preacher, followed by a depiction of the Jews wishing to cast Him from the moun—[fol. 4v.]—tain; and He escaped from their hands, as the Gospel says: Jhesus autem transiens per medium illorum ibat.

Above the centre compartment is the Passion, Entombment and Resurrection when Our Lord showed himself to be true man and true God the Creator. Thus through knowledge of all creatures, one comes to knowledge of God the Creator. And through knowledge of the humanity of Our Lord Jesus Christ one knows His miracles; and through His miracles, His divinity.

Next, above, in the last segment on the left is a representation of the glory of Our Lord Jesus Christ which God the Father bestowed on Him after the Ascension. He crowned Him with honour and glory, and made Him judge and lord over every creature, and made Him sit at His right hand for ever.

VIII *Sapientia and the Disciple behold Christ Carrying the Cross* Fol. 24

[Fol. 4v.] La viij^e hystoire est en ce mesmes secont chapitre a l'endroit du parafe qui commence: *De tant comme je semble*; et est la cause de l'ystoire pour ce que le disciple demanda ung pou devant et pria a sa tres bien amee dame Sapience, que elle se monstrat a luy en forme et en maniere tres piteuse et tres amoureuse qu'elle acquist de sa tres amere passion. Et elle lui apparut en la plus piteuse forme et figure que on ne puet considerer et penser, la charité et amour de Nostre Seigneur Jhesucrist, c'est assavoir tout nu, tout playé, decraché, debouté, portant la croix a ses espaules. De quoy, quant le disciple l'ouyt, s'en esbaït, disant que ce n'estoit pas forme ne figure amoureuse. Et Sapience lui respondit par plusieurs similitudes, car les vrays amans ne quierent pas ce qu'il voyent par dehors de couverture mais ce qui est contenu dedens, a la semblance de deux coffres, l'ung vieil et vermolu, plain de joyaulx et l'autre couvert de drap d'or et plain de charoigne. Et aussi les mondains ne recraingnent pas la pointure de l'espine pour la souefve odeur de la rose. Et ce est qui est contenu et figuré en ceste hystoire. Ycy est le doulx Jhesus tiré, bouté, playé et decrachié, tout nu, portant la croix, et dame Sapience monstrant au disciple deux coffres[24], comme dit est, et tenent une branche de rosier a toutes les espines et roses.

The eighth illustration occurs in this same second chapter where the paragraph De tant comme je semble *begins. The reason for the picture is as follows: a little earlier, the Disciple asked and begged his dearly beloved Sapientia to show herself to him in the quite pitiful and endearing likeness and manner she assumed at His most unmerciful Passion. And she appeared to the Disciple in the most compassionate likeness and form possible to imagine, the charity and love of Our Lord*

[24] MS. *costes.*

Jesus Christ: namely, completely naked and wounded, spat upon and buffeted, and carrying the Cross on His shoulders. When the Disciple heard about all this, he was in dismay, saying that such was not a loveable likeness or form. Then Sapientia replied, using several similitudes: true lovers do not desire what is visible on the covering outside but rather what is within, like the two chests, one being old and worm-eaten, yet filled with jewels, the other covered with a cloth of gold and full of rotting flesh.

Furthermore, because of the rose's sweet scent, the worldly wise do not fear the prick of the thorn. And this is what is contained and represented in this illustration. Behold sweet Jesus, dragged, buffeted, wounded and spat upon, totally naked, bearing the Cross; and Sapientia shows the Disciple two chests, as is said, while holding a briar with all its thorns and roses.

IX (1) *Agony in the Garden* Fol. 25
(2) *Christ before Annas*
(3) *Christ before Caiaphas*
(4) *Christ before Pilate*
(5) *Christ before Herod*
(6) *Buffeting of Christ*
(7) *David's Soldiers at the Well*
(8) *Soldiers bring Water to David*
(9) *Sapientia and the Disciple*

[Fol. 4v.] La ix^{e} hystoire est au commencement du tiers chapitre qui se commence: *Ante diem festum.* Ou quel chapitre recite le commencement de la passion Nostre Seigneur Jhesucrist, commencent a la priere qu'il fit ou jardin, presens troys de ses disciples, c'est assavoir: Pierre, Jehan et Jacques, en laquelle sua sang, et pria Dieu le pere que il se povoit faire que il ne souffrit point ce que il avoit a souffrir. Et ce est figuré ou premier parquet d'en hault a la dextre.

A l'autre parquet est comment Nostre Seigneur Jhesucrist fut presenté a Anne, prince des prestres de la loy, la ou il fut democqué et frappé.

Et en l'autre parquet a la senestre en hault est comment il fut presenté a Caÿphe, prestre de la loy, et compaignon de Anne, lequel prophetisa que il estoit expedient que il mourut ung homme pour le peuple.

En l'autre parquet de la moyenne renge a la dextre est comment il fut presenté a Pylate et lui demanda se il estoit roy, et il respondit que vrayement il estoit, et les gens de Pylate le saluoient comme par derision.

Aprés ou milieu de la page est comment [fol. 5] il fut mené a Herode et il se mocqua de lui en le vestant d'une robe blanche.

A la senestre ou milieu est comment il fut ramené à Pylate et couronné d'espines, les Juifz crians que on le crucifiast, et lors fut jugé a estre pendu a la croix.

Et pour ce que le disciple avoit demandé a sa tres amee dame Sapience que elle lui monstrast son amour, sa doulceur et sa divinité, elle lui monstre par le contraire, amertumes, douloureuses enfermetez et passions esquelles veult que son disciple et son amy se conforme; pour ce le disciple se complaint, et elle l'enhorte que il prengne couraige vigoureux a la semblance des troys fors chevaliers de David qui alerent parmy l'ost de leurs ennemis puiser de l'eaue de la fontaine ou cisterne de Bethleem pour faire plaisir au roy David, lequel, quant il vit que ses chevaliers s'estoient mis en peril pour acomplir le desir de leur roy, il n'en voulut point boire, mais l'offrit a Dieu en sacrifice; et c'est ce qui est figuré en deux parqués bas, a la dextre.

Et a la senestre ensuivant est dame Sapience qui monstre a son disciple tout ce qui est dit et figuré par cy devant, concluent que ainsi fault il que le vray aimant de Sapience se conforme aux passions de Nostre Seigneur Jhesucrist, et que il n'ait pas le cuer failly, mes que il prengne en soy couraige vigoureux.

The ninth illustration is at the opening of the third chapter Ante diem festum. *In the chapter is rehearsed the Passion of our Lord Jesus Christ, beginning with the prayer He formulated in the Garden when three disciples were present, namely, Peter, John and James. He sweated blood and prayed to God the Father that it would come about that He would not suffer what He was destined to suffer. This is the subject of the first segment at the upper right.*

In the second compartment Our Lord Jesus Christ is brought into the presence of Annas, the High Priest of the Faith; on this occasion He was mocked and smitten.

In the other segment, upper left, Christ is brought in front of Caiaphas, a priest of the Faith

Pour mettre en ton cuer et p^r souvent
ramener a ta memoire et donner art
et forme a toy de moy amer te vueil re
traire z raconter une partie de celle saincte
passion Coment le bon disciple de sapi
ence se doit conformer a la passion de Jhu
crist / et pour quoy il nous voulut par
telle mort racheter. Troisiesme chapitre.

La. iv^e. hyst

and colleague of Annas who prophesied that it would be expedient that a man die for the people.

The next picture in the centre row on the right, represents Christ before Pilate who asked Him if He was king; He said He was in truth, whereupon Pilate's men greeted Him with derision.

Then, in the middle of the page, we see how [fol. 5] Christ was led to Herod who mocked Him by clothing Him with a white robe.

On the centre left, Christ is shown taken back to Pilate and crowned with thorns. The Jews cried out that He be crucified; and He was then condemned to be hung on the Cross.

Because the Disciple had asked his dearly beloved Sapientia to show him her love, her tenderness and her divinity, she shows him, on the contrary, piteous things, painful infirmities and suffering. This was because she wanted her Disciple and friend to mould himself on all these. Whereupon, the Disciple complains, so Sapientia exhorts him to be courageous and of stout heart in the manner of David's three valiant knights, who ventured into the midst of the enemy's army to draw water from the cistern or well of Bethlehem in order to please their king. When David saw that his knights had put themselves in danger to carry out his wishes, he did not want to drink, but offered the water to God as a sacrifice. And this is what is depicted in two compartments, bottom right.

Next on the left, Sapientia shows her Disciple what is represented and stated above, and concludes that thus it behoves the true devotee of Sapientia to conform to the sufferings of Our Lord Jesus Christ, and not have faint heart, but show resolute courage within himself.

X (1) *Christ Nailed to the Cross* Fol. 27v.
(2) *Christ Offered the Sop*

[Fol. 5] La xe hystoire est en ce mesmes tiers chapitre, a l'endroit du parafe: *Quant je fuz amené ou lieu etc.* Ceste hystoire despent de la precedent et est accomplissement d'elle. Pour ce que le disciple c'estoit espouenté d'avoir ouÿ le commencement de la passion et bataille de son amé seigneur, et sa tres bien amee dame Sapience l'avoit conforté et alumé par plusieurs exemples et induit a suivre son Seigneur, maistre et roy, comme loyal chevalier et bon campion, maintenant il poursuit son hystoire et racompte le Crucifiement de Nostre Seigneur Jhesucrist en lieu deshonneste et entre deux malfaitteurs larrons. Et c'est ce qui est figuré clerement en ceste hystoire.

The tenth illustration is in this same third chapter, at the place of the paragraph Quant je fuz amené ou lieu etc. *The illustration is related to the preceding one and is its conclusion. Because the Disciple had become fearful on hearing about the beginning of his beloved Lord's Conflict and Passion, and his very dear Sapientia had comforted him and enlightened him by several* exempla *and had induced him to follow his Lord, Master and King as a loyal knight and doughty champion, now he continues the narrative and relates the Crucifixion of Our Lord Jesus Christ on an ignoble site and between two malefactors. And this is what is clearly depicted in this illustration.*

XI (1) *Jael Welcoming Sisera* Fol. 30v.
(2) *Jael Slaying Sisera*
(3) *Samson and Delilah*

[Fol. 5] La xj^{e} hystoire est ou quart chapitre qui parle tout de l'amertume de peché et de penitence, et est a l'endroit du parafe qui se commence: *O peché, que m'as tu procuré.* Ceste hystoire signifie la faulceté decevable du monde. Et met ycy deux figures, la premiere est de Jahel, femme de Abel Cineus qui deceut Cisaram, le prince de la chevalerie de Chanaan, lequel avoit guerré contre Barach qui estoit juge et capitaine des filz de Israel. Ce Cisara fut vaincu de Barach et s'enfuit de pié, cuidant se sauver avecq ce Cineus qui lors avoit paix avec le roy de Chanaan. Et quant la femme de se dit Cineus le vist, elle le receut en tromperie et deception et lui dist: "Entrez, monseigneur, entrez, n'ayez paour." Et quant il lui demanda de l'eaue a boire, elle lui donna du lait. Et puis elle le mist dormir en sa tente, et en dormant, elle lui ficha ung [cloe] en la temple, et le monstra ainsi mort a Barach, capitaine des filz de Israel, son ennemy. Et ceste hystoire est escripte ou quart chapitre du livre des Juges[25].

L'autre figure ou tiers parquet est de Sanson le fort qui fut trahy de son amie Dalila[26] qui lui feist crever les yeulx par les Philistiins qui estoient ses ennemis. Et est cecy escript ou xvj^{e} chapitre du livre des Juges[27].

Ces deux figures signifient la faul —[fol. 5v]—ceté decevable du monde. Qui promet amitiez et presente douleurs[28], a la fin deçoit a mort et a mort pardurable.

The eleventh illustration is in the fourth chapter which treats of the bitterness of sin and penitence; it is at the place where the paragraph O peché que m'as tu procuré *begins. This painting illustrates the falseness and deception of the world. And he provides here two representations, the first one is of Cinaeus's wife, Jael, who deceives Sisera, the prince of the knights of Canaan, who had made war on Barak, the judge and leader of the sons of Israel. This Sisera was defeated by Barak and fled on foot thinking he would escape with the said Cinaeus who then was at peace with the King of Canaan. And when the wife of Cinaeus saw him, she received him with treachery and deception, saying: "Enter, my lord, enter, fear not." And when he asked her for water to drink, she gave him milk. Then she allowed him to sleep in her tent, and while he slumbered, she drove a nail into his temples. She then showed the corpse to Barak, his enemy and leader of the sons of Israel. And this story is written down in the fourth chapter of the Book of Judges.*

The next representation in the third section of the painting is that of the valiant Samson who was betrayed by his mistress Delilah. She had his eyes put out by the Philistines who were his enemies. And this story is written in the sixteenth chapter of the Book of Judges.

These two representations depict the falseness [fol. 5v.] and deception of the world. He who promises friendship and presents sorrow, in the long run is a deadly deceiver and earns everlasting death.

[25] Judges, IV, 17-22.
[26] MS. *dalida.*
[27] Judges, XVI, 13-31.
[28] MS. *promet et presente douleurs et amitiez.*

hystoire.

O pechie que mas tu procure. O
maleureuse coulpe. O voulente
desordonnee en quel estat mas tu mis
O monde vil ⁊ ort qui par ta mauuai
stie ⁊ par tes fallaces decois les ames
Bien est mauldit qui sert a toy et qui
a ta voulente se soubsmet Qui en toy
se fie il est rempli dangoisses de miseres
et dauersitez. Se tu appers par dehors
amables si as tu les dens agues et
affillees par dedens pour mordre et po
deuorer Se tu es debonnaire en apparence
tu es felon ⁊ despiteux en ton essence.
Las las faulx traictre que me
promis tu au commencement quant
tu me voulz deceuoir Tu en eus bien
la maniere. Tu estriuoies a moy doul
cement tu me donnoies de petis dons
souuent tu affaitoyes tes parolles ⁊
ton langaige pour moy attraire. Et
si comme la seraine attrait en mer
les personnes par son chant Tout ainsi
par tes raisons dorees ⁊ emmellees
mas tu oste mon sens ⁊ mon entende
ment. O mondaine vanite O mon
daine plaisance tu mas monstre vng
viaire charnel ⁊ attrayant vne gorge
polie vngs yeulx riants Et par ainsi

faire tu as nauire ⁊ perce mon poure
dolent cuer Je demandoye leaue et
tu mas donne le lait Je demandoie a
boire et tu mas enyure de poisons et
de venin Tu mas presente belles pommes
et doulces mais elles contendrent la
mort Tu mas donne pour mon loyer
les roses de ton rosier Tu mauoyes
promis que mes [illegible] richesses mon
auoir et mes richesses jamais ne me
fauldroient O malheureux retart.
O face enuenimee veci le tempz que
tes promesses mont villainement
barretee et tout a vne foys failli Et pis
encore car tu mas pris ⁊ enlacie et
transmue apres toy jusques a la mort
Tu mas fait comme on seult faire
a loysel ou a la beste mue a qui len
monstre vng rain sel vert pour la
conduire a la mort. Certes mau
uaise beste sauuaige O beste cruelle.
tu poins maintenant cruelment. celui
que tu souloies oindre doulcement.
Tu laisses cheoir durement qui sou
loyes releuer courtoisement Tu faic
que tu ayes voix de seraine tu fais
entendant que tu es plain de joyes.
et je scay tout le contraire quant

XII (1) *Apostles* Fol. 33
(2) *Martyrs*
(3) *Confessors*
(4) *Female Saints*
(5) *City of Religion in Ruins*
(6) *Money-Changer*
(7) *Charity and Envy*
(8) *Chastity and Lechery*
(9) *Obedience and Sloth*
(10) *Chastisement and Pride*
(11) *Poverty and Avarice*

[Fol. 5v.] La xij^{e} hystoire est au commencement du cinquiesme chapitre lequel commence: *Quomodo sedet sola civitas.* Ceste hystoire signifie l'estat de l'Esglise et de religion. Ou milieu en hault est une cité qui signifie l'Esglise et religion. De ceste cité sont saillis tous les sains de paradiz qui sont au costé dextre: les apostres, les martirs, les confesseurs, et les vierges et sainctes femmes, et ce est l'estat de saincte Esglise passé, ou quel temps Jhesucrist estoit prince de la cité et y regnoit paisiblement.

Depuis la cité est venue en grant ruine par le mauvais gouvernement des habitans. Et est ceste cité toute plaine de monstres et de bestes sauvaiges en guise d'ommes qui signifient que ja çoi ce que les[29] gens ayent habit de cristien ou de religieux, si portent ilz le cuer et la voulenté de beste mue. Et Jhesucrist, jadiz prince et roy de la cité, est a la porte comme pelerin, voulent entrer en la cité et desirent regner et dominer en ycelle pour le salut de son peuple, mais il est rebouté par ces monstres et gens de mauvaise vie.

Et y a deux femmes a l'entree de la porte qui signifient deux vices qui fort impugnent Jhesucrist et l'empeschent fort a entrer et dominer en la cité de l'Esglise et de religion et en la maison de nostre cuer, ce sont Ire et Oyseuze. Ire est vieille rechignee qui tient deux caillous desquelz fait saillir feu de noises et de riotes. Et Oyseuse est jouvencelle qui ne veult que jouer et perdre son temps et ce demonstre elle qui tient sa main en son sain et de l'autre se joue a son gan.

Au pié d'une tour sont ycy deux homes, l'ung aiant le visaige enflé et l'autre une gibesiere, qui minent au pié d'une tour et hotent la pierre quarree accouplant deux murs, qui signifie paix, pour faire cheoir l'edifice.

Au costé senestre sont cinq dames de noble attour et grande signification qui sont bouteez hors de l'Esglise et de religion. Et derriere elles sont plusieurs vices qui persecutent et poursuivent ces nobles dames par lesquelles sont vertus signifiez.

La premiere dame est Charité qui porte a sa main dextre une pierre quarree ou y a troys lettres, c'est assavoir P. A. X. qui assembleez font *pax*: "paix"; et a son espaule porte une auge a mortier qui signifie que c'est le ciment qui fait tenir les autres pierres et vertus ensemble. Et a sa sainture pent une bouette a lettres ou a escript *Testamentum pacis*: le testament de paix.

Derriere ceste noble dame Charité, paisible et portant Paix et Amour, s'ensuit une vieille maigre a quatre piez de laquelle saillent de ses yeulx deux sagettes qui signifient faulx regart et despit d'onneur[30]. Celle a nom Envie, portant sur son doz deux de ses filles, quelle est la mere telles ses filles: la premiere a nom Trayson, elle a ung faulx visaige comme riant, et a sa main senestre porte une bouette plaine d'ongnement qui signifient flaterie et beau semblant par dehors, et si a son propre et naturel visaige cruel et aïreux[31], plain de felonnie et de despit. Et a sa dextre derriere tenant une dague [fol. 6] aigue pour poindre et frapper quant temps sera et que son opportunité vendra.

[29] MS. *les les.*

[30] MS. *regard despit et lonneur*

[31] MS. *eureux.*

La .xvi^e. hystoi

Derriere est la seconde fille d'Envie nommee Detraction, tenent a sa dextre ung os de char vive et le rongent, et en la senestre, une broche enhastee de oreilles des escoutans sans lesquelles ne vauldroit riens parler en derriere d'autruy, se il n'y avoit qui l'escoutast et a cui ne pleust. Ces deux desloyales filles avecq leur mere sont qui boutent hors de toute bonne compaignie de l'Esglise et de religion Charité, Paix et Amour.

Aprés est la seconde dame Chasteté qui a les mains armeez de gantelés de fer contre tous atouchemens deshonnestes. Elle tient a sa dextre ung baton long a faire lit, par ce signifient que elle doit estre dortouriere et garde du dortouoir[32] de saincte religion et n'y doit laissier entrer quelconque creature qui ne veulle chasteté garder. A sa senestre tient une lampe ardant et si est seinte d'ung beau tissu vert a ses reins, en signifiant que toute creature religieuse doit avoit en cuer et cogitation, chasteté et pureté, bonnes oeuvres en operation.

Derriere ceste dame pure et nette vient une laide vieille et orde qui couvrir sa laidure se targe d'une face de pucelle que nous povons appeller Farderie. Et pour plus monstrer son ordure est montee sur une truie qui ne prent plaisir fors que se voultrer en fange et en boue. Elle porte aussi ung dart de quoy elle navre tous ses contraires ou par dedens en cuer ou par dehors a l'ueil. Et celle est ditte selon le commun langaige Luxure et selon les anciens poetes estoit appellee Venus. Elle boute Chasteté hors de ceste religion.

Aprés ceste belle vertu de Chasteté est une autre dame et maistresse de vertus qui est nommee Obedience, laquelle porte en ses mains de beaux gés et lasses de soye qui signifient les commandemens et consaulx et observances de saincte religion par lesquelz sont lïez les piez et les mains des bons religieux et bien obeissans auxquelz n'est pas de la cité aler[33] ne de rien faire sans obeyssance.

Derriere ceste noble dame [et] maistresse est une vieille, laide et boyteuse, qui porte une congnee a son col et ung fardeau de cordes noires et en la main tient une grosse corde noire. Ceste vieille est ditte Paresse qui de sa congnee assomme les negligens, et est ditte ceste congnee Ennuy de bien faire. Et quant sont ainsi assommez, elle les lie de ses cordiaux, c'est assavoir Fetardie, Negligence et Lacheté et puis, quant elle a eu ainsi son homme lïé de ses cordiaux, elle lui met la corde ou col de quoy Judas se pendit: c'est Desesperation.

Aprés ceste dame et maistresse de religion Obeyssance vient sa prieuse nommee Correction qui tout corrige et reprent. Quan que de transgression elle treuve, elle fourbit et escure des verges que elle tient en sa main. Et affin qu'elle ne excede en ses corrections, elle porte une targe qui est ditte Prudence.

Derriere ceste prieuse viennent deux autres vieilles, l'une portant l'autre. Celle qui est portee est enflee oultre mesure tellement que aler ne pourroit, se portee n'estoit. C'est Orgueil qui est enflee, plaine de vent de propre reputation; elle porte une corne ou front qui signifie Fierté. Elle tient ung cornet en sa main [fol. 6v.] qui signifie Ventance. Elle a ung souflet en escharpe qui signifie Vaine Gloire; elle porte en sa dextre ung baton de cornelier qui signifie Obstination. Elle a ung manteau affublé qui signifie Ypocrisie de quoy cueuvre ses iniquitez. Celle qui la porte, c'est Flaterie; sens elle, ne vivre ne aler pourroit. Sa maistresse qui ditte est porte deux esperons dont l'ung a nom Rebellion, l'autre Inobedience dont point et stimule fort sa porteresse. Ceste cy porte ung mirouoir ou continuellement se mire Orgueil, et se ce n'estoit qu'elle ne se veïst, elle vouldroit tout homme hurter de sa corne de fierté. Ce sont ycy les filles ou operations d'Orgueil qui boutent hors de religion toute subjection et correction.

Aprés toutes ces nobles dames et leurs vices opposez et les deboutans hors de saincte Esglise et de religion, est la souveraine et la plus glorieuse de toutes, Povreté voluntaire, a laquelle Nostre Seigneur donne en ce monde desja le royaume de paradis. Et pour ce, porte elle la couronne comme possessoire de l'eritaige de paradiz. Rien ne porte en ses mains, et pour ce partout peut elle aler

[32] MS. *dortouoire.*

[33] MS. *pas la cite de aler.*

ne rien ne doibt doubter, ne vivre ne mourir. Elle porte sur son dos une enclume qui est ditte Pacience, car communement sont ceulx lesquelz, amateurs du monde, n'en tiennent[34] compte, et sont de tous persecutez.

Sa droitte contraire est Avarice qui derriere elle est pour la bouter hors de saincte Esglise et de religion. Ceste cy, laide vieille, bossue et boyteuse, n'a talent de voler au ciel, car sur elle porte ung sac de metail et a son cueur a tousjours soin et sollicitude de le garder et de le multiplier. Pour ce elle a vj mains et deux mongnons. Son pere de l'infernal pallut luy couppa les mains desquelles elle labouroit pour acquerir son vivre et desquelles elle donnoit, et depuis ce dy n'eust le cuer de riens faire, ne de donner par ausmone a ceulx qui en ont mestier. Elle a vj mains autres, troys dextres et troys senestres, et a la dextre a premierement une main a ungles de griffon, c'est la gentille qui est ditte Pillerie[35]; de ceste main n'a povre laboureur, ne coq[36] ne poule ne blez ne avaine qui lui puisse demourer. L'autre main derriere porte pareillement a griffes qu'elle porte derriere, car ce qu'elle prent est en secret et est ditte Larrecin. L'autre est tenant ung baston crochu duquel elle houe et mine et fouit, et ceste appartient aux prelatz de l'Eglise qui gratent et minent et prennent les biens de l'Eglise a tort et a travers, ce que peuvent sans riens y mettre. Et pour ce est ce que les vieilles esglises s'en vont maintenant tout a neant. A la senestre sont troys autres mains; la premiere porte une lime: ceste appartient aux faulx marchans et aux changeurs qui rongnent les escus, appeticent le poiz et les mesures pour decepvoir le povre pueple. En l'autre main porte unes[37] balances esquelles poise le souleil et le zodiac[38]. Ceste cy est ditte Usure car elle vent le Temps. L'autre main ditte est Truanderie qui par questes et quaimenderies multiplie avoir.

Ceste vieille cy devant ditte trait la langue enflee et meselee par quoy est signifié Mençonge, car bonnement ne puet on acquerir ou multiplier richesses et avoir par les manieres dessus dittes sans mentir.

Aprés en ung parquet bas ou milieu [fol. 7] de la page est ung changeur assis a son bang qui reçoit des gens d'eglise joyaulx et vaisselle d'eglise et en baille des escus et monnoye, laquelle chose fait moult a la ruine de l'Eglise, car quant les prelas ou gouverneurs de l'Eglise ont despendu en leurs excés les revenues de l'Eglise, ilz vendent les joyaulx et vaisselles consacrez et donnez au service et a l'onneur de l'Eglise. Ainsi leurs esglises vont en ruine et ne treuve on maintenant homme qui veulle mettre ung denier a la reparation de leurs esglises.

The twelfth illustration is at the head of the fifth chapter which commences Quomodo sedet sola civitas. *It represents the state of the Church and Religion. In the upper centre is a walled city which depicts the Church and Religion. From the city have emerged all the saints of Paradise who are on the right-hand side: apostles, martyrs, confessors, virgins and women saints; and this was the condition of the Holy Church in the past at a time when Jesus Christ was prince of the city and reigned there in peace.*

Since then the city has fallen into great ruin because of the bad government of the inhabitants. It is filled with monsters and wild animals disguised as humans, which means that, although the people wear the garb of Christians or of religious, they have the hearts and minds of dumb animals. And Jesus Christ, formerly prince and king of the city, is at the gate in the guise of a pilgrim, wishing to enter and desirous of reigning over and ruling the city for the salvation of its people, but He is turned away by these monsters and people of evil life. And there are two women at the entrance-gate who represent two Vices who oppose Jesus Christ and prevent Him outright from entering and ruling the city of the Church and Religion and the abode of our hearts; these are Wrath and Idleness. Wrath is an old, grimacing hag who holds two pebbles with which she sparks

[34] MS. *ne tiennent.*

[35] MS. *pilliere.*

[36] MS. *cog.*

[37] MS. *une balances*, the scribe no longer recalls that *unes* conveys the idea of a pair.

[38] MS. *zodiacal.*

off disputes and disturbances. Idleness is a young maid who wishes only to play and idle away the hours, and this she makes clear with one hand in her bosom while the other plays with her glove.

At the foot of the tower are two men, one with a puffed face, the other holding a game-bag. They mine under the tower and carry off the square stone joining the two walls together, which means peace. This is to make the structure collapse.

On the left side are five women, nobly attired and of great importance; they are thrust out of the Church and Religion. Behind them are several Vices who hound and pursue the noble women who represent Virtues.

The first lady is Charity who holds in her right hand a square stone bearing the three letters P.A.X., which together constitute pax, *meaning 'peace'; on her shoulder is a mortar tray representing the cement that binds the other stones and virtues together. From her belt hangs a box for letters, bearing the name* Testamentum pacis *'the Last Will of Peace.'*

Behind this noble lady Charity, who is peace-loving and bearer of Love and Peace, comes an old thin hag on all fours; from her eyes dart two arrows which represent False Glance and Contempt for Honour. Her own name is Envy and she carries on her back two daughters. Like mother, like daughter. The first daughter, named Treachery, has a false face which seems to wear a smile, and her left hand holds a box full of pomade. They represent Flattery and Fair Countenance on the outside, yet her own natural face is cruel and fearful, and is filled with contempt and spite. In her right hand, held behind her, is a pointed dagger [fol. 6] to stab and strike with when the moment is right and the occasion presents itself. Behind Envy is the second daughter, Denigration, grasping in her right hand a bone with living flesh which she gnaws at. In her left hand she holds a brooch encrusted with the ears of listeners; without the ears it would not be worth speaking behind peoples' backs, if there was not someone listening and someone who was disgruntled. These two deceitful daughters and their mother are the ones who cast out Charity, Peace and Love from every good assembly of the Church and Religion.

Next comes the second lady, Chastity, whose hands wear iron gauntlets against dishonourable contacts. She has in her right hand a long rod for making beds, signifying that she should be the dormitory-maid and keeper of the dormitory of Holy Religion, and should not permit any person to enter who does not wish to preserve chastity. In her left hand she holds a lighted lamp; around her waist is a beautiful green cloth, signifying that every person in Religion should be chaste and pure in thought and heart and do good works.

Behind this stainless and pure lady is a filthy old woman who, to cover her ugliness, shields herself with the face of a young maid which can be called Disguise. And to underscore even more her filth, she is riding a sow, which derives its pleasure solely from wallowing in mire and slime. She carries a dart with which she wounds her enemies, either inside in the heart or outside in the eye. And this woman is commonly known as Lechery, whom the ancient poets called Venus. She drives Chastity out of Religion.

Next, after this fair virtue named Chastity, comes another lady and mistress of the Virtues, by name Obedience. In her hands are the fine thongs and threads of silk which represent the counsels, commandments and observances of Holy Religion which bind the hands and feet of the good religious and of the truly obedient, all of whom may not depart the city or do anything without compliance.

Behind this noble lady and mistress is an old, limping harridan with an axe at her neck and a bundle of black cords; in her hand is a very thick black rope. Her name is Sloth; she fells the negligent with her axe which has the name Bored-with-doing-well. And when they are thus smitten, she binds them with her ropes which are named Laziness, Negligence and Cowardice. When she has thus bound her captive, she puts around his neck the rope which hanged Judas, and its name is Despair.

Next, after this noble lady and mistress of Religion, Obedience, comes her prioress, by name, Correction, who takes everything in hand and corrects it. Whatever she finds that is in error, she furbishes and cleanses with rods which she holds in her hand. To ensure that she is not excessive, she carries a shield called Prudence.

Behind this prioress are two other old women, one carrying the other. The one being carried is puffed up beyond measure, to the point that she could not move unless she were carried. Her name is Pride, who is puffed up with the hot air of her own importance; the horn on her forehead represents Haughtiness. In her hand is a small horn [fol. 6v.] signifying Boasting. Across her body are bellows standing for Vain Glory. Her right hand grasps a dog-wood stick, the representation of Obduracy. The cloak she wears fastened up is the Hypocrisy with which she hides her iniquities. The woman carrying her is Flattery, and without her Pride could not exist or move. Flattery's mistress bears two spurs, Revolt and Disobedience, with which she pricks and incites her bearer. The bearer holds a mirror in which Pride admires herself continually; if she did not see herself, she would wish to strike everyone with her horn of Pride. These are the daughters or workings of Pride, who drive out all subordination and redress from Religion.

After all these noble ladies and their opposing Vices that eject them from the Holy Church and Religion comes the sovereign and most glorious of them all, Voluntary Poverty, to whom Our Lord gives already in this world the Kingdom of Paradise. Because of this, she wears the crown as holder of the heritage of Paradise. She has nothing in her hands, and therefore, she can go everywhere and fear nothing, neither life nor death. On her back she bears an anvil called Patience, for it is a commonplace that lovers of the secular life pay no attention to Patience and are persecuted by everyone.

Avarice is the true opponent of Poverty and is stationed behind her to eject her from the Holy Church and Religion. Avarice is an ugly old hag, hunchbacked and limping. She has no wish to go heavenwards, since she carries on her back a sack of metal; in her heart she is desirous and solicitous of protecting and enlarging it. Because of this, she has six hands and two stumps. Her father from the Slough of Hell cut off her hands with which she toiled to earn a living and with which she used to give alms; and from that day she has had no heart to do anything nor to give alms to the needy. She has six other hands, three right and three left. On the right there is first of all a hand with griffin claws; it is the noble one and is called Plunder. No corn or wheat, no rooster or hen, no poor toiler in the fields can stand up to this hand. The second right hand behind the other has claws also; it is placed behind, because what it takes, it does so stealthily, and its name is Robber. The third right hand holds a crooked stick with which it hoes, mines and digs, and this hand belongs to the prelates of the Church who scrape away, undermine and steal the wealth of the Church for no rhyme or reason, something they can do without replacing anything. And because of this, the old churches are now on their way to oblivion.

On the left are three other hands; the first holds a file and belongs to the treacherous traders and money-changers who file off the corners of the écus, *thus reducing the weight and measurements to deceive the poor. The second left hand holds a pair of scales in which to weigh the sun and the zodiac; its name is Usury, because it sells Time. The third left hand, by name, Beggary, is so called because it seeks to increase wealth by begging and importunate requests.*

This old hag Avarice pokes out a swollen and leprous tongue by which is represented Mendacity, for one simply cannot acquire or increase wealth and riches by the above means without telling lies.

Next, in a section in the lower centre part [fol. 7] of the page, a money-changer sits on his bench and receives jewels and plate from persons of the Church, and in exchange he gives them écus *and money. This activity causes much harm to the Church, because, when the prelates or rulers of the Church have spent the revenues of the Church on their own excesses, they sell jewels and consecrated vessels that were donated for use in the offices and for the honour of the Church. So it is that their churches are in a ruinous state and one cannot find anyone willing to give a penny for the restoration of the edifices.*

XIII *Ram with the Iron Crown* Fol. 37v.

[Fol. 7] La xiij^{e} hystoire est en ce mesme chapitre bien avant a l'endroit du parafe: *En celle cité devant ditte*, et est consequent a l'ystoire devant escripte, car comme l'Eglise soit ainsi tendent a ruine, par ce que elle est toute plaine de monstres et de vices et les vertus en sont fors boutez, consequent est quant elle sueffre de bas scismes et dissentions. Et est ceste hystoire du temps temporel de l'Eglise de Romme qui a esté devant mil cccc et xviij, c'est assavoir devant que le pape Martin fut du conseil general esleu de Constances. Ou quel temps l'Eglise fut en tres abhominable scisme, tellement que a Romme avoit ung pape et en Avignon ung autre, et chascun d'iceulx deffendoit sa partie et son oppinion par force de seigneurs seculiers et non par raison et bonnes vies. Et c'est ce que est figuré en ceste hystoire sur la methaphore du mouton, qui signifie le pape, qui soubz l'ombre de simple habit et belles paroles ottroioit les princes temporelz et avoit aprés lui les ensuivans, plusieurs regnars qui signifient les flateurs, couvoitans et ambicieux. Et pour ce que pour le temps le roy de France estoit simple innocent, avecq se puissant tant de puissance seculiere comme de prelas de l'Eglise et de clers et docteurs des universitez de toutes facultez, chascun[e] des parties se retraihoit vers lui; et aussi il laboura[39] bien fort et tant feist que il appaisa tout et mist bonne paix en l'Esglise, et est qui est figuré en ceste hystoire qui est de soy clere.

The thirteenth illustration is in the same chapter, further on, at the paragraph beginning En celle cité devant ditte, *and it is consequential to the illustration just described. For, since the Church falls into ruin because it is filled with monsters and Vices, and the Virtues are cast out, a consequence is that the Church suffers villainous schisms and dissensions. And the period for this illustration is that of the Church of Rome before one thousand four hundred and eighteen, that is, before Pope Martin was elected by the General Council of Constance. At the time, the Church was in a most abominable schism so that there was a Pope at Rome and another at Avignon, and each defended his side and his aims with the aid of secular lords and not through reason and exemplary life. And this is what is depicted in this illustration through the metaphor of the ram which represents the Pope, who, by simple garb and fair words, favoured the temporal princes; after the ram come several foxes who represent the flatterers, the greedy and the ambitious. And each of the parties turned to the King of France who at this time was simple, yet powerful, as much because of secular strength as because of strength from the prelates, clerks and doctors from all university faculties. He toiled so hard and did so much that he appeased everyone and brought back a lasting peace to the Church. And this is what is depicted in this illustration, which is of itself explicit.*

[39] MS. *il laboura il bien.*

En celle cite devant dite avoit
une rue qui sembloit estre une
des principaulx de la cite ou habitoient
hommes de grant sapience ⁊ de bonne
doctrine Mais la plusgrant partie des
docteurs dycelle en brief tempz mouru
rent ⁊ trespassèrent La avoit un homme
de dieu qui tousiours prioit nrẽ seign ⁊
que il voulsist garder ⁊ sauver celle cite
Et vit une nuyt en dormant un mou
ton cornu qui sapparut par devers
occident et de partie dutropel Il avoit
deux cornes en son chief ⁊ une couronne
de fer sur sa teste et avoit puissance
de regner sur ycelle cite Apres lui
aloient regnars ⁊ goupilz qui tous
estoient de sa route et avoit chascun
une couronne Avec eulx avoit grant
multitude de toutes manieres de be
stes masles ⁊ femelles les autres
par paour les autres par convoitise ⁊
les autres par ygnorance ⁊ par sim
plesse se tenoient en celle multitude.
Le mouton qui vouloit regner
et usurper pour lui ⁊ pour les siens
la dominacion de la cite sen aloit cer
chant toute la region comme ung
tirant a grant force Et tous ceulx

XIV (1) *The Prior's Investigation* Fol. 39
(2) *Susanna*
(3) *Ahasuerus, Esther and Haman*
(4) *Haman*
(5) *Mordecai receiving the Signet Ring*
(6) *Saul Threatening David*
(7) *Saul's Death*

[Fol. 7] La xiiij^{e} hystoire est en ce mesme chapitre v^{e} a l'endroit du parafe *Mes les delivrés*. Ceste hystoire despent des deux precedentes. Pour ce que en tel estat de l'Eglise si ruineux, plain de vices, vuidé de vertus et tout sismatique, plain de couvoitises et d'ambissions, ne pevent les bons subgietz voulens tendre a perfection estre en paix sans tribulation, dame Sapience les enhorte a souffrir paciemment toute tribulation et a avoir en Dieu foy et refuge. Et met en ceste hystoire la forme de religion en laquelle par bonne coustume on tient tous les jours ou souvent, ou quel pour garder la discipline de l'Ordre et ung chascun de mal faire, puet l'ung l'autre accuser pour emendation et en charité. Et pour ce que chascun n'a pas ceste vertus mais, advient souvent que on accuse son frere par vengance, rancune ou maltalent, et est aucune foiz le juste condampné, et le hayneux emporte le droit. Il met en ce texte plusieurs exemples esquelz les justes estoient condampnez, et par le jugement de Dieu, le jugement du monde a esté retourné et les mauvais [punis].

Ou premier parquet est la forme de chapitre[40] et sont ycy les accusans tous drois et le accusé prosterné aux piez de son prelat accoutant et attendant ce que on [fol. 7v.] lui vouldra dire et portant la correction ou reprehension.

Aprés est l'ystoire de saincte Susanne qui fut accusee et jugee a mort injustement, mais la bonne dame mist tellement son cuer a Dieu que Nostre Seigneur lui envoya Daniel le prophete. Ja çoi[41] que jeusne fut, toutesvoyes il jugea les deux faulx vieillars par la mensonge de leur propre bouche, et ilz furent condampnez a mort par le peuple, et la bonne dame innocente delivree[42].

Aprés est l'ystoire de Aman et de Mardocheus. Comme ledit Aman fut honnouré envers le roy Assuerus, il abusa de l'amour et l'onneur que le roy lui avoit faitte, et voulut faire mourir le roy par deux portiers[43] de l'ostel du roy. Quant Mardocheus sceut, il le denunça a la royne Hester qui estoit sa niepce, et elle le denunça au roy, et il les fit pendre au gibet, pour laquelle chose ledit Aman print en hayne ledit Mardocheus, tellement que il s'esforça de destruire tous les Juifz qui estoient ou royaume de Assuerus en ung jour. Quant Mardocheus le sceut, il le denunça a la royne Hester et elle recourut a Dieu en grant fiance et puis s'en complaingnist au roy en la presence dudit Aman. Et le roy lui fit oster son aneau qu'i lui avoit baillié par grant honneur et le feist pendre en son hostel a une poultre ou il avoit proposé de faire pendre Mardocheus. Et le jour de devant que il avoit proposé de faire mourir tous les Juifz, ledit Mardocheus par la puissance et commandement du roy feist mourir tous les amis dudit Aman qui estoient apprestez a tuer les Juifz. Et le roy bailla audit Mardocheus son aneau qu'il avoit osté a Aman. L'istoire est clere.

Aprés est l'istoire de Saül roy qui persecutoit son loyal serviteur et esleu de Dieu a estre roy aprés lui, mais par envie et le mauvais esperit qui le tenoit et tourmentoit, il persecutoit David a le faire mourir: comme une foyz que ledit David jouoit de la herpe devant lui, il le cuida perser de sa lance contre la paroir, et il en eschappa par la grace de Dieu le gardent; et plusieurs autre[s] foiz que Saül cheut es mains de David, mais il le espergna ne ne osa mettre la main a son seigneur roy. Et a la fin par la vengance de Dieu, ledit roy Saül avecq le peuple de Israel descendirent en bataille

[40] MS. *forme cha de chapitre.*
[41] MS. *coiz.*
[42] After *delivree* the scribe copies *macedoine iuif*, which do not belong to the context.
[43] MS. *portieres.*

Susanne
aman
Assuerus
Hester
aman
aman
Assuerus
Hester
le roy saul
Dauid
le roy saul

La.xiiij.l

et la fut vaincu avecq le peuple et lui mesmes fut tué de son propre glaive, ainsi que il avoit persecuté son innocent chevalier et successeur. Nostre Seigneur, vray juge, [est] vengent les tribulations de ses innocens serviteurs.

Et par ceste hystoire appert que les accusations ou tribulations, se ilz sont justes, ilz[44] sont purgeez en ce monde. Se ilz sont injustes, la vengence leur tourne sur la teste de ceulx qui les font; ainsi, par tout fault avoir pacience.

The fourteenth illustration is in the same fifth chapter at the paragraph beginning Mes les delivrés. *The picture is related to the two preceding ones. The condition of the Church is so ruinous, vice-ridden, bereft of virtue, totally schismatic, full of covetousness and ambitions, that honest subjects wishing to strive for perfection cannot be at peace without sorrow. Because of this, Sapientia exhorts them to suffer patiently every tribulation and to have faith and refuge in God. He depicts in this illustration the religious practice customarily observed daily or at intervals, by which, in order to preserve the discipline of the Order and keep a brother from causing harm, one may accuse him with a view to correction and in* caritas. *And because no one has this virtue any more, it often happens that one accuses one's brother out of vengeance, bitterness or anger; sometimes the innocent is condemned and the malicious wins. He includes in this text several* exempla *in which the just are condemned; but through God's judgment people's condemnation has been reversed and the wicked [punished].*

The first compartment shows the practice in chapter, and here the accusers stand upright and the accused is prostrate at the feet of his prior who listens and waits for what [fol. 7v.] will be stated and who then imposes the correction or reprimand.

Next comes the picture about St. Susanna who was accused and condemned to death unjustly; but the good woman set her heart on God in such a manner that Our Lord sent the prophet Daniel to her. Although Daniel was young, nevertheless, he judged the two elders by the lies they mouthed; they were condemned to death by the people and the innocent good woman was freed.

Then comes the illustration about Haman and Mordecai. Whereas the said Haman was honoured by King Ahasuerus, he abused the love and trust shown him by the monarch, and was intent on having him murdered by two door-keepers of the royal household. When Mordecai discovered this, he revealed it to Queen Esther, his niece. She told the king and he had the two men hung on the gallows. For this reason Haman conceived a great hatred for the said Mordecai to the point that he strove to annihilate in one day all the Jews of Ahasuerus's kingdom. On learning this, Mordecai denounced him to Queen Esther; she trustingly turned to God and then complained to the monarch in the presence of Haman. And the king made him take off the ring he had given him by way of high trust and had him hanged in his household from a beam destined for Mordecai. The day before the one that was set down for slaying the Jews, the said Mordecai, acting with the might and authority of King Ahasuerus, had put to death all the friends of the said Haman who were ready to kill the Jews. And the king gave to the said Mordecai the ring he had taken back from Haman. The illustration is unambiguous.

Next in order is the picture about King Saul who persecuted his loyal servant David, chosen by God to succeed him. But, out of envy and because of the evil spirit which gripped and tormented him, Saul persecuted David to death. Once, when David played the harp in his presence, Saul thought to run him through against the wall with his lance, but David escaped, thanks to God's protection. And on several occasions when Saul fell into David's hands, the latter spared him, not daring to lay a hand on the lord his king. In the end, through God's vengeance, the said King Saul

[44] The pronoun *ilz* may represent a fem. pl. noun in Medieval French *scripta*, see A. Tobler and E. Lommatzsch, *Altfranzösisches Wörterbuch*, Berlin and Wiesbaden, IV, 1958, col. 1311.

went down with the people of Israel in battle and all were slain, the monarch himself by his own sword, just as he had persecuted his innocent knight and successor. Our Lord, the true Judge, avenges the tribulations of His innocent servants.

And by this illustration it is manifest that if tribulations or accusations are well-founded, they are atoned for in this world. If they are unjust, vengeance falls on the heads of those who perpetrate them. Thus, in all things, we must be patient.

XV (1) *Carole* Fol. 40
(2) *Sick and Dead Youths*
(3) *Sapientia with Lilies and the Disciple*

[Fol. 7v.] La xv^{e} hystoire est au commencement du vj^{e} chapitre ou quel il interrogue et demande a la plus belle des femmes, c'est a son propos dame Sapience: "Quel est son bel amy?" Et en ce chapitre le disciple medite et enquiert quel est vraye amour, et fait comparation de l'amour mondaine et temporelle en recordant les amours, joyes et plaisances que il avoit eues et prinses en sa jeus—[fol. 8]—nesse. Et quant il eust bien consideré que ces plaisances terriennes et mondaines tost[45] viennent et tost passent, au jour d'ui fleuré et demain flaittriez, au jour d'ui sain et lié et demain en mal aise et mourant, il considera que sa tres desiree amie, la eternelle Sapience de Dieu le pere, ne peust estre telle, si muable, qui est eternelle et perpetuelle. En ce considerant, il lui fut en advis que il veoit a l'encontre de lui vers le ciel une fleur singuliere, plus belle, plus blanche et plus resplendissant que oncques en sa vie n'avoit veue. Et comme il[46] approuchoit de prendre ceste belle fleur, elle se esvanuit, et en lieu d'elle, apparut une belle et excellent dame, plaine de toute beauté, de toute doulceur. Ses paroles estoient toutes plaines de beau langaige et de grant subtilité, contenant en soy la somme de parfection de toutes choses que on pourroit souhaider. Et ceste dame estoit dame Sapience que le bon disciple avoit couvoité a aimer des sa jeusnesse et l'avoit quise pour estre son espouse. Et c'est ce que est figuré en ceste histoire, car le premier parquet monstre les plaisances mondaines transitoires, et le secont monstre les fleurs flaitries et jeusnes malades et mourans. Et le tiers dame Sapience belle, fle[u]rissent et perpetuelle.

The fifteenth picture is at the opening of the sixth chapter in which he interrogates and asks the most beautiful of women, that is, in his words, Sapientia: "Who is her fair friend?" And in that chapter the Disciple meditates and inquires what is true love, and makes comparisons with worldly and temporal love by recalling the loves, joys and pleasures he partook of in his youth. [fol. 8] And after reflecting that earthly and worldly pleasures arrive quickly and depart equally fast, today blooming and tomorrow faded, today hale and hearty then tomorrow sick and dying, he came to the view that his greatly desired friend, the eternal Wisdom of God the Father, could not be so changeable, since it is eternal and perpetual. As he pondered all this, he thought he saw, high up in front of him, a wondrous flower, whiter, more beautiful, and more resplendent than any he had ever seen in his life. And as he drew near to seize this striking bloom, it disappeared from view; in its place stood a fair and noble woman of extraordinary beauty and gentleness. Her speech was that of a refined tongue and filled with great subtlety; it contained the sum of perfection for everything one could wish for. And this woman was Sapientia whom the good Disciple had longed to love since his youth and whom he had sought as his spouse. This is the subject matter of this illustration, for the first segment shows the transitory pleasures of the world, the second depicts faded flowers and youths sick and dying; and the third portrays Sapientia with flowers, beautiful and eternal.

[45] MS. *tout.*

[46] MS. *sil.*

O pardurable sapience qui toutes
choses conuertis en bien pour tes amis
il me semble que jay a faire vne seule
chose Cest assauoir que en tous cas
et a toute heure je doye retourner a
sainte z deuote oroison. et ton ayde req
rir de cuer deuot Doncques o tu
pere de misericorde z dieu de toute
consolation je te prie treshumblemet
que tu ayes pitie de ton peuple et de
ton heritaige et nous donne paix de
cuer et paix de temps en nos jours
Ayes aussi par la vertu de to saint
nom pitie de ceulx qui heent paix
et leur donne bon esperit pour eulx
traire a paix z a bonne vnite A
fin que nous tous ensemble puissos
seruir a sainte eglise dun cuer et
dune voulente Et que finablemet
nous puissions venir a la cite glori
euse du pays celeste Amen ainsi
soit il.

Cy commence le sixiesme chapitre
qui parle de la beaulte de la diuine
sapience. et quelle est son amour
et sa charite.

Qualis est dilectus tuus
O pulcherrima mulieru
O la plus belle des femes
Dy moy quel est ton ami
Qui vouldra oyr
la response de ceste demande si tende ses
oreilles de lamour diuine. Qui veult
entendre ceste matiere damours il doit
estre vray amant desirant z de fin cu
er amour ardant. Et tel pourra en
tendre ce comme je le dy Car aucu
parle bien damours qui nest pour ce
enamoure Et tel parle damours
bien auant qui na assauoure point en
son cuer la plaisance qui est en amours
Nous nous tairons de ceste ma
tiere car nous sommes encore trop
loing de assauourer z de sentir la doul
ceur damours Mais nous desirons que
nous en soions pres mais viengne auant le
courtois amoureux z nous die coment il
a este seurpris de lamour de son ami z comet
son ami la tient a lui Si porrons sauoir
la condicion z auoir la congnoissance de lamant
et de lamie Et si ne couient la que il soit
trop secret ne trop couuert car vraie amour
ne se peut celer mais se reuelle vueille ou no

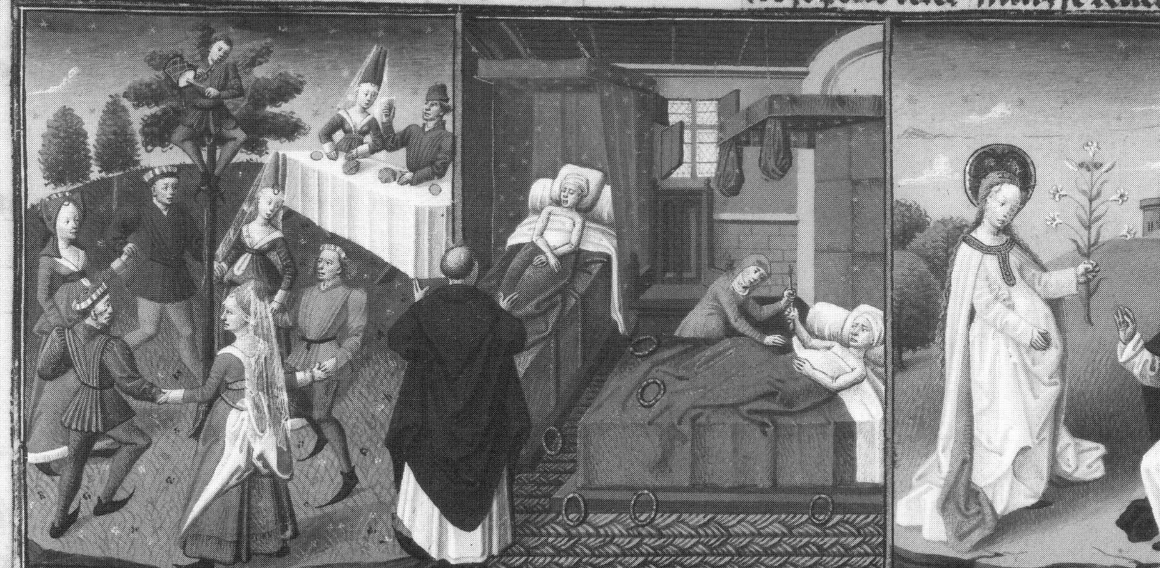

la. vi

XVIa *Last Judgment* Fol. 46v.

XVIb *Shipwreck* Fol. 47

[Fol. 8] La xvj^{e} hystoire est a l'endroit du vij^{e} chapitre ou quel il parle de la redoubtance de la Divine Sapience et du jugement de Dieu. Et met une similitude de l'ame troublee et paoureuse ou temps esmeu et tempestez de vens, de pluiez, de naiges et de gresles; et d'une nef enmy la mer cassee et brisee par fortune et tempeste. Comme ung chascun, voient sa nef cassee et presque effondree, ha paour et craint a mourir, aussi une chascune creature, considerent sa vie plaine de offenses et de pechez, et estre en brief presentee au jugement de Dieu qui est et sera si destroit que il n'y ara si juste qui ne doubte a estre condampné. Par raison doit estre bien paoureuse et bien craindre a mourir et estre jugee et condampnee a feu pardurable, chascune creature plaine de pechez et ne ayent nul bien en soy par quoy elle puist rapaiser son juge.

Pour ce est en ceste hystoire le disciple en une nef cassee, rompue et brisee et presque effondree en la mer, et tout entour tempestes, naiges, vens et gresles, et deux anges dessus vers le ciel a deux trompes, crians et denuncens que on viengne au jugement de Dieu, du quel sault de sa bouche ung fleuve de feu ardant le monde et descendant sur les dampnez. Et le disciple est devant le juge paoureux et tremblent. Et le dyable est devant lui, tenant ung livre ou quel sont ses pechez escripts, et ung angel a l'opposite est le aidant et suppliant pour lui.

The sixteenth illustration is positioned at the seventh chapter in which he speaks of fear of Divine Sapientia and of God's Judgment. And he adduces the similitude of the troubled and fearful soul in stormy weather with wind, rain, snow and hail, and of a ship at sea that is split open and broken by storm and disaster. Just as anyone, beholding his ship rent asunder and almost sunk, becomes afraid and fears death, so every creature, contemplating his life filled with crime and sin, and on the point of being presented to the Judgment of God is, and will be, so distraught that none will be so innocent as not to fear damnation. Thus, every creature that is full of sin and has no good in himself with which to assuage his Judge should be well and truly terrified, and afraid to die and be judged and damned to eternal fire.

And because of this, the Disciple in this illustration is in a ship that is split open and broken, disintegrating and almost sinking under the waters; all around are storms, snow, winds and hail, while heavenwards are two angels with trumpets, proclaiming and summoning him to God's Judgment. From God's mouth flows a river of fire that burns the world and descends over the Damned. And the Disciple is fearful and trembling in front of the Judge, while the Devil is in front of the Disciple and holds a book in which his sins are written. On the opposite side is an angel who assists the Disciple and supplicates for him.

XVII (1) *The Disciple Offering his Heart to Sapientia* Fol. 49v.
(2) *Sapientia with Faith, Hope and Charity*
(3) *The Disciple Shutting out Sapientia*

[Fol. 8] La xvij^{e} hystoire est ou viij^{e} chapitre a l'endroit du parafe: *O mon cuer*, ou quel chapitre il parle comment la Divine Sapience visite l'ame de son loyal amy. Et premierement le inspire de sa grace...[47] en lui donnant grace de recourir a Dieu, de se donner a devocion, et de se enamourer de la Divine Sapience en lui donnant son cuer. Aprés elle vient par grace cooperant ou par son aide et frappe a l'uys de son ame, temptant et aysaient se il lui ouvrera l'uys de son cuer par consentement. Et lors ne vient elle [fol. 8v.] pas seulement mais amaine avecques elle trois de ses filles, les troys vertus theologicales: Foy, Esperance, Charité (*MS.* Foy, Charité, Esperance). La premiere dit *Credo* "je croy"; la seconde, ce que il a conceu et croist par vraye foy, il a esperance de obtenir, et ceste dist *Spero* "je ay esperance"; la tierce demourra aprés toutes, c'est Charité qui est parfaitte et sera en la vie perpetuelle. En cest estat est l'ame devote en grant consolation, en grant joye et devocion. Mais n'en puet pas demourer longuement en ceste presente vie pour la corrupcion du corps qui aggreve et appesantit l'ame qu'elle ne puet penser a ce que elle ayme, mais se descent a penser les choses terriennes avecques lesquelles elle et son corps habitent. Et pour ce le disciple de Divine Sapience, qui par avant lui avoit donné son cuer et puis luy avoit ouvert l'uys par grant devocion, par les cogitations terriennes et mondaines survenans et umbragens son entendement, il clost la fenestre a la lumiere et ne seuffre le souleil de vraye lumiere entrer en la chambre de son cuer. Ainsi son amie, a qui il estoit fiencé par foy, il avoit donné erres par esperance, il c'estoit fermé en amour par charité, se depart de luy, portant sa lumiere absconse et mucee entre ses mains. Et demeure le povre mescheant en tenebres et desolé. Et c'est ce qui est figuré es troys parqués de ceste hystoire.

The seventeenth illustration is in the eighth chapter at the place of the paragraph O mon cuer. *In the chapter he relates how Divine Sapientia visits the soul of her faithful friend. First of all, she fills him with her grace... by giving him the grace of recourse to God, of commitment to devotion and of love for Divine Sapientia when he surrenders his heart. Then she arrives, with the concurrence or assistance of grace, at the door of his soul and knocks, trying and testing him to open the door to her by consent. She does not arrive [fol. 8v.] alone but in the company of three of her daughters, the theological Virtues, Faith, Hope and Charity. The first says* Credo *"I believe"; the second says* Spero *"I have hope"—what he has conceived and believes in true faith, he hopes to obtain. The third will endure after all others, this is Charity, which is perfect and will continue in everlasting life. The devout soul in this state is filled with great solace, joy and piety. Yet it cannot remain so for very long in this present life because of the body's corruption which distresses and burdens the soul so that it can no longer think of what it loves; rather, it sinks to a consideration of earthly things with which it and its body dwell. Because earthly and wordly thoughts arise and darken his understanding, the Disciple of Divine Sapientia who had earlier given her his heart and opened the door for her in great devotion, now closes the window on the light and does not allow the sun's true rays to penetrate the abode of his heart. Thus his friend, to whom he was affianced in faith, to whom he had given pledges of hope and for whom he had become strengthened in love because of Charity, now leaves him, carrying her light sheltered and hidden between her hands. And the hapless Disciple remains forsaken and in darkness. And this is what is depicted in the three compartments of this illustration.*

[47] An illegible word follows *grace*.

querray je helas pourquoy me suis je
endormy Que nay je tousiours veille
Que ne lay je tenu pres de moy. O
mon doulz amy ou es tu. monstre toy
a moy. Vens toy a moy. Ton absence
me blesce autant comme ta presence
me consoloit Ta venue z ton alee. ton
aler z ton venir que tu fais si souvent
me tourmente malement. Ore me
monstre lombraige ou tu te rafreschis
et umbroies a midi. a fin que je ne perde
mon temps en toy querant. a fin que
je ne men voise hors de moy mesmes
qui autreffois avoie acoustume de
toy trouver en moy. Laz laz seray
je longuement privee de ta presence.
Quant je tavoye javoye tous les
biens que cuer humain peut avoir
ne souhaidier Et encores avoye je
telle richesce sil te plaisoit. Si te
supplie mon tresdoulz amy encline
toy a la priere de mon cuer. Je suis
tant constraint de ton amour qui
me assault par telle voulente et violence que je suis hors de moy et suis
privee z estrange de moy mesmes
Je ay un faissel au col qui me foule
et me quesse efforcement Doncques
mon chier amy. retarde moy de tes
yeulx piteux Je deffaulz tout et ne
puis plus porter Donne toy a moy
a fin que me puisse herbergier en toi.
qui tay perdu par moy. Quelle
merveilleuse vertu damours Je
dy et puis bien dire en verite que
vraye amour est forte comme la
mort. O amours amours tu mas
oste mon cuer z las donne a mon amy
Tu las tellement conjoinct z cole
a lui sans departir quil semble
aucuneffois que je soie sans cuer.
et quil me ait du tout relenqui. et
ne scay a la fois avecques le quel
ou avec toy. ou avec moy et si ne
sauroie dire avecques lequel il
est mieulx conjoinct ou avec mon
corpz auquel il donne vie ou avec
son amy quil aime tant ardamment.

XVIII (1) *The Disciple Offering his Soul to the Trinity* Fol. 56v.
(2) *Life's Pilgrim*
(3) *The Disciple's Contemplation*

[Fol. 8v] La xviij^{e} hystoire est ou ixe chapitre a la fin, a l'endroit du parafe: *Lieve toy doncques.* Pour ce que en ce chapitre par avant le disciple de Sapience c'estoit esmerveillié des jugemens de Dieu, comment ceulx, qui se mettent a Dieu servir, ont a la foys plus de tribulations que n'ont ceulx qui ensuivent le monde et ne voient pas ce que il receveront pour leurs merites, ou payens pour les maulx ou gloire pour les biens faiz, dame Sapience l'enhorte de considerer non pas les choses presentes qui sont temporelz, mais les choses a avenir qui sont perpetuelz ou payens pour les maulx ou gloire pour les biens faiz. Et lui dit que il se lieve de ceste consideration terrienne et mondaine et considere soy mesmes qui il est, car il est fait a la semblance de Dieu, la saincte Trinité et une essence. Car il a en une ame troys puissances, c'est assavoir: entendement, memoire et voulenté. Et si est exemplaire de la eternité de Dieu, car il est, quant a l'ame, de perpetuelle durance et de infinie capacité. Et est ce qu'il met ycy ou premier parquet.

Aprés, doit considerer ou il est, car il n'est pas en son pays, c'est assavoir, en paradis du quel il est bouté hors et banni et exili[é], mais est pelerin en une valee de lermes, plaine de joyes transitoires qui au jour d'uy sont joyes et demain pleurs, plaine de las et de temptations de l'ennemy, de tempestes et de tribulations, biens et maulx meslez ensemble, ou il n'y a point de stabilité. Et c'est ce qui est figuré ou secont parquet ou quel le disciple est comme pelerin a my le monde ou il y a de joyes et plaisances mondaines, de tristesses, de tempestes, de las, de temptations de l'ennemy, de biens et maulx meslez ensemble, ou il n'y a ne stabilité ne asseurance.

Aprés, doit considerer ou il va, car selon ce que il ara fait, il yra ou pour mal en enfer, ou pour bien en paradiz, entre lesquelz n'a point de moyen; car qui meurt a tous [fol. 9] ses pechez, en enfer descendra, et qui meurt en bonne voulenté, contrit et confés de ses pechez, il est en bonne voye, en voye de salut et de paradiz. Et c'est ce qui est figuré ou tiers parquet, ou quel est le disciple, est tout seul en chambre considerant et regardant le temps a avenir et les lieus ou il doit estre logé pour ses merites aprés la mort: ou enfer ou paradiz.

The eighteenth illustration is in the ninth chapter, at the end, where the paragraph is Lieve toy doncques. *Sapientia's Disciple had earlier in this chapter wondered at God's Judgments, and how those who devote themselves to His service have at the same time more tribulations than those who adhere to the secular world and do not see what they will receive for their deserts either pains for their misdeeds or glory for their good works. Therefore Sapientia exhorts the Disciple to contemplate, not present things which are temporal, but things to come which are everlasting, whether pains for misdeeds or glory for good works.*

And she instructs him to rise above this consideration of earthly and wordly things and consider himself who he is, for he is formed in the image of God, the Trinity and one essence. For in one soul there are three powers, namely, understanding, memory and will. And he exemplifies God's eternity, for he is, as far as the soul is concerned, of perpetual continuance and infinite capacity. And this is what he depicts here in the first segment.

Next, the Disciple should consider where he is, for he is not in his own land, that is, Paradise, from which he was expelled, banished and cast out, rather, he is a pilgrim in a vale of tears, replete with transitory joys that are today joys and tomorrow tears. It is a vale filled with the Devil's snares and temptations, with storms and tribulations, good and evil interwoven, where there is no calm. And this is what is depicted in the second compartment in which the Disciple resembles a pilgrim in a world where there are joys and mundane pleasures, sorrows, storms, snares and temptations of the Devil, good and bad intertwined, and where stability and certainty are lacking.

Then, the Disciple must consider where he is going, for, according to his deeds, he will go either for worse to Hell or for good to Paradise. Between the two there is no middle way. He who dies with all [fol. 9] his sins, will go to Hell; and he who dies with steadfast will, being contrite and confesses of his sins, is on the right path, the path to salvation and Paradise. This is what is depicted in the third segment in which the Disciple is seen all alone in his cell, contemplating and beholding the future and the places where after death he may have his abode for his merits: either Hell or Paradise.

XIX (1) *Devils of Hell* Fol. 57 missing
(2) *Cold of Hell*
(3) *Fires of Hell*

[Fol. 9] La xixe hystoire est au commencement du xe chapitre qui se commence: *Aprés ces choses*, ou quel il parle des payens de enfer. Pour ce que en l'ystoire precedent estoit le disciple tout seul pensif, en considerant ou il yroit, il eust en consideration que il veoit une region et pays obscur, plain de tenebres, de fumee et de punaisie, et celle ou on ne veoit riens, mais on y oyoit coups de marteaulx et de maillés, et amez braire et plourer, cris et horribles voiz getter par desesperation. Et y avoit horribles faces de dyables, et c'est ce qui est figuré ou premier parquet.

Aprés y avoit d'autres tormentez d'autre et diverse maniere, les ungs estoient en naige et eaue de glace, les autres en feu horrible. Et les dyables les gettoient de l'ung en l'autre, de chaut en froit, de froit en chaut, pour accroistre les tourmens. Et c'est ce qui est figuré en ces deux parqués ensuivans.

The nineteenth illustration is at the opening of the tenth chapter which begins Aprés ces choses. *He speaks here about the pains of Hell. Because in the preceding picture the Disciple, all alone and pensive, was considering where he would go, he had in his mind the view of an obscure region covered in darkness, smoke and filth. It was a place where one could see nothing, but one could hear hammers and mallets, and weeping and wailing souls uttering horrible sounds out of desperation. And hideous faces of demons were everywhere. This is what is depicted in the first segment.*

Then, there were tortures of one kind or another, some were in snow and icy water, others in terrifying fires. The devils cast the souls from one to the other: from hot to cold, from cold to hot, so as to augment the torture. And this is what is shown in these two compartments that come next.

XX (1) *Gallows of Hell*
(2) *Pains of Hell*
(3) *Pains of Hell*

Fol. 57v. missing

[Fol. 9] La xxe hystoire est en ce mesmes chapitre a l'endroit du parafe[48] *Ceulx qui avoient esté etc.*, et figure divers tormens de enfer dont[49] le premier parquet monstre et signifie le gibet d'enfer ou sont penduz plusieurs seloncq leurs desertes, ou par les piez, ou par le col, ou par la langue; et autres divers tormens selon les desertes de chascun pecheur. Sont les payens plus grandes en enfer que on ne saroit descripre ne estimer.

The twentieth illustration is in this same chapter where the paragraph begins Ceulx qui avoient esté etc., *and it depicts the divers torments of Hell. The first segment shows and represents the gallows of Hell where many hang according to their deserts: by the feet, by the neck, or by the tongue. There are various other tortures according to the deserts of each sinner. The pains of Hell are greater than one can describe.*

[48] MS. *parquet.*

[49] MS. *doncq.*

XXI (1) *The Slothful, Hedonists and Wasters in Hell* Fol. 58
(2) *The Fornicators in Hell*
(3) *The Avaricious in Hell*

XXII (1) *False Judges in Hell* Fol. 58v.
(2) *Wicked Women in Hell*
(3) *Pit of Hell*

[Fol. 9] Et semblablement aux autres parqués ensuivans[50], et aux deux hystoires ensuivans, c'est assavoir, en la xxj^e^ et xxij^e^, sont autres divers tormens, fornaises et gehinnes, et a la fin, est le puis d'enfer, lesquelles pevent apparoir en les considerant et pensant.

[50] The first six words refer still to the triptych on lost fol. 57v.

qui les mordoient et devouroient
sans cesser. Puis regarda et vit
les orgueilleux qui estoient en serre
lis en ordure & punaisie et en vilte
telle quelle ne fait a raconter Et
si comme charongne puant et pour

rissoit mondaine Ilz estoient en celui
enfer foulez et charmez de fais pesans
et horribles tant come pourroit estre
charge ung homme seul avoit la
plus grosse tour du monde mise
et chargee sur son ventre.

Les yvrongnes & ceulx qui avoient
tout leur temps despendu en
gloutonnie et en servir a leur ventre
estoient tous enraigiez de faim et vou
loient comme loupz affamez et lan
guissoient de soif en telle maniere que

gettoient dedens le corps celui an
groisseux buvraige. Apres veoit les
luxurieux et ceulx qui delicieusement
avoient leur char nourrie Et qui
tout le temps de leur vie avoient de
moure en leur obstinacion et en le

La.xx

And likewise in the two other compartments that follow, and in the two illustrations that come next, that is, in the twenty-first and twenty-second ones, are still seen torments and tortures, and lastly, the Pit of Hell. They may all appear as one ponders and meditates about them.

estes enferrees de flambe ardant et
les navroient douloureusement.
Ceulx et celles qui avoient este au
monde compaignons en pechie et
en coulpe estoient en enfer compai
gnons en tourmens. Sur tous

faulx questeurs. Car ilz estoient
en fosses parfondes plaines de
metal boullant et la dedens se bai
gnoient et souvent sefforcoiet dissir
hors du baing mais les ennemis
estoiet la qui tatost les y rebouttoiet.

Ceulx aussi qui avoient este
justiciers z souverains au
monde estoient moult rigoureusemet
pugnis en meshaine z a tourment Et celui
qui avoit este plus puissant au mode
estoit plus puissamment tourmente.

et plusieurs autres faulx crestiens estoi
ent laidement tourmentez et aspremet
pugnis z travaillez qui tous ensemble
ulloient z avoient comme bestes mues
a haulte voix par telle maniere et par
telle horreur que cestoit grant affictio

XXIII *Court of Heaven* Fol. 62

[Fol. 9] La xxiij^{e} hystoire est en xj^{e} chapitre a l'endroit du parafe *Or regarde.* Pour ce que en la xviij^{e} hystoire dame Sapience avoit enhorté son filz et disciple entre les autres choses de considerer ou il yroit et depuis il se occupe a considerer les paines d'enfer, maintenant en ce chapitre il considere la gloire et la beauté de la cité et gloire de paradiz, laquelle est essencialement en la vision de Dieu, laquelle tous les sains et sainctes de paradiz aront comme le denier de la retribution eternelle different accidentalement, plus ou moins selon les differences des merites, come une estoile est differente a une autre en clarté. Et ce est qui est figuré en ceste hystoire. Selon ce que on peut mieulx comprendre de la gloire de paradiz pour la diversité des sains et sainctes, icy est premierement Dieu en magesté, les ordres des angelz et archangelz etc., la benoite Vierge Marie prouchaine de la saincte Trinité, sur tous les ordres saint Jehan Baptiste, les apostres et evangelistes, les martirs, confesseurs, vierges et sainctes matrones et la tourbe populaire innumerable. Et entre eulx tous, sains et sainctes, angelz desquelz merites ont esté glorieux[51].

The twenty-third illustration is in the eleventh chapter at the place of the paragraph Or regarde. *In the eighteenth illustration, Sapientia had exhorted her son and Disciple among other things to consider whither he was going, and he had considered the pains of Hell; now in this chapter he contemplates the beauty and refulgence of the city and glory of Paradise. It is essentially in the vision of God, which all the saints of Paradise will have as the coin of external reward; it will vary adventitiously more or less according to the differences of individual merits, just as one star varies from another in brightness. And this is what is depicted in this picture. Because one can better comprehend the glory of Paradise through the diversity of the saints, here we see first of all God in Majesty, the Orders of angels and archangels etc., the Blessed Virgin close to the Holy Trinity, St. John the Baptist above all the Orders, then the Apostles and Evangelists, the martyrs, confessors, virgins and holy matrons, and the countless numbers of the populace. And among all the male and female saints are angels whose merits have been illustrious.*

[51] The words *sains et sainctes* have been transcribed after *glorieux*.

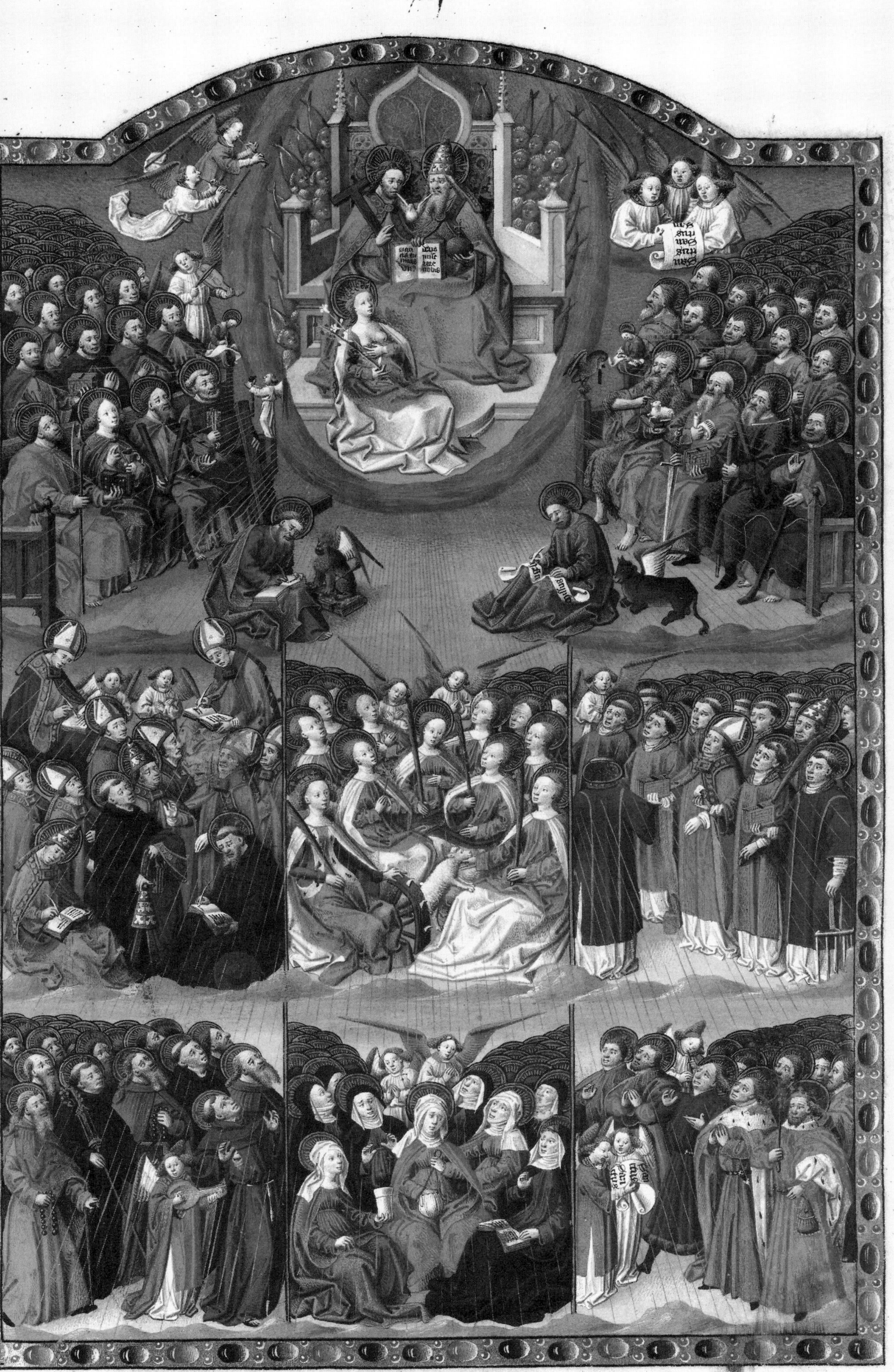

La.xxvi.

XXIVa (1) *Absalom Becoming King* Fol. 72
(2) *David Fleeing from Absalom*
(3) *Absalom Slain by Joab*
(4) *Sons of Jocab*
(5) *Isaiah, Jeremiah, Habakkuk and Daniel*

XXIVb (1) *Three Youths in the Furnace* Fol. 72v.
(2) *Job on the Dunghill*
(3) *Antiochus Torturing the Maccabees*

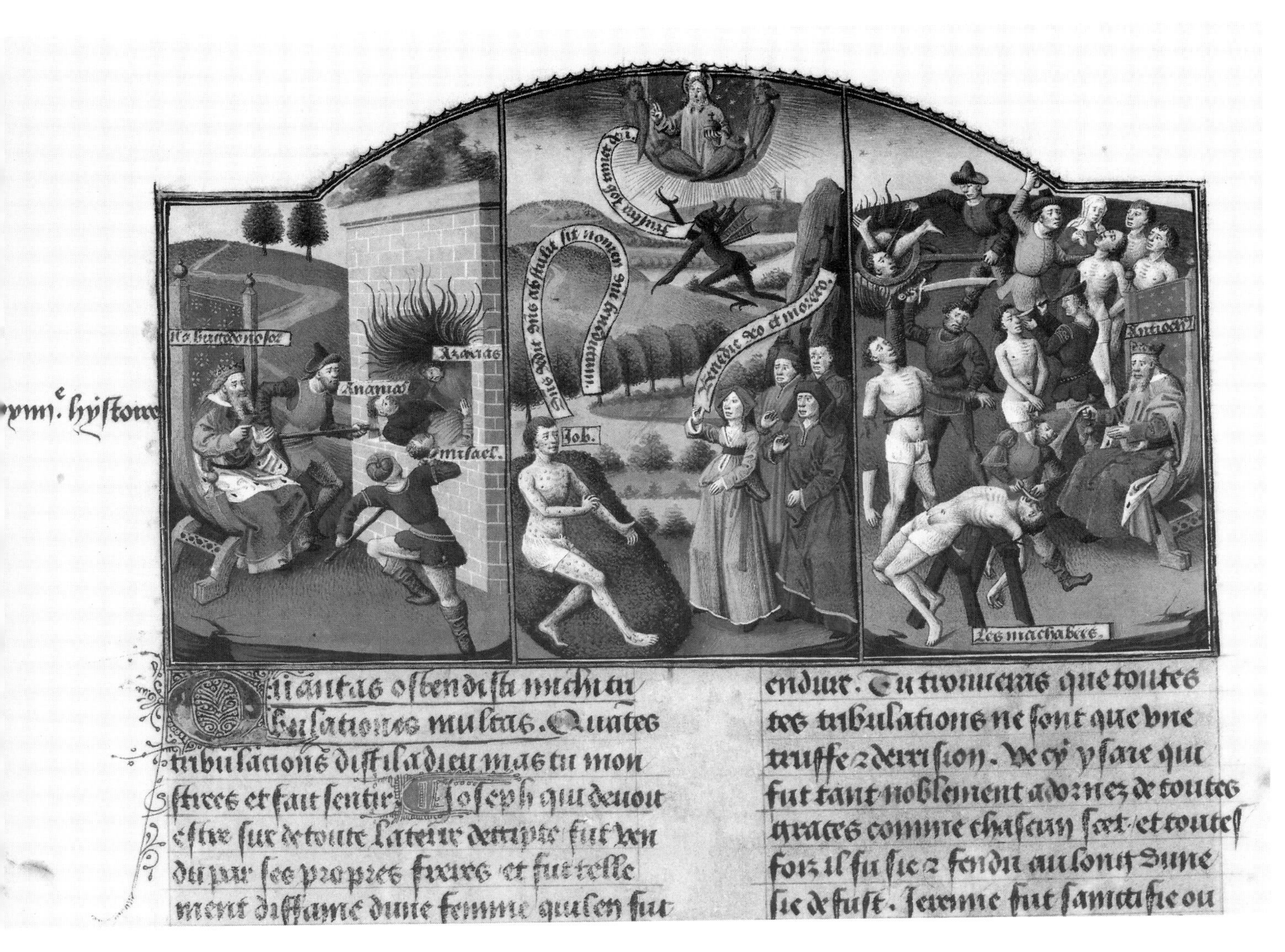

[Fol. 9] La xxiiij^e hystoire est ou xiij^e chapitre a l'endroit du parafe *Quantas ostendisti michi*, ou quel chapitre est enhorté le disciple de endurer paciemment [fol. 9v.] toutes tribulations a l'exemple de plusieurs sains du temps passé, desquelz il en figure aucuns en ceste hystoire, car tous ne pourroit on pas figurer. Premierement recite et figure l'istoire de David qui fut persecuté de son filz Absalon, laquelle est recitee ou xve chapitre du secont livre des Roys[52]. Absalon vint en Jherusalem et se feist dire roy, et David s'en fouit a pié. Et toutesvoyes, ainsi que Absalon poursuivoit son pere et le persecutoit, par la vengence de Dieu il fut pendu a ung arbre par les cheveux, et fut tué par Joab, le connestable de David, et David s'en retourna victorieux et paisible roy de tout le peuple de Israel. Et cecy est figuré en la premiere renge de l'istoire et au premier parquet de la renge ensuivant.

Aprés est figuré de Joseph que ses freres getterent en ung puis sec et faignerent a leur pere que une beste mauvaise l'avoit devoré, et puis le vendirent. Et toutesvoyes, il fut aprés seigneur de toute la terre d'Egypte et ses freres le adorerent comme il leur avoit dit par ung songe. Et cecy est figuré ou secont parquet de la seconde renge de ceste hystoire.

Aprés est figuré de troys prophetes qui furent tormentez pour le nom de Dieu qu'il avoient denuncé et adoré[53], le premier Ysaye fut sié, le secont Jeremie fut lapidé, le tiers Daniel fut mis

[52] II Sam. XV, 1-37.

[53] MS. *adorer*.

en une fosse a lyons pour le devourer. C'est ce qui est figuré en ce tiers parquet de la derreniere renge de ceste page.

[Fol. 9v.] En l'autre page ensuivant est figuré comment Nabugodonosor, roy de Babiloine, feist mettre troys enfans en une fournaise ardent, lesquelz meurent mal par la puissance de Dieu. Et ce est figuré ou premier parquet de ceste page.

Aprés recite de la pacience de Job qui perdit toute sa substance, et fut navré par le corps par l'operation de l'ennemy, et en toutes ses tribulations beneissoit Nostre Seigneur. Et ce est figuré ou secont parquet.

Aprés est figuré des tormens que[54] souffrirent les Machabeez par le commandement du roy Anthyoche. Et ce est figuré ou derrenier parquet de ceste hystoire.

The twenty-fourth illustration is in the thirteenth chapter where the paragraph begins Quantas ostendisti michi. *In the chapter, the Disciple is exhorted to endure patiently [fol. 9v.] every tribulation as did several saints of yore. He portrays only some of them in this illustration, for he could not portray them all. First, he rehearses and depicts the story of David who was persecuted by his son Absalom; it is told in the fifteenth chapter of the Second Book of Kings. Absalom came to Jerusalem, and had himself proclaimed king, whereupon David fled on foot. Yet, while Absalom pursued his father and harrassed him, he was, through God's vengeance, hung by his hair from a tree. He was then slain by David's Constable, Joab. And David returned, a victorious and peaceable king of all the people of Israel. And this is what is depicted in the upper register of the illustration and in the first segment of the next row.*

Next is the portrayal of Joseph whom his brothers threw into a dry cistern, pretending to their father that a wild beast had devoured him; they then sold him. And yet, Joseph was afterwards lord of all Egypt and his brothers worshipped him just as he had told them through a dream. And this is depicted in the second compartment of the second register of this illustration.

After this is a representation of three prophets who were tortured for the name of God whom they had proclaimed and worshipped: the first, Isaiah, was sawn in two; the second was Jeremiah who was stoned; and the third was Daniel, cast into a lion-pit to be eaten. This is what is depicted in this third segment of the last row on the page.

On the page which follows is a portrayal of Nebuchadnezzar, King of Babylon, having three youths cast into a burning oven. Thanks to God's power, they did not die. And this is depicted in the first segment of the page.

Next comes an account of Job's patience: he lost all his possessions, he was wounded in the body by the Devil's workings, yet in all his tribulations he blessed Our Lord. And this is represented in the second compartment.

Then comes a depiction of the torture the Maccabees suffered at the command of King Antiochus. And this is represented in the last segment of this illustration.

[54] MS. *qui.*

XXV *Christ Offered the Sop* Fol. 80

[Fol. 9v.] La xxv^{e} hystoire est au commencement du xv^{e} chapitre ou quel il parle que le disciple et vray amant de la Divine Sapience de Nostre Seigneur Jhesucrist se doibt conformer aux passions de Nostre Seigneur Jhesucrist son amy, qui a tant souffert pour lui; ainsi, doibt il souffrir et endurer pour l'amour de lui toutes paines, souffrances et tribulations et s'i conformer de tout son povoir. Et pour en avoir memoire, recite il ycy une partie de sa passion, laquelle est figuree en ceste hystoire.

The twenty-fifth illustration is at the opening of the fifteenth chapter in which he states that the Disciple and true votary of the Divine Sapientia of Our Lord Jesus Christ should conform to the agonies of Our Lord Jesus Christ, his friend, who suffered so much for him. Thus, for his love of Him, the Disciple must endure all manner of pain, distress and sorrow, and be completely like Him. And as a reminder, he rehearses here part of His Passion, which is depicted in this illustration.

XXVI *Virgin Suckling the Holy Child* Fol. 83v.

[Fol. 9v.] La xxvj^e hystoire est au commencement du xvj^e chapitre ou quel il parle singulierement et grandement de la louange de la glorieuse Vierge Marie, et specialement de la louange des precieuses mammelles et du precieux laict virginal duquel fut aletté et nourri nostre glorieux sauveur Jhesucrist, et du precieulx ventre qui le porta. Et pour ce en ceste hystoire est figuree la benoite matrone qui par l'inspiration du saint Esperit eust grace de dire ceste benoite parole: *Beatus venter qui te portavit et ube*—[fol. 10]—*ra que suxisti*[55].

The twenty-sixth illustration is at the beginning of chapter sixteen in which he pronounces grandiloquently and at length the praises of the glorious Virgin Mary; and in particular, he praises her precious breasts and the precious virginal milk with which Our Glorious Saviour Jesus Christ was suckled and nourished; and he speaks of the precious womb which bore Him. Therefore, in this illustration is portrayed the blessed matron who, under the inspiration of the Holy Spirit, was favoured to utter these blessed words: Beatus venter qui te portavit et ube—[*fol. 10*]—ra que suxisti.

[55] Luke, XI, 27.

fait le coulon qui se loge ou pertuis de
la pierre. Car en celui lieu trouueras
tu a toute heure pleine remission de
tes pechiez plenitude de grace pro
tection defense et sauuegarde contre
toute male aduenture qui te pourroit
aduenir. Le disciple Encore te vueil
je faire une petite demande de la ma
tiere de ta tres doulce et tres piteuse
passion de laquelle tu as ja parle
cy deuant mais tu ten es passe bien
briefment. Je te demande comment
se porta et comment se maintint
celle reuerend dame celle bieneureuse
ta mere quant elle estoit deuant
la croix et elle te veist pendre et cru
cifier. Sapience Quant a ceste dema
de je te donne licence de en aler parler
a ta mere et de lui en demander la
verite quelle ten saura bien respon

La xxvj
toire.

XXVII (1) Stabat Mater — Fol. 87
(2) *Deposition*
(3) *Christ Appearing to His Mother*

[Fol. 10] La xxvij^e hystoire est au commencement de la seconde partie du xvj^e chapitre qui se commence *Stabat iuxta crucem Jhesu. Marie mere de Jhesu.* En laquelle partie il traitte[56] des grandes douleurs que Nostre Dame mere de Jhesu endura ou temps de la passion Nostre Seigneur Jhesucrist et consequemment de la grande joye que elle eust a la resurrection quant il se apparut a elle. Et c'est ce qui est figuré en ceste hystoire. Ou premier parquet est de la passion Nostre Seigneur, presente sa douleureuse mere; le second, de la deposicion de la Croix, le recepvent sa doulce mere; le tiers est de l'apparicion a sa tres excellemment amee mere.

The twenty-seventh illustration is at the beginning of the second part of the sixteenth chapter which commences Stabat iuxta crucem Jhesu. Marie mere de Jhesu. *In this part he treats of the great sorrows which Our Lady, mother of Jesus, endured at the time of the Passion of Our Lord Jesus Christ; later, he speaks of the great joy she had at the Resurrection when He appeared to her. And this is what is depicted in the illustration. The first segment is about the Passion and His grieving mother is present; the second compartment concerns the Deposition, with Jesus being received by His sweet mother; the third segment shows Him appearing to His most exceptionally beloved mother.*

[56] MS. *traittie.*

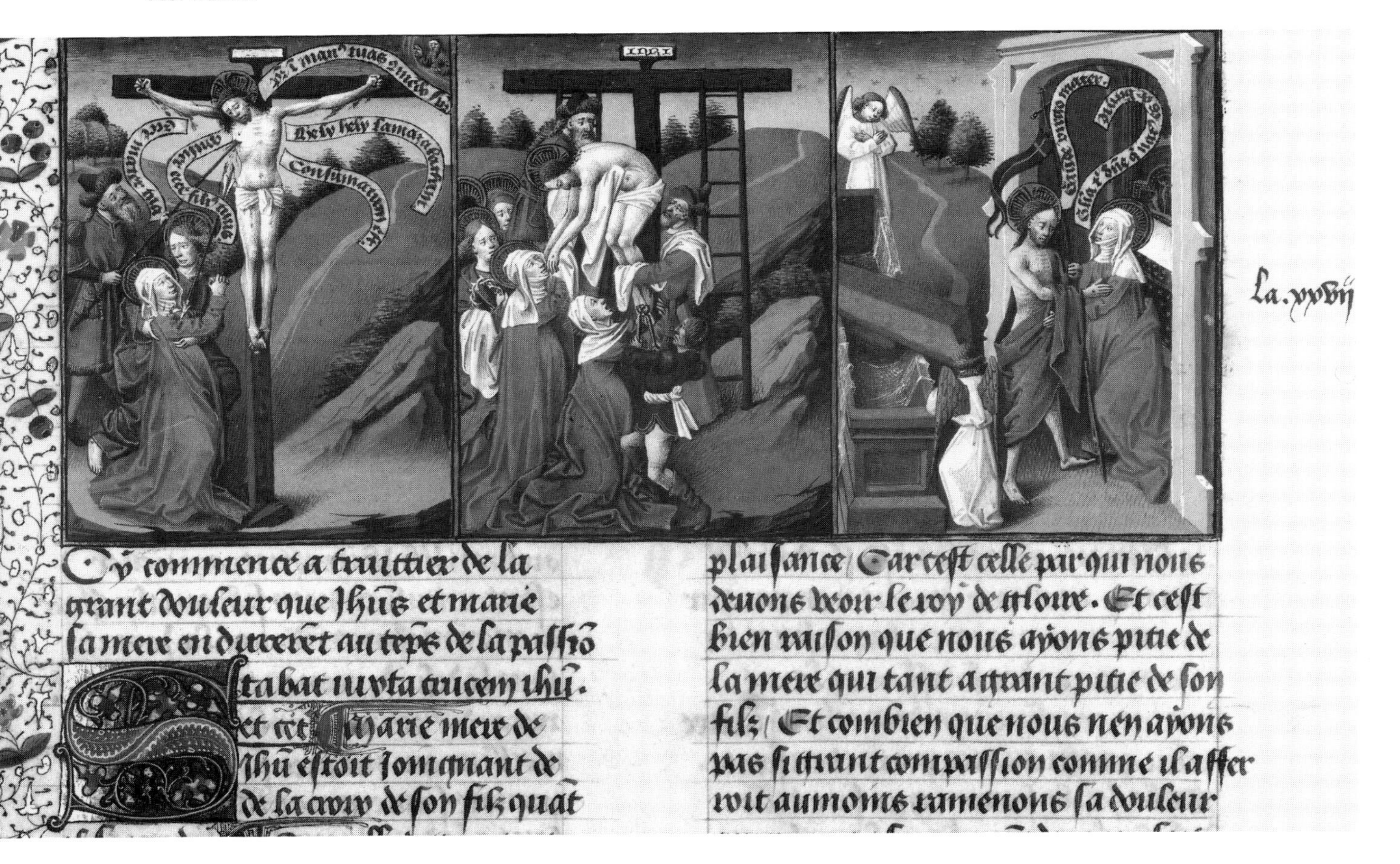

XXVIII (1) *School of Theology* Fol. 90v.
(2) *Seven Liberal Arts*

[Fol. 10] La xxviiij[e] hystoire est au commencement du secont livre qui se commence *Sapientia antiquorum*, ou quel premier chapitre il traitte de la diversité des doctrines et des escoliers, dont il en y a de purs seculiers et mondains qui ne estudient fors pour sçavoir et pour l'onneur du monde. Ceulx cy estudient les sept ars liberaulx ou pour acquerir richesses ou pour nom ou pour honneur. Et ceulx cy ne sont pas de l'escole ne de l'estude de vraye sapience, car ne leur science ne leur entention n'est mie espirituelle. Autres y a doctrines de soy spirituelles, esquelles preside dame Sapience. Qui s'apelle la doctrine de verité theologique, ceste doctrine est de soy bonne et saincte. Mais les escoliers sont en troys manieres selonc leur fin et leur entencion, les ungs veulent estre honnourez de leur science et ceulx cy se debattent ensemble par argumens, cuidens et s'esforcens de oster le nom et la gloire l'ung a l'autre, qui est signifié par l'esteuf d'argent qui premierement vient de sapience. Cest esteuf est la saincte escripture et doctrine de sapience laquelle vient du ciel a terre, et qui plus laboure et diligemment s'estudie a entendre, cil en est honnouré. Mais ceulx qui, voient l'onneur d'autrui, en ont envie et lui cuident oster et le vaincre et confondre par argumens, par cauteles et subtilitez. Ceulx cy n'ont pas bonne fin ne bonne entention, mais estudient pour avoir honneurs et vanitez du monde. Les dyables leur presentent en leur entention bonnés, chapperons fourrez, chaires haultes et pareez, et a ceste entention prennent ilz le labour et l'estude que ilz en ont.

Les autres escoliers sont a la dextre assis en sieges, estudiens et escoutans diligemment, affin de sçavoir, et ne leur chaut de autre honneur ou promocion, contens de leur simple estat.

Et les autres sont aux piez de dame Sapience, tous entenduz a sa doctrine, eslevez en contemplation et en amour d'ycelle, qui ne attendent ne ne acquierent fruit d'ycelle fors que l'amour de Dieu et la vie pardurable; ceulx cy sont dignes de presider en saincte Esglise.

C'est ce qui est figuré en ceste hystoire: en la moitié d'embas sont les sept ars liberales, qui sont vers terre ne ne quierent fors que terre et vie terrienne, qui sont signifieez par ces sept dames.

En la moitié d'en hault est l'estude de dame Sapience, l'estude de theologie qui pour divers fins et entencions se estudie, les ungs a senestre pour l'onneur mondaine affin qu'ilz soient sceuz et honnourez et prisez, les autres a dextre qui estudient [*fol. 10v.*] pour sçavoir et se occuper et y employer leur temps, les autres ou milieu estudient pour Dieu aimer et pour acquerir la gloire pardurable.

The twenty-eighth illustration is located at the beginning of Book II which opens Sapientia antiquorum. *In the first chapter he treats of the diversity of teachings and of pupils among whom there are the purely secular and worldly who study only to acquire knowledge and for worldly honour. They are the ones who study the Seven Liberal Arts either to acquire riches or for renown or for honour; and they do not belong to the School or study of true wisdom, for neither their intent nor knowledge is spiritual. There are other teachings, in themselves spiritual, over which Sapientia holds sway. The one which is called the Doctrine of Theological Truth is of itself sound and holy. But the pupils are of three kinds according to their aims and intent; some wish to be honoured for their knowledge; they debate among themselves, thinking and striving to remove by argument, each other's name and glory. This is represented by the silver ball which comes first from wisdom. This ball is holy writ and teachings about wisdom which descend from heaven to earth. He who works hard and strives earnestly to comprehend, is thereby honoured, but those who, seeing the honour of others, are envious of it and think they will usurp it, and overcome and confound it by arguments, ruses and cunning, they are the ones who are bereft of sound aims or intentions;*

xxviii^e. hystoire

they merely study to acquire the honours and vanities of the world. The devils offer them fur-lined hats, high-backed chairs with adornments; and for these objectives they undertake the study and work they do.

The other students are seated on the right in stalls, studying and listening diligently in order to learn; honour or preferment is not important to them; they are content with their simple state.

And still others are at Sapientia's feet, all attentive to her teachings, exalted in contemplation, and in love of her; they do not expect and they do not receive from the teaching any benefit except love of God and everlasting life. These are the ones who are worthy of holding responsible positions in the Holy Church.

This is what is depicted in this illustration: in the lower half we see the Seven Liberal Arts which are earth-bound and only research the earth and life on earth; they are represented by these seven ladies.

In the upper half is the studium *of Sapientia, the studium of theology. It is studied for various aims and objectives: students on the left do so for wordly honours so that they are known, respected and appreciated; others on the right study to acquire knowledge [fol. 10v.] and to busy themselves and fill their time; still others in the centre study in order to love God and acquire everlasting glory.*

XXIX (1) *Death-Bed Scene with Death Personified* Fol. 95v.
(2) *Battle for the Soul*

Fol. 10v.] La xxixe hystoire est ou secont chapitre du secont livre a l'endroit du parafe *O mort a telles gens.* Pour ce que ou premier chapitre il avoit parlé generalement des diversitez de doctrines, en ce chapitre il descent a monstrer de singulieres sciences pourfitables a ceulx qui quierent Dieu et le salut de leurs ames et la gloire de paradiz. Et premierement en ce chapitre il parle de la science de bien mourir. Et la demonstre par ung exemple sensuel et materiel selon ce que on voit tous les jours avenir que une personne qui a mené sa vie en joyes, en plaisances et delices de son corps sans se preparer a bien mourir, meurt en grant douleur, en douloureux regrez et en grant paour. Les dyables sont attendans a ravir l'ame pour les meffais, et la maleureuse personne ne pense que a ce qu'elle aime: ou son corps ou ses richesses. Et l'ung et l'autre, fault qu'elle lesse: les richesses a ceulx qui demeurent, car riens n'en emportera; son corps que elle laissera a menger a vers. Ors ne demeure que l'ame, laquelle a a passer ung estroit pont, lequel ne peut on passer legierement, qui n'est pur et net comme de baptesme. Et qui est celui qui tellement a vescu qui n'et rien a purger? Sur ce pont sont les angelz qui gardent, defendent et conduisent les ames a eulx commises. Et dessoubz est ung gouffre et fleuve de souffre, plain de couleuvres, de serpens et de dyables lesquelz, se ilz voient que l'ame passant trebuche ne choppe tant soit pou, ilz la aggrafent et tirent pour plungier en ce gouffre. Et c'est ce qui est figuré en ceste hystoire qui est clere de soy et de commun cours.

The twenty-ninth illustration is in the second chapter of Book II where the paragraph begins O mort a telles gens. *Since he had spoken in general terms in chapter I about the diversity of teachings, in the present chapter he finds time to detail particular information that will be advantageous to those seeking God, the salvation of their souls and the glory of Paradise.*

First of all in this chapter, he speaks of the Art of Dying well and he demonstrates it by a sensual and material exemplum *the like of which one sees acted out all the time: a person who has led a*

sauuement. Et si te fera acquerir
habundance de bonnes vertus qui te
seront enracinees en ton cuer sans
jamais partir. Ores donques pẽse
et te souuiengne dun homme qui est
au lit et a leure de la mort et fay
tout ainsi comme sil parlast a toy
tout sur le point de mourir. Quant
le disciple ouy cest exemple il print
a soubztraire son cuer et son entende
ment de toutes choses mondaines
et reprinst sa memoire en soy mes
mes et prist a considerer moult dili
gemment la semblance de lomme
qui vouloit mourir en la maniere
que dame sapience lauoit dit et de
uise. Lors luy vint une vision que
il auoit deuant luy ung tresbel jou
uencel qui estoit seurpris du mal
de la mort et le conuenoit mourir
soudainement z tantost et si nauoit
quelconques ordonnance faicte
pour son sauuement. Et se com
plaingnoit moult piteusement en
disant. Les pleurs et les douleurs
de la mort mont assailli et enuironne
et la paine denfer ma fait assault.
helaz mon dieu z mon createur pour
quoy fus je oncques ne au monde
et pour quoy ne mourus je la jour
nee que je fus ne. Laz laz le comme
ncement de ma vie fut en larmes z
en pleurs et la fin de ma vie est et
sera en tresgriefues complaintes et
en peine et en douleur. O mort cõe
ta memoire et la souuenance de toy
est amere comme dure chose est
de attendre ta venue especialment
a ceulz qui sont gais z jolis et qui
sont souef et delicieusement nourris

La. xxi
hystou

life filled with joy, pleasure and carnal delights without preparing to die well, dies in great grief, painful remorse and sheer terror.

The devils wait to carry off his soul because of his misdeeds, and the wretch only thinks of what he loves, his body or his wealth. Yet, he must give up both; riches to those who remain behind for he can take nothing with him; and his body he must leave to be eaten by worms. Thus, only the soul remains; it has to traverse a narrow bridge which cannot be crossed easily unless the soul is as pure and unblemished as at baptism. And who is there who has lived in such a way that he has nothing to cleanse? On the bridge are the angels which guard, defend and lead the souls entrusted to them. Below is an abyss and river of sulphur, filled with snakes, serpents and devils. The demons, on seeing a soul slightly falter and stumble as it crosses, seize it and drag it so that it plunges into the abyss. And this is what is depicted in this illustration which is in itself clear and of common currency.

XXX *St. Arsenius* Fol. 104v.

[Fol. 10v.] La xxxe hystoire est ou iij[e] chapitre du secont livre a l'endroit du parafe: *Vien cy et escoute.* En ce chapitre enseigne dame Sapience son disciple a bien vivre. Pour ce que ou chapitre precedent elle le avoit enseigné a bien mourir, et qui veult bien mourir, si[57] est necessaire que bonne vie precede; pour ce enseigne elle ycy a bien vivre. Et pour ce que toutes les sainctes escriptures sont plaines de ceste doctrine et tant y en a de livres que ce n'est que une confusion a simples gens qui ne sont pas acoustumez des livres, pour ce le disciple demande une briefve doctrine. Et Sapience le renvoye a ung saint heremite, de qui sa vie et doctrine est entre les *Vies des sains peres*, qui est appellé[58] Arseny, qui en troys mos enseigne la somme de toute parfection, c'est assavoir: Fuy les hommes ou compaigniez mondaines, parle pou, et te repose en Dieu. Et c'est ce qui est figuré en ceste hystoire en laquelle est le docteur, le disciple et l'ange denuncent ceste doctrine en une eschelle de degré en degré.

The thirtieth illustration is in the third chapter of the second Book where the paragraph begins Vien cy et escoute. *In this chapter Sapientia teaches her Disciple to live well. She had in the previous chapter instructed him how to die well; and whoever wishes to die well, an unblemished life must go before. For this reason, she gives instruction here on the art of living well. And because all of Holy Scripture is filled with this teaching, and because there are so many books on the subject that confusion reigns among simple folk unaccustomed to books, the Disciple asks for a succinct doctrine. And Sapientia refers him to a holy hermit whose life and teaching are found among the* Vies des sains peres. *He is called Arsenius. In three words he teaches the sum total of perfection, namely: shun men and worldly company; speak little; rest in God. And this is what is depicted in this illustration where we see the teacher, the Disciple and the angel proclaiming this doctrine step by step on a ladder.*

[57] MS. *qui bien veult mourir sil.*

[58] MS. *appellee.*

Vien cy et escoute la lecon que
je vueil lire. Le disciple en
ot grant joye. et fut moult desirant
quil peust veoir le livre dont le jouven
cel vouloit lire et tantost mist en
sa main un petit livret moult ancien

le lecon et il le fist voulentiers et
lui dist la lecon si comme devant de
mot a mot. Lors le disciple lui dist
en grant liesse desperit. O comme
ceste doctrine est sainte et prouffita
ble et lui demanda qui est celle sainte
creature qui ma apris ceste science
et sainte parole et bonne doctrine.
Le jouvencel lui respondi. Cest ar
seni le souverain philosophe. Le
disciple lui demanda sil avoit plus
ou petit livre. Adoncques il reprint
a lire et disoit. La fontaine et le fon
dement de tous maulx est aler et ve
nir sans prouffit. Quant le disci
ple louyt il en fut un pou troublé.
ou cuer et sefforcoit de monstrer
que il nestoit mie ainsi et que le livre
ne disoit mie verite et le promoit

XXXI (1) *The Last Supper* Fol. 105v.
(2) *Elevation of the Host*
(3) *Tantalus*

[Fol. 10v.] La xxxj^{e} hystoire est au commencement du iiij^{e} chapitre du secont livre qui se commence: *Sapience pardurable.* Pour ce que aux deux precedens chapitres dame Sapience a enseigné son amé disciple a bien mourir et a bien vivre, et a ces deux sciences acquerir et parfaire y a plusieurs dangiers et soussis, et pour ce, il demande a sa maistresse aucuns remedes contre ces dangiers, laquelle respont que generalement les sept sacremens de l'Esglise sont remedes souffisans pour [fol. 11] transmuer l'omme en spirituelle creature, mais plus especialement le sacrement de l'autel qui est proprement dit sacrement d'amour duquel il parle tout au long de ce present chapitre. Et pour ce que ce saint sacrement fut institué par Nostre Seigneur Jhesucrist en la derreniere cene qu'il feist avecques ses disciples devant sa passion, en ceste presente hystoire est figuree celle cene ou premier parquet.

Et l'autre parquet ensuivant est la commune maniere de celebrer la messe selon la tradicion et coustume de saincte Esglise. Aprés doncques que le disciple, devant qu'il fut enseigné et introduit de la verité et de la haultesse et doulceur de ce saint sacrement, ne sentoit pas les delices qui sont contenues en ce saint sacrement, il se compare a Tantalus, qui selon les poetes estoit ung homme qui estoit logé entre une fontaine et ung arbre et quant il vouloit boire, il ne se povoit besser, et quant il vouloit menger, il n'y povoit avoindre. Ainsi sont tous couvoiteux de ce monde que il ne pevent entreprendre et aussi estoit ce disciple devant qu'il eust congnoissance de la vertus de ce saint sacrement, car il estoit a l'eglise en la presence de ce saint sacrement et n'aparcevoit point la noblesse et la doulceur de ce saint sacrement. Et ce est figuré ou darrenier parquet ou quel est ung homme en ung traveil, lïé et mains et col, entre une fontaine et une table chargee de viandes, et n'y peut avenir. Et le disciple est la, contemplant les choses dessus dictes.

The thirty-first illustration is at the opening of the fourth chapter of Book II, which begins Sapience pardurable. *In the two preceding chapters, Sapientia taught her beloved Disciple to die well and to live well, pointing out that in acquiring and perfecting these two arts there were many dangers and troubles. He therefore asks his mentor for remedies against these dangers; she replies that the Seven Sacraments of the Church generally suffice [fol. 11] to transform man into a spiritual being, but in particular, the Sacrament of the Altar, properly called Sacrament of Love. He speaks about it throughout the present chapter. And because this Sacrament was instituted by Our Lord Jesus Christ at the Last Supper He took with His disciples before the Passion, the very same Last Supper is depicted in the first segment of this illustration.*

The next compartment shows the usual manner of celebrating mass according to the tradition and custom of Holy Church. Then since the Disciple, before he was taught and introduced to the truth, majesty and tenderness of this Holy Sacrament, did not experience the spiritual joys contained in it, he compares himself to Tantalus, who, according to the poets, was a man placed between a fountain and a tree; when he wanted to drink, he could not lower himself; when he wished to eat, he could not reach. In the same mould are those who covet this world but cannot lay their hands on it. The Disciple was in a similar state before he was cognizant of the powers of this Holy Sacrament: he was in church in the presence of this Holy Sacrament and did not perceive its majesty or tenderness. And this is what is depicted in the last compartment, in which we see a man with hands and neck tied in a tripalium, *stationed between a fountain and a table laden with food, and he is unable to reach any of it. And the Disciple is there contemplating the afore-mentioned things.*

Sapience tres humble et
souueraine vueilles
auoir aucune com
passion de mon desir
et te haste de moy se
courir. Ta doctrine et la doulceur de
tes parolles me traient a ton amour
quant je me aduise que je treuue en
ta doctrine comment je doy viure et
comment je doy mourir. Et je my
entens et my ocupe et je y pense vng
pou parfondement je deffaulx tout.
Et pour ce je retourne a ta maieste
et a ta souueraine puissance en sup
pliant que par ta pitie tu me vueilles
monstrer et enseignier a quel remede
je puis et doye recourir quant je me sen
tiray en aduersite. Sapience. Les sept
sacremens de saincte eglise sont les
sept principaulx remedes pour toy.
car par eulx est lomme retenue
et renouuelle en vne creature espiri
tuelle et est garde et nourri et souue
rainement pourueu a estre de perfec
tion par la grace de dieu. Mais a
voir dire le sacrement de lautel est
le principal car de lui principalment
est venu et yssu le ray et le fonde
ment de lamour diuine. Cest le
fleuue de grace qui arrouse et enyure
de bon eur les ames et donne grace
et vertus plus habundamment
que nul autre. Car tout ainsi et
encore mieulx comme la busche
seiche fait legierement feu alumer
et la flambe en hault esleuer et
espandre tout a lenuiron. Ainsi vo
ray le sacrement de lautel est cau
se de la chaleur de lamour diuine
et donne matiere et nourrissement
de vraye charite a ceulx qui le recoi
uent deuotement. Et la cause est
car entre tous les signes damours
le principal et le plus delitable.
et qui plus tost reprend le cuer et
le courage de lamant cest la pre
sence de ce quil aime. pour ceste
cause je me donnay je aimee tresa
mez disciples le soir de ma derre
niere cene et leur donnay puissance
et auctorite et a tous ceulx qui de
leur auctorite et puissance seroient
prestrez ordonnez que en la vertu
de mes parolles ilz puissent con

XXXII *The Spiritual Marriage of Sapientia and the Disciple* Fol. 127v.

[Fol. 11] La xxxij^{e} et darreniere hystoire est ou vij^{e} chapitre du secont livre a l'endroit du parafe: *Comment peut ce estre,* ou quel chapitre il traitte comment les loyaulx amés de la Divine Sapience la pevent espouser et estre conjoings a elle par vray et loyal amour. Et c'est ce qui est figuré en ceste hystoire, ou est Dieu le pere comme souverain prestre conjoingnent le disciple amant a la Divine Sapience qui est vray filz de Dieu. Et ce est fait par la conjonction du saint Esperit. Et puis est la Divine Sapience en chambre et en secret avecq son loyal amy lequel se humilie et reputé indigne de si grant don, et elle luy impose ung nouveau nom, c'est assavoir: frere amant.

The thirty-second and last illustration is in the seventh chapter of the second Book at the location of the paragraph Comment peut ce estre. *In the chapter he discusses how the faithful devotees of Divine Sapientia are able to marry her and be joined to her in true and loyal love. And this is what is represented in this illustration where God the Father is shown as a High Priest joining together the loving Disciple to Divine Sapientia who is the true Son of God. And this is carried out with the conjunction of the Holy Spirit. And then divine Sapientia is in a chamber and alone with her loyal devotee who kneels humbly, thinking himself unworthy of such a great gift; she bestows on him a new name:* Frater Amandus.

xxxvije hystoire.

Comment peut ce estre que vn
homme de nulle valeur vil.
et ort. z indigne pecheur sans meri
tes peut estre z est a si hault destre
eleue. Qui pourroit priser la valeur
de tel don qui tant est digne et pre
cieux. Encore ne sest pas tenu a tat
le roy trespuissant. mais a encore
aggrandis ces dons. Il a multiplie
ses benefices. et plus fort que deuat
espendue sa grace. et distribuee tres
largement. Car celui disciple de
sapience qui ca deuant est nomme
tant de foiz. en celle vision qui lui
fu monstree fu appelle par vng nou
uel nom de sa chiere espouse quat
ilz furent en leurs secrez retrais.
en la chambre des nopces en silence
et en secret. et il fut endormy entre
ses bras et par grant fiance et par
douleeur. mais il veilloit de cuer
en traittant de son sauuement et
du sauuement des autres creatures

Son espouse lappella frere amat
et lui dist secrettes parolles. saiges
subtilles z fiables z moult espiri
tuelles lesquelles trespassoient
engin humain et lui disoit. De toy
vendra chose par la quelle toutes gens
seront beneis et receuront la diuine be
neicon. Quant je louy ainsi parler
il me vint vne souueraine plaisance
au cuer et fus forment esmerueille.
Adoncques je me retourne vers elle
et lui dis. O lumiere de mes yeulx. O
desir de mon ame. O joye de ma pensee.
telle z si grande que je ne la pourroie
dire. O ma tresbelle z ma tresdoulce
amie noble de cuer z de lignee et de
toutes graces souuerainement douee
et garnie. Tu scez que la maniere.
la coustume la condicion. z la pro
priete de bonne amour est telle.
que quant vne personne aime vne
autre. Il vouldroit que celle personne
en qui il a mis son cuer pleust a tout
le monde aussi bien comme elle lui
plaist. et quelle fust amee de toutes
gens pour sans preiudice de son amour.
et sauf son droit. Et pour tant tu es
celle toute seule qui as en toy perfection
et grande assemblee de tous biens.
Et as ceste condicion que de tant coe.
vne personne te aimera mieulx et
comme plus vouldra que tu soies a
tous amie z de tous amee de tant la
aimeras tu plus et de tant sera elle
mieulx amee de ton cuer et de ton amour.
Et pour ce o tresbenigne dame qui
as la diligence du sauuement
de toute creature. Donne moy ma
mere z voye compectant par quoy
lamour dont tu me aimes soit
chascun jour z de jour en jour plus
grant z plus forte et quelle soit
renouuellee es cuers de ceulz qui
te ont oubliee. z mise en non chaloir

[Fol. 11] S'ensuivent les Declarations de troys hystoires qui ne sont pas de l'Orloge de Sapience et est la premiere des x commandemens

En ce mesmes volume y a troys autres hystoires au commencement de aucuns livres.

Here follow the Expositions of three illustrations which are not part of the Horloge de Sapience, *and the first one concerns the Ten Commandments.*

In this same volume are three other illustrations at the beginning of certain books.

XXXIII (1) *God in Majesty on Mount Sinai* Fol. 157
(2) *Moses climbs the Mountain*
(3) *Moses descends with the Tables*
(4) *Moses casts down the Tables on seeing the Worship of the Golden Calf*

[Fol. 11] En ceste hystoire est la montaigne de Synaÿ auquel sommet est Dieu en magesté, environné de angelz, et tout rouge comme feu, et la montaigne toute ardant et fument. Et ceste montaigne est enclose d'une haye. Et dessus la montaigne est Moÿse avecq son disciple Josué en diverses contenances. A la dextre est Moÿse descendant de la montaigne et gettent les tables a terre[59] par courroux; et devant luy dehors la closture de la montaigne est le peuple de Israel et Aaron entre eulx adorans le veau d'or a menetriez et dances et chançons, de quoy Moÿse fut moult courroucé, car ilz avoient laissé Dieu et adoroient les ydoles [fol. 11v.]. Et pour ce par courroux Moÿse getta les tables que Dieu lui avoit bailleez ou estoient les x commandemans la loy escrips du doit de Dieu, et les despeça et cassa en pieces.

Aprés, ou milieu, est Moÿse et son disciple montant en la montaigne, et derriere hors de la closture, est le peuple regardant Moÿse monter a grant reverence.

Aprés est Moÿse et son disciple descendens de la montaigne, apportant les tables que Dieu lui avoit bailleez contenans les x commandemens escrips du droit de Dieu. Et dehors la closture est le peuple le regardant et attendant en grant reverence. Et est la dicte hystoire ou clvij[e] feullet.

In this illustration we see Mount Sinai with God in Majesty at the summit; He is surrounded by angels and everything is fiery red, the mountain being all alight and smoke-covered. And this mountain is enclosed all around by a hedge. On the mountain one sees Moses and his disciple Joshua doing various things. On the right Moses descends the mountain and casts down in wrath the Tables of the Law. In front of him, and outside the mountain's enclosure are the people of Israel and Aaron adoring the Golden Calf with minstrels present as they sing and dance. Moses was greatly incensed because they had abandoned God and were adoring idols [fol. 11v.] And because of this, Moses angrily cast down the Tables which God had given him and on which were written by God's finger the Ten Commandments of the Law; he broke up the Tables and reduced them to fragments.

Next, in the centre, we see Moses and his disciple climbing the mountain, and behind, beyond the enclosure, the people watch in awe as Moses ascends.

Then, Moses and his disciple are depicted descending the mountain, carrying the Tables God had given him which contain the Ten Commandments written by God's finger. And outside the enclosure, the people watch him and wait showing great reverence. And the said illustration is on the 157th folio.

[59] MS. *de a terre.*

Gloire soit a dieu en qui nom pour le salut des ames du simple peuple crestien. et pour eulx monstrer en brief la teneur de nostre foy. et pour savoir ce que dieu nous commande et defend. est ordonnee ceste briefve escripture. pour lexposer a ceulx qui ne pevent oir souvent autres sermons. Et contient xviij chapitres. qui se pourront convenablement diviser en trois. ou en quatre parties. Coment le monde fu cree de dieu qui est un en trinite. et en especial pour quoy homme fu fait

Fermement nous devons croire et tenir quil est un dieu en trois personnes. qui sont le pere. le filz. et le saint esperit. par qui a este cree tout le monde. Car il peut tout. et scet tout. et gouverne tout. selon sa voulente. et divine bonte. En especial il gouverne les creatures humaines hommes et femmes ausquelz il a donne ames immortelles crees a sa semblance pour lui congnoistre amer. servir et honnorer. a fin que par bien vivre en ce siecle elles viennent a la gloire pardurable de paradis en lautre siecle apres ceste vie mortelle avec les bons angles. es lieux des angeles mauvais qui par leur pechie devindrent ennemis. et chairent des cieulx. avec lesquelz seront dampnez tous ceulx qui vivent ainsi comme se ilz feussent bestes sans ame. ou ennemis en char humaine.

Comment pour le pechie des premiers parens nous sommes en telles miseres

Vray est que nos premiers parens adam et eve qui a ceste fin davoir paradis furent noblement creez. donc par leur pechie et desobeissance lentree

XXXIV *Court of Heaven* Fol. 199v.

[Fol. 11v.] Une autre hystoire est au commencement du livre de la *Mandicité Spirituelle.* En ce livre traitte comment la povre ame malade et pecherresse qui a mestier de garison, de delivrance et de aumosne spirituelle, comment elle se doit pourchasser. Et en ce signifient est en ceste hystoire la figure de paradiz, Dieu, Nostre Dame et les angelz et les sains, et les sains et S. Pierre a la porte de paradiz; et dehors le paradiz est le docteur qui ensaigne l'ame estant en une litiere, couchee, malade, povre et chetive, comment elle doit demander l'aumosne spirituelle aux seigneurs, roys et princes de la cité de paradiz. Et est la dicte hystoire au c.iiij^{xx} et xixe[60] feullet.

A second illustration is found at the beginning of the work La Mendicité Spirituelle. *In it he discusses how the poor, sick and sinful soul should conduct itself when in need of protection, deliverance and spiritual succour. And representing this in the illustration is the portrayal of Paradise, God, Our Lady, angels and saints, and saints with St. Peter at the gate of Paradise. Outside Paradise is the doctor instructing the sick, wretched and miserable soul, recumbent on a litter, how it should seek spiritual succour from the lords, kings and princes of the City of Paradise. And this particular illustration is on the 199th folio.*

[60] MS. *et xij^{e}.*

Complainte de lomme a son ame
et lennorte a mendier et demander
espirituelment.

Ma povre . ma malade .
ma chartriere . ma mi
serable ame . hors mise
en hostaige loing de
ton pays . Tu qui nas riens . et pour
ton labour ne scez . et ne peus quelcon
ques chose acquerir . Croy mon con
seil . apren le mestier de mendier
et truander . et que ton pourchas te
soit en lieu de rente . Pourquoy mor
roies tu de faim . de soif . et de froidure .
A blasmer seroit une telle honte et
paresse .

La responce de lame desconfortee

Comme mon hostellain qui
avecques moy es gettee hors
de ton premier pays . et sommes
ensemble en la chartre obscure et
douloureuse de ce present exil . Je
congnois bien helas que je suis po
vre malade . emprisonnee . blecee . et
navree . nue sans vesteure . et si nay
riens . Trop male fu leure . quant
le premier pere commist envers dieu
le souverain roy telle trayson . tel crime
de leze maieste que tout nostre heritaige
fu forfait et oste . et que du lieu de
plaisance . de joye . et dexcellence nous
feusmes dechaciez . en ce lieu de pleur
de tribulacion . dangoisse . et de desola
cion . de richesse en povrete . de noblesse
en vilte . de vie en mortalite . de lieu
seur en lieu hors de toute seurte . Et
a brief dire de tout bien en toute ma
leurte . Si ne scay que je face . car
a labourer suis impotent . enferme
et non sachant . Mon mestier deue
roit estre congnoistre dieu . le veoir .
lamer . le loer et honnorer . mais lasse
moy trop loing en suis eslongnee .
trop ma gettee arriere de lui . Je voy

XXXV (1) *Agony in the Garden* Fol. 229
(2) *The Betrayal*
(3) *Christ Nailed to the Cross*
(4) *Crucifixion*
(5) *Deposition*
(6) *Entombment*

[Fol. 11v.] Une autre hystoire est au commencement du sermon de la Passion de maistre Jehan Gerson. En ce sermon est preschee la passion de Nostre Seigneur Jhesucrist selon le texte des quatre evvangelistes. Et est ce qui est figuré en ceste hystoire, car au commencement, au parquet d'en hault a dextre, est le jardin et Nostre Seigneur en ce jardin priant Dieu le pere. Dessoubz est la capcion comment Nostre Seigneur fut prins et trahy de Judas son disciple. Enaprés, est ou milieu, embas, le crucifiement; et dessuz, le crucifix et la teneur de la passion. A la senestre en hault est la deposicion de la croix et comment Nostre Dame le receut devotement et reveremment en son geron. Dessoubz en ce mesmes costé est le sepulchre de Nostre Seigneur et le service et obseques, et les sains et sainctes dames lui firent sa sepulture. Et est ceste hystoire ou ccxxix^{e} feullet.

Another illustration is at the opening of Master Jean Gerson's Sermon de la Passion. *In this sermon the Passion of Our Lord Jesus Christ is preached according to the text of the four Evangelists. And this is what is represented in this illustration: to begin with, in the upper right segment is the Garden, where Our Lord prays to God the Father. Below, is the Arrest showing how Our Lord was captured and betrayed by His disciple, Judas. Next, in the lower centre, is the Nailing to the Cross, while above, appears the Crucifix and the subject matter of the Passion. In the upper left occurs the Deposition and how Our Lady received Him on her lap with piety and devotion. Below, on the same side, is shown Our Lord's Sepulchre with the ministrations and obsequies. The saints and holy women carried out the entombment. And this illustration is on the 229th folio.*

A Dieu sen va par mort amere. Jhus devant sa doulce mere. Si de vons bien par penitance de ce dueil avoir remembrance. Car ainsi est il de vostre dueil doncement de vons nous bien avoir remembrance tousiours. O benoist ihs seul et vray sauveur de tout le monde bien nous en doit souvenir. Quant apres ce que vous no avez fait en vng seul moment de vostre vouloir et de neant vous nous avez voullu reparer par ceste labour par angoisse

et par douleur et telle douleur que oncques en ce monde ne fut la pareille et tout pour nous. O devotes ames pour nous dy je miserables et povres creatures subiectz aux po lui navoit il nous deservi. C Et commencea des son enfance en povrete en douleurs en fain et en soif. en plours en froit et en chault. en pelerinaige en egypte en veilles et en reponces . avis au iour dhui au iour de sa gloriouse passion. toute de plaintes de miseres et de pleurs toute sa vie fut acomplie et consommee. C Car a dieu va et mort amere. Jhs devant sa doulce mere. C O tres doulce mere je esliève a present mes yeulx

Glossary

The *lemmata* are referenced by the serial number in Roman capitals at the head of the sections of the French *Déclaration*.

absconse XVII, *sheltered*
accoutant XIV, *listening*
accroistre XIX, *augment*
affublé XII, *fastened up*
aggreve XVII, *distress*
aigue XII, *pointed*
alumé X, *enlightened, inspired*
amere VIII, *unmerciful*
amertume IX, XI, *piteous thing, bitterness*
appesantit XVII, *burden*
appeticent XII, *reduce*
apprestez XIV, *prepared, made ready*
arbre de la Croix III, IVB, *Holy Rood*
ars liberales, arts liberaux XXVIII, *Liberal Arts*
asseurance XVIII, *certainty*
assome, assommez XII, *fell*
atouchemens XII, *contacts*
attour XII, *attire*
auctorité *passim, saying*
auge XII, *hod, tray*
avenir XXXI, *reach*
avoindre XXXI, *reach*
avoir XII, *wealth*
aysaient XVII, *trying, testing*

baille XII, baillent IVB, bailleez XXXIII, *hand over*
balances XII, *scales*
bang XII, *bench*
baulme VI, *balm*
beste mue VB, XII, *dumb animal*
bonnement XII, *simply*
bonnés XXVIII, *academic bonnets*
bouette VI, XII, *box*
bouté, VIII *buffeted*—bouter hors XII, *eject, push out*
boyteuse XII, *limping*
brain VI, *sprig*

campion X, *champion*
cauteles XXVIII, *ruses*
cedules IVB, *small scrolls*
cene XXXI, *Last Supper*
chaires XXVIII, *academic chairs*—en chaire, I, *in a pulpit*
chambre IVB, *monastic cell*
changeur XII, *money-changer*
chapitre IVB, XIV, *chapter (held by Dominicans)*
charité XII, *charity*; XIV, *caritas*
charoigne VIII, *rotting flesh*
cheoir XII, cheut XIV, *fall*
choppe XXIX, *stumble*
cloe XI, *nail*
clost XVII, *shut, close*

closture XXXIII, *enclosure*
coffre VIII, *chest*
cogitation XII, XVII, *thought, reflection*
comparation XV, *comparison*
composeur I, *author of a text*
conforme (se) IX, *mould oneself*
congnee XII, *axe*
consaulx XII, *counsels*
consentement XVII, *consent*
consequemment XXVII, *later*
conservation de memoire IVB, *retention in memory*
consolation VII, XVII, *solace, delight*
contenir (se) VI, *restrain oneself*
contraires XII, *opponents*
cooperant XVII, *concurring*
cordiaux XII, *ropes*
cornelier XII, *dog-wood*
cornet XII, *small horn*
cours, de commun c. XXIX, *of common currency, usage*
coustume, a c. III, *customarily*
couvenablement II, *suitably*
couvoitans XIII, *greedy*
couvoité XV, *desired*
couvoiteux XXXI, *covetous*
couvoitise XIV, *covetousness*
crucifiement III, X, XXXV, *nailing to the Cross*
cuidant XI, *thinking*

dague XII, *dagger*
darnier, darrenier, darreniere, derrenier, derreniere IVB, VII, XXIVA/XXIVB, XXXII, *last, final*
debouté VIII, *buffeted*
decevable XI, *deceiving, treacherous*
declaration I, XXXIII, *exposition*
deçoit, deceut XI, *deceive*
decraché, decrachié VIII, *spat upon*
democqué IX, *mocked*
demonstre XII, *make clear*
denunça XIV, denuncé XXIVA, denuncens XVIB, *reveal, proclaim*
depart (se) XVII, *leave*
desertes XX, *rewards, deserts*
desesperation XII, XIX, *despair*
deshonneste X, XII, *ignoble, dishonourable*
desloyales XII, *deceitful*
desolé XVII, *forsaken*
despent de X, XIV, *to be related to*
despit XII, *contempt*
doctrine *passim, teaching, doctrine*
dortouoir XII, *dormitory*
dortouriere XII, *dormitory-maid*
drap d'or VIII, *cloth of gold*
durance XVIII, *duration*
dy XII, *day*

effondree XVIA/XVIB, *sinking*
emendation XIV, *correction*
emprés VII, *afterwards*
enamourer (s') XVII, *be loved*
enclume XII, *anvil*
enhastee XII, *encrusted*
enhorte IX, XIV, *exhort*
enquiert XV, *inquire*
ensuivans XIX, XXII, *following, next*; ensuivent *XVIII, adhere to, follow*
erres XVII, *pledges*
esbaït VIII, *be dismayed, be in dismay*
escoliers XXVIII, *pupils*
escure XII, *cleanse*
escus XII, *crowns* (*coin*)
esmouvent IVB, *moving, inciting*; — esmeu VI, *moved*
espouenté X, *fearful*
estains II, *dead at heart*
estas du monde I, *classes of society*
esteuf XXVIII, *ball*
estude XXVIII, *studium*
et = ait XXIX, *pr. sub. of* avoir
excede XII, *be excessive*
excellemment XXVII, *outstandingly, exceptionally*
excellent XV, *noble*

failly IX, *faint*
fange XII, *slime*
fardeau XII, *bundle*
fermé XVII, *strengthened*
ficha XI, *drive in*
fiencé XVII, *affianced*
figuré *passim, depicted, represented*
figure *passim, likeness, representation, picture*
flaitries, flaittriez XV, *faded*

flateurs XIII, *flatterers*
fleur, en f. d'aage IVA, *in the prime of life*
fleuré XV, blooming
fleurissent XV, *with flowers*
forme I, *image*—XIV, *practice, procedure*
fornaise, fournaise XXI/XXII, XXIVB, *oven*
fors, fort IX, *valiant*—XI, strong
fort (*adv.*) XII, *outright*
fouit XII, *dig*
frappé IX, *smitten*

gan XII, *glove*
gantelés XII, *gauntlets*
gehinnes XXI/XXII, *tortures*
geron XXXV, *lap*
gés XII, *lanyards, thongs*
gibesiere XII, *game-bag*
gratent XII, *scrape away*
gresle XVIA/XVIB, *hail*
guerré XI, *made war*

habit IVB, *monk's habit*
hayne, prendre en h. XIV, *conceive a hatred for*
hayneux XIV, *hateful*
hostel, *see* ostel
hotent XII, *carry off*
houe XII, *hoe*
hystoire *passim, story, account, illustration*

impugnent XII, *oppose*
induit X, *induced*
iniquitez XII, *iniquities*
inspiration IVA, *revelation*
instigation VB, *incitement*
interrogant VI, *questioning*—interrogue XV, *interrogate*

jouvencel, jouvencelle IVA, XII, *juvenile*

laboura XIII, laboure XXVIII, labouroit XII, *toil*
laboureur XII, *agricultural labourer*
laidure XII, *ugliness*
las XVIII, *snares*
liberales, liberaux, *see* ars
lié XV, *happy*
lie XII, lié *passim, bind, bound*
lime XII, *file*
loy, IX, *faith*
loyal XVII, faithful

maillés XIX, *mallets*
maintien II, *conduct*
maistresse II, *mentor*
mal XXIVB, *not at all*
malfaitteurs X, *malefactors*
maltalent XIV, *anger*
matines III, *matins*
medite XV, *meditate*
meselee XII, *leprous*
mestier XII, *need*
mire (se) XII, *admire oneself*
monde I, *orb of the world — passim, secular life*
mongnons XII, *stumps.*
mortier XII, *mortar*
moyen, moyenne VII, IX, *middle, central*
mucee XVII, *hidden*
mue, *see* beste

noises XII, *disputes*

odeur VIII, *scent*
ongnement VI, XII, *balm, ointment*
operations XII, XXIVB, *workings*
oppinion VB, *aims, resolve*
opportunité XII, *occasion*
opposite, a l'o. XVIB, *on the other side facing*
orde XII, *filthy*
ordre IVB, XIV, *religious Order* — XXIII *Order of angels*
ost IX, *army*
ostel, hostel XIV, *household*
ottroioit XIII, *favour, encourage*

paines XXV *sufferings*—XXIII, payens *passim, pains of Hell*
pallut infernal XII, *slough of Hell*
parafe *passim, paragraph*
pareillement XII, *similarly*
paroir XIV, *wall*
parquet *passim, segment, compartment*
partie XIII, *side*
payens, *see* paines
perpetuellement IVB, VII, *for ever, everlastingly*

perser XIV, *run through, pierce*
piteuse VIII, *piteous, compassionate*
plaisances *passim, pleasures*
playé VIII, *wounded*
poindre XII, point XII, *stab, prick*
pointure VIII, *prick*
poise XII, *weigh*
portant XIV, *imposing*
porteresse XII, *bearer who is a female*
portier XIV, *doorkeeper*
possessoire XII, *holder*
poultre XIV, *beam*
pourfitable XXIX, *advantageous*
poursuivent XII, poursuivoit XXIVA, poursuit X, *pursue*
praiau III, *stretch of grass*
premiere XXIVA, *upper, starting from the top*
presenté IX, *brought into the presence of*
prestres IX, *priests*
pretoire III, *praetorium*
prieuse XII, *prioress*
principe I, *essence*
procession III, *perambulation, procession* — faire une p. III, *perambulate*
prosterné XIV, *prostrate*
punaisie XIX, *filth*
purger XXIX, purgeez XIV, *atone for, purge*

quaimenderies XII, *importunate requests*
quarree XII, *squared*
quart *passim, fourth*
quierent I, VIII, *seek out*

racompte X, *relate*
ralume II, *revive*
rapaiser XVIA, *assuage*
rebouté XII, *repelled, turned away*
rechignee XII, *grimacing*
reciter XXIVA *rehearse, retell*
recraingnent VIII, *fear*
reduisant a soy (se) VA, *turning inward*
refectouoir, refectuoir IVB, VI, *refectory*
refroidiz II, *benumbed*
regnars XIII, *foxes*
renge VII, IX *row, register of pictures*
reparation XII, *restoration, repair*
reparé I, *reconstituted*
reprehension XIV, *reprimand*
reputation XII, *importance*
resveil II, *alarm*
retraihoit (se) XIII, *turn*
riotes XII, *disturbances*
rongent XII, *gnawing, file off*
rosier, branche de r. VIII, *briar*

sagettes XII, *arrows*
saillir III, IVA, *emerge* — saillis XII, *emerged, abandoned*
sain XII, *bosom*
sainture XII, *belt, girdle*
scismes XIII, *schisms*
seinte XII, *girded around*
semblance IX, *manner* — XVIII, *likeness*
sepulture VII, XXXV, *entombment*
serpente VII, *serpent (of the Garden of Eden)*
signification XII, *import, significance*
signifie *passim*, signifient, signifiant I, XII, *represent, mean*
sismatique XIV, *schismatic*
soigneuse, estre s. de II, *take pains to*
souef, souefve VI, VIII, *fragrant, sweet*
souffre XXIX, *sulphur*
souflet XII, *bellows*
souhaider XV, *wish*
soussis XXXI, *cares, worries*
stabilité XVIII, *calm, stability*
subjection XII, *subordination*
substance XXIVB, *possessions, wealth*
subtilité XII, XV, XXVIII, *subtlety, cunning*
suppliant XVIB, *supplicating*

tables IVB, *writing tablets* — XXXIII, *Tables of the Law*
talent XII, *desire*
targe (se) XII, *shield oneself*
terriennes XV, XVII, *earthly, of the world*
therebint VI, *terebinth tree*
theume I, *subject matter, theme*
tourbe XXIII, *crowd*
traitte XXVII, XXVIII, XXXII, *discuss, treat of*
transmuer XXXI, *transform*
trebucher VII, *cast down*
truie XII, *sow*

ueil XII, *eye*

umbragens XVII, *darkening*
unes XII, *a pair of*
ungles XII, *nails*
uys XVII, *door*

vent XII, *hot air*
verges XII, *rods*
vermolu VIII, *worm-eaten*
vers XXIX, *worms*
vertueuse VI, *with powerful properties*
voultrer (se) XII, *wallow*
vuidé XIV, *bereft*
vuit III, *empty*

ymagination VB, *mental picture*
ysseue de I, *emanated from*

Selected Bibliography

Articles such as *L'*, *La*, *Le*, *Les* or *The*, which serve as the first words of titles, are omitted; thus, *The Hours* is listed as *Hours*, *Le Livre* as *Livre*, *Les Heures* as *Heures*, etc.

Ancelet-Hustache, Jeanne, "Quelques indications sur les manuscrits de l'*Horloge de Sapience*," in *Heinrich Seuse. Studien zum 600 Todestag (1366-1966)*, ed. E. M. Filthaut, Cologne, 1966, pp. 161-70.

Ancona P. d', and Aeschlimann, E., *Dictionnaire des miniaturistes du moyen âge et de la Renaissance*, Milan, Hoepli, 2nd ed. 1949.

Argan, G. C. and Lassaigne, J., *The Great Centuries of Painting. The Fifteenth-Century from Van Eyck to Botticelli*, Lausanne, Skira, 1955.

Arts libéraux et philosophie au moyen âge. Actes du quatrième congrès international de philosophie médiévale, Montreal, Institut d'études médiévales, 1969.

Avril, F., *Manuscript Painting at the Court of France. The Fourteenth Century (1310-1380)*, London, Chatto and Windus, 1978.

——, "Le destinataire des Heures 'Vie à mon desir': Simon de Varie," *Revue de l'art*, LXVII (1985), pp. 33-44.

Backhouse, Janet, *The Illuminated Manuscript*, Oxford, Phaidon, 1979.

——, "A Reappraisal of the Bedford Hours," *British Library Journal*, VII (1981), pp. 47-69.

——, "French Manuscript Illumination 1450-1530," in T. Kren (ed.), *Renaissance Painting in Manuscripts. Treasures from the British Library*, London, British Library, 1983., pp. 145-74.

——, *Books of Hours*, London, British Library, 1985.

Beatson, *see* Meiss.

Belles Heures, see Meiss and Beatson.

Besserman, L. L., *The Legend of Job in the Middle Ages*, Cambridge, Mass., Harvard University Press, 1979.

Biblia Sacra iuxta Vulgatam Clementinam, ed. A. Colunga O. P. and L. Turrado, Madrid, Editorial Catolica, 1965.

Bigarne, C., *Etude historique sur le chancelier Rolin et sur sa famille*, Beaune, Lambert, and Dijon, Lamarche, 1860.

——, "Documents sur la famille du chancelier Rolin," *Mémoires de la Société Eduenne*, n.s. VI (1877), pp. 481-502.

Blum, A. and Lauer, P., *La Miniature française du XVe et XVIe siècle*, Paris, Van Oest, 1930.

Boase, T. S. R., *Death in the Middle Ages. Mortality, Judgment and Remembrance*, London, Thames and Hudson, 1972.

Bradley, J. W., *A Dictionary of Miniaturists, Illuminators, Calligraphers and Copyists*, London, Quaritch, 1887-1889, 3 vols.

Brandis, T. *et al.* (eds.), *Zimelien. Abendländische Handschriften. Ausstellung 13. Dezember 1975 — 1. Februar 1976*, Wiesbaden, Reichert, 1975.

Breckenridge, J., "*Et prima vidit*: The Iconography of the Appearance of Christ to His Mother," *Art Bulletin*, XXXIX (1957), pp. 9-32.

Buren, Anne H. van, "New Evidence for Jean Wauquelin's Activity in the *Chroniques de Hainaut* and for the Date of the Miniatures," *Scriptorium*, XXVI (1972), pp. 249-68.

——, "The Canonical Office in Renaissance Painting. Part II: More about the *Rolin Madonna*," *Art Bulletin*, LX (1978), pp. 617-33.

Byvanck, A. W., *Les principaux manuscrits à peintures de la Bibliothèque Royale des Pays-Bas et du Musée Meermanno-Westreenianum à La Haye*, Paris, Société Française de Reproduction de Manuscrits à Peintures, 1924, 2 vols.

Calkins, R. G. (ed.), *A Medieval Treasury*, Ithaca, Cornell University, 1968.

——, "Medieval and Renaissance Illuminated Manuscripts in the Cornell University Library," *Cornell University Library Journal*, XIII (1972), pp. 3-95.

——, *Illuminated Books of the Middle Ages*, London, Thames and Hudson, 1980.

Cazelles, *see* Longnon.

Chavannes-Mazel, C. A. *et al.* (eds.), *Schatten van de Koninklijke Bibliotheek. Acht eeuwen verluchte handschriften*, The Hague, Koninklijke Bibliotheek, 1980.

Cinq années d'acquisitions 1969-1973. Exposition organisée à la Bibliothèque Royale Albert Ier du 18 janvier au Ier mars 1975, Brussels, Bibliothèque Royale, 1975.

Cinq années d'acquisitions 1974-1978. Exposition organisée à la Bibliothèque Royale Albert Ier du 24 septembre au 31 octobre 1979, Brussels, Bibliothèque Royale, 1979.

Cockshaw, P., *Les Secrétaires de la Chancellerie de Flandre-Bourgogne sous Philippe le Bon*, Brussels, Université de Bruxelles, 1964, Unpublished Thesis.

——, *Les Miniatures des Chroniques de Hainaut (15e siècle)*, Mons, Service Provincial du Tourisme du Hainaut, 1979.

Cotton, Françoise, "Les manuscrits à peintures de la Bibliothèque de Lyon," *Gazette des Beaux-Arts*, LXV (1965), pp. 265-320.

Davies, M., *Rogier van der Weyden. An Essay, with a Critical Catalogue of Paintings assigned to him and to Robert Campin*, London, Phaidon, 1972.

Debae, Marguerite *et al.* (eds.), *De Librije van Bourgondien en enkele recente aanwinsten van de Koninklijke Bibliotheek Albert I. Vijftig Miniaturen*, Brussels, Cultura, 1970.

——, *see also* Dogaer.

Delaissé, L. M. J., "Les *Chroniques de Hainaut* et l'atelier de Jean Wauquelin à Mons, dans l'histoire de la miniature flamande," *Bulletin des Musées royaux des Beaux-Arts*, 1955, pp. 21-56.

——, "Les principaux centres de production de manuscrits enluminés dans les Etats de Philippe le Bon," *Cahiers de l'Association Internationale des Etudes Françaises*, VIII (1956), pp. 11-34.

——, (ed.) *Miniatures médiévales de la librairie de Bourgogne au Cabinet des manuscrits de la Bibliothèque Royale de Belgique*, Brussels, Bibliothèque Royale, 1959.

Dhanens, Elisabeth, *Hubert and Jan van Eyck*, New York, Alpine Fine Arts Collection, 1980.

Dogaer, G., "Handschriften over de kruistochten in de librije der hertogen van Bourgondië," *Spiegel Historiael*, II (1967), pp. 457-65.

——, "L'Ecole ganto-brugeoise: une fausse appellation," in *Miscellanea codicologica F. Masai dicata*, Ghent, 1979, II, pp. 511-18.

—— and Debae, Marguerite, *La Librairie de Philippe le Bon. Exposition organisée à l'occasion du 500e anniversaire de la mort du duc,* Brussels, Bibliothèque Royale, 1967.

Durrieu, P., *Un grand enlumineur parisien au XVe siècle: Jacques de Besançon et son oeuvre*, Paris, Champion, 1892.

——, "Les *Antiquités judaïques* de Josèphe à la Bibliothèque Nationale," *Gazette des Beaux-Arts*, 1906, part II, pp. 5-13.

——, *Les Antiquités judaïques et le peintre Jean Fouquet*, Paris, Plon, 1908.

——, *Les Manuscrits à peintures de la Cité de Dieu*, Paris, Leclerc, 1910.

Enluminure, see Gagnebin.

Farquhar, J. D., *Creation and Imitation. The Work of a Fifteenth-Century Manuscript Illuminator*, Fort Lauderdale, Nova / NYIT University Press, 1976.

——, *see also* Hindman.

Francis, W. N., *The Book of Vices and Virtues. A Fourteenth-Century English Translation of the Somme le Roy of Lorens d'Orleans*, London, Oxford University Press for the Early English Text Society, 1942.

Gagnebin, B. *et al.* (eds.), *L'Enluminure de Charlemagne à François Ier*, Geneva, Bibliothèque Publique et Universitaire, 1976.

Galle, L., *Note sur le Missel d'Autun de la Bibliothèque de la Ville de Lyon*, Paris, Plon-Nourrit, 1901.

Gaspar, C. and Lyna, F., *Les principaux manuscrits à peintures de la Bibliothèque Royale de Belgique*, Paris, Société Française de Reproduction de Manuscrits à Peintures, 1937 and 1949, 4 vols.

——, *Philippe le Bon et ses beaux livres*, Brussels, Editions du Cercle d'Art, 1944.

Gerson, Jean, *Oeuvres complètes*, ed. P. Glorieux, Tournai, Desclée, 1961-1973, 10 vols.

Glorieux, P., *see*: Gerson, Jean.

Guratzsch, H., *Dutch and Flemish Painting*, New York, Vilo, 1981.

Harthan, J., *The Book of Hours*, London, Thames and Hudson, 1977.

Hindman, Sandra, *Text and Image in Fifteenth-Century Illustrated Dutch Bibles*, Leiden, Brill, 1977.

—— and Farquhar, J. D., *Pen to Press. Illustrated Manuscripts and Printed Books in the First Century of Printing*, College Park, University of Maryland Art Department, and Baltimore, Johns Hopkins University Department of the History of Art, 1977.

Hours of Etienne Chevalier, see Sterling and Schaefer.

Huttois, Isabelle (ed.), *L'Iconographie musicale dans les manuscrits de la Bibliothèque Royale Albert Ier*, Brussels, Bibliothèque Royale, 1982.

Jenni, *see* Pächt.

König, E., *Französische Buchmalerei um 1450. Der Jouvenel-Maler, der Maler des Genfer Boccaccio und die Anfänge Jean Fouquets*, Berlin, Mann, 1982.

Korteweg, A. S. (ed.), *Liturgische Handschriften uit de Koninklijke Bibliotheek*, The Hague, Rijksmuseum Meermanno-Westreenianum, 1983.

Kren, T. (ed.), *Renaissance Painting in Manuscripts. Treasures from the British Library*, London, British Library, 1983.
Künzle, P. (ed.), *Heinrich Seuses Horologium Sapientiae. Erste kritische Ausgabe unter Benützung der Vorarbeiten von Dominikus Planzer O.P.*, Freiburg, Universitätsverlag, 1977.
Lassaigne, *see* Argan.
Lauer, *see* Blum.
Leroquais, V., *Les Sacramentaires et les Missels Manuscrits des Bibliothèques Publiques de France*, Paris, Imprimerie Nationale, 1924, 4 vols.
——, *Les Livres d'heures manuscrits de la Bibliothèque Nationale*, Paris, Imprimerie Nationale, 1927-1940, 3 vols.
——, *Les Bréviaires manuscrits des Bibliothèques Publiques de France*, Mâcon, Protat, 1934, 5 vols.
——, *Les Psautiers manuscrits latins*, Mâcon, Protat, 1940-1941, 2 vols.
Liturgische Handschriften, *see* Korteweg.
Longnon, J. and Cazelles, R., *Les Très Riches Heures du duc de Berry*, Paris, Draeger, 3rd ed. 1975.
Lyna, *see* Gaspar.
Mâle, E., *L'Art religieux à la fin du moyen âge en France*, Paris, Colin, 6th ed. 1969.
Mandonnet, P., "Order of Preachers," in *The Catholic Encyclopedia*, New York, XII, pp. 354-70.
Manion, Margaret M., *A Study of the Wharncliffe Horae*, Master's Thesis, Melbourne University, 1962.
——, *The Wharncliffe Hours. A Study of a Fifteenth-Century Prayerbook*, Canberra, Australian Academy of the Humanities, 1972.
——, *The Wharncliffe Hours. Facsimile*, London, Thames and Hudson, 1981.
—— and Vines, Vera F., *Medieval and Renaissance Illuminated Manuscripts in Australian Collections*, London, Thames and Hudson, 1984.
Marrow, J. H., "Miniatures inédites de Jean Fouquet: Les Heures de Simon de Varie," *Revue de l'Art*, LXVII (1985), pp. 7-32.
Meer, F. van der, *Maiestas Domini: théophanies de l'Apocalypse dans l'art chrétien*, Paris, Belles-Lettres, 1938.
Meiss, M., *French Painting in the Time of Jean de Berry. The Late XIVth Century and the Patronage of the Duke*, London, Phaidon, 1967, 2 vols.
——, *French Painting in the Time of the Duke de Berry. The Boucicaut Master*, London, Phaidon, 1968.
——, *French Painting in the Time of Jean de Berry. The Limbourgs and their Contemporaries*, London, Thames and Hudson, 1974.
——, and Beatson, Elizabeth H., *Les Belles Heures de Jean duc de Berry*, London, Thames and Hudson, 1974.
——, and Thomas, M., *The Rohan Book of Hours*, London, Thames and Hudson, 1973.
Mély, F. de, *Les Primitifs et leurs signatures: les miniaturistes*, Paris, Geuthner, 1913.
Meurgey, J., *Les principaux manuscrits à peintures du Musée Condé à Chantilly*, Paris, Société Française de Reproduction de Manuscrits à Peintures, 1930, 2 vols.
Michel, H., "L'*Horloge de Sapience* et l'histoire de l'horlogerie," *Physis*, II (1960), pp. 291-8.
Miniature flamande. Le Mécénat de Philippe le Bon, Brussels, Palais des Beaux-Arts, 1959.
Monks, P. R., *A Study of the Compositional Elements in Jean Fouquet's Paintings for the Antiquités Judaïques*, Master's Thesis, Northwestern University, 1980.
——, "The Influence of the Master of Jean Rolin II on a Master of the Vienna *Girart de Roussillon*," *Codices Manuscripti. Zeitschrift für Handschriftenkunde*, XI (1985), pp. 1-8.
——, "Reading Fifteenth-Century Miniatures; the Experience of the *Horloge de Sapience* in Brussels, Bibliothèque Royale, MS. IV.111," *Scriptorium*, XL (1986), pp. 242-8, pls. 13-16.
——, *A Study of the Art Work of the Rolin Master in the Horloge de Sapience, Brussels, Bibliothèque Royale, MS. IV.111, and of his other Known Surviving Works*, PhD Thesis (unpublished), Melbourne University, 1986, 813 pp. and 122 pls.
——, "Pictorial Programmes in Manuscripts of the French Version of Suso's *Horologium Sapientiae*," *Archivum Fratrum Praedicatorum*, LVII (1987), pp. 31-43, 2 pls.
——, "Two Parisian Artists of the Dunois Hours and a Flemish Motif," *Gazette des Beaux-Arts*, CXII (1988), pp. 61-8, 3 figs.
——, "Master of Jean Rolin II," in *Dictionary of Art*, ed. H. Brigstocke London, Macmillan.
Pächt, O. and Thoss, Dagmar, *Die illuminierten Handschriften und Inkunabeln der Oesterreichischen Nationalbibliothek: Französische Schule I und II*, Vienna, Verlag der Oesterreichischen Akademie der Wissenschaften, 1974-1977, 4 vols.
——, Jenni, Ulrike and Thoss, Dagmar, *Die illuminierten Handschriften und Inkunabeln der Oesterreichischen Nationalbibliothek: Flämische Schule I*, Vienna, Verlag der Oesterreichischen Akademie der Wissenschaften, 1983, 2 vols.
Panofsky, E., *Early Netherlandish Painting*, Cambridge, Mass., Harvard University Press, 1953, 2 vols.
Perdrizet, P., *Le Calendrier parisien à la fin du moyen âge d'après le Bréviaire et les Livres d'heures*, Paris, Belles-Lettres, 1933.
Perls, K. G., *Jean Fouquet*, London, Hyperion, 1940.
Perrat, Marie-Josette, *Autun, Bibliothèque Municipale. Le Livre au siècle des Rolin*, Autun, Marcellin, 1985.
Plummer, J., *The Hours of Catherine of Cleves*, London, Barrie and Rockliff, 1966.
——, *The Last Flowering. French Painting in Manuscripts 1420-1530*, New York, Pierpont Morgan Library, 1982.
Pope-Hennessy, J., *Fra Angelico*, London, Phaidon, 2nd ed. 1974.
Porcher, J. *et al.* (eds.), *Les Manuscrits à peintures en France du XIIIe au XVIe siècle*, Paris, Bibliothèque Nationale, 1955.

Purtle, Carol J., *The Marian Paintings of Jan van Eyck*, Princeton, Princeton University Press, 1982.
Quinze années d'acquisitions de la pose de la première pierre à l'inauguration officielle de la Bibliothèque, Brussels, Bibliothèque Royale, 1969.
Réau, L., *Iconographie de l'art chrétien*, Paris, Presses Universitaires de France, 1955-1959, 6 vols.
Renaissance Painting in Manuscripts, see Kren.
Robb, D. M., "Iconography of the *Annunciation* in the 14th and 15th Centuries," *Art Bulletin*, XVIII (1936), pp. 480-526.
Rohan Book of Hours, see Meiss and Thomas.
Samaran, C. and Marichal, R., *Catalogue des manuscrits en écriture latine portant des indications de date, de lieu ou de copiste*, Paris, C.N.R.S., 1959-1981, 6 vols.
Sandler, Lucy F., "Jean Pucelle and the Lost Miniatures of the Belleville Breviary," *Art Bulletin*, LXVI (1984), pp. 73-96.
Schaefer, C., *Recherches sur l'iconographie et la stylistique de l'art de Jean Fouquet*, Lille, Service des Reproductions de Thèses, 1971, 3 vols.
—–, *see also* Sterling.
Shapiro, M., *Words and Pictures. On the Literal and the Symbolic in the Illustration of a Text*, The Hague, Mouton, 1973.
Schatten, see Chavannes-Mazel.
Schiller, Gertrud, *Iconography of Christian Art*, Greenwich, Conn., New York Graphic Society, 1971-1972, 2 vols.
Spencer, Eleanor P., *The Maitre François and his Atelier*, PhD Thesis (unpublished), Harvard University, 1931.
—–, "Les Heures de Diane de Croy attribuées à Jean Fouquet," *Gazette des Beaux-Arts*, 1931, part II, pp. 329-339.
—–, "The International Style and Fifteenth-Century Illuminations," *Parnassus*, XII (1940), pp. 30-1.
—–, "L'*Horloge de Sapience*, Bruxelles, Bibliothèque Royale, MS. IV. 111. *Scriptorium*, XVII (1963), pp. 277-99 and pls. 21-30.
—–, "Gerson, Ciboule and the Bedford Master's Shop," *Scriptorium*, XIX (1965), pp. 104-8.
—–, "The Master of the Duke of Bedford: the Bedford Hours," *Burlington Magazine*, CVII (1965), pp. 495-502.
—–, "The Master of the Duke of Bedford: the Salisbury Breviary," *Burlington Magazine*, CVIII (1966), pp. 607-12.
—–, "The First Patron of the *Très Belles Heures de Notre Dame*," *Scriptorium*, XXIII (1969), pp. 145-9.
—–, "Dom Louis de Busco's Psalter," in *Gatherings in Honor of Dorothy E. Miner*, Baltimore, Walters Art Gallery, 1974, pp. 227-40.
—–, "Le Lectionnaire du cardinal Charles II de Bourbon," *Dossiers de l'archéologie*, XVI (1976), pp. 124-9.
—–, "The Hours of Anne de Neufville," *Burlington Magazine*, CXIX (1977), pp. 704-7.
—–, *The Sobieski Hours*, London, Academic Press for the Roxburghe Club, 1977.
Sterling, C., "Jan van Eyck avant 1432," *Revue de l'Art*, XXIII (1976), pp. 7-82.
—–, and Schaefer, C., *The Hours of Etienne Chevalier. Jean Fouquet. Musée Condé, Chantilly*, London, Thames and Hudson, 1972.
Suso, *see* Künzle.
Thomas, *see* Meiss.
Thoss, Dagmar, *Französische Gotik und Renaissance in Meisterwerken der Buchmalerei*, Graz, Akademische Druck- und Verlagsanstalt, 1978.
—–, *see also* Pächt.
Très Riches Heures, see Longnon and Cazelles.
Tuve, Rosemond, "Notes on the Virtues and Vices," *Journal of the Warburg and Courtauld Institutes*, XXVI (1963), pp. 264-303 and XXVII (1964), pp. 42-72.
—–, *Allegorical Imagery. Some Mediaeval Books and their Posterity*, Princeton, Princeton University Press, 1966.
Van der Meer, *see* Meer.
Verdier, P., "L'Iconographie des Arts Libéraux dans l'art du Moyen Age jusqu'à la fin du quinzième siècle," in *Arts Libéraux et Philosophie au Moyen Age*, Montreal, 1969, pp. 305-55.
Véronée-Verhaegen, Nicole, *L'Hôtel-Dieu de Beaune*, Brussels, Centre National de Recherches, Primitifs Flamands, 1973.
Villain-Gandossi, Christiane, *Le Navire médiéval à travers les miniatures*, Paris, C.N.R.S., 1985.
Vines, *see* Manion.
Wardrop, J., "Egregius Pictor Franciscus," *Apollo*, XV (1932), pp. 76-82.
Wescher, P., *Jean Fouquet and his Time*, London, Pleiades Books, 1947.
Wharncliffe Hours, see Manion, 1972.
White, L. T. Jr., "The Iconography of Temperantia and the Virtuousness of Technology," in *Action and Conviction in Early Modern Europe. Essays in Memory of E. Harbison* (ed. T. K. Rabb and J. E. Seigel), Princeton, Princeton University Press, 1969, pp. 197-219.
Wieck, R. S. (ed.), *Late Medieval and Renaissance Illuminated Manuscripts 1350-1525 in the Houghton Library, exhibited March 15 - June 3, 1983,* Cambridge, Mass., Harvard College, 1983.
Winter, P. M. de, "Copistes, éditeurs en enlumineurs de la fin du XIVe siècle. La production à Paris de manuscrits à miniatures," in *Actes du 100e Congrès des Sociétés Savantes (1975)*, Pais, 1978, pp. 173-98.
—–, "Manuscrits à peintures produits pour le mécénat lillois sous les règnes de Jean sans Peur et de Philippe le Bon," in *Actes du 101e Congrès des Sociétés Savantes (1976)*, Paris, 1978, pp. 233-56.
—–, "The Illustrations of the *Chroniques de Hainaut* in the Fifteenth Century," *Scriptorium*, XXXVI (1982), pp. 139-40.
—–, *La Bibliothèque de Philippe le Hardi, duc de Bourgogne*, Paris, C.N.R.S., 1985.
Wittek, M., *Le Livre illustré en Occident du haut moyen âge à nos jours*, Brussels, Bibliothèque Royale, 1977.
Zimelien, see Brandis.

Index

Numbers in italics refer to pages with plates.